Lecture Notes in Artificial Intelligence

Edited by R. Goebel, J. Siekmann, and W. Wahl

Subseries of Lecture Notes in Computer Science

Kevin Korb   Marcus Randall
Tim Hendtlass (Eds.)

Artificial Life: Borrowing from Biology

4th Australian Conference, ACAL 2009
Melbourne, Australia, December 1-4, 2009
Proceedings

 Springer

Series Editors

Randy Goebel, University of Alberta, Edmonton, Canada
Jörg Siekmann, University of Saarland, Saarbrücken, Germany
Wolfgang Wahlster, DFKI and University of Saarland, Saarbrücken, Germany

Volume Editors

Kevin Korb
Monash University, School of Information Technology
Clayton, Victoria 3800, Australia
E-mail: kevin.korb@infotech.monash.edu.au

Marcus Randall
Bond University, School of Information Technology
University Drive, Gold Coast, Queensland 4229, Australia
E-mail: mrandall@bond.edu.au

Tim Hendtlass
Swinburne University of Technology
Faculty of Information and Communication Technologies
Melbourne, Victoria 3112, Australia
E-mail: thendtlass@swin.edu.au

Library of Congress Control Number: 2009938816

CR Subject Classification (1998): I.2, J.3, G.2, F.1, I.2.11, I.2.9

LNCS Sublibrary: SL 7 – Artificial Intelligence

ISSN	0302-9743
ISBN-10	3-642-10426-6 Springer Berlin Heidelberg New York
ISBN-13	978-3-642-10426-8 Springer Berlin Heidelberg New York

springer.com

© Springer-Verlag Berlin Heidelberg 2009
Printed in Germany

Typesetting: Camera-ready by author, data conversion by Scientific Publishing Services, Chennai, India
Printed on acid-free paper   SPIN: 12798056   06/3180   5 4 3 2 1 0

Preface

As we approach the limits and capabilities of machines, we find that the principle of diminishing returns is forcing scientists to turn their attention toward biology to provide practical solutions to problems in our increasingly complex world. An emerging field of science that studies the systems related to life, its processes and evolution is known as Artficial Life (ALife). It draws on the skills and talents of scientists from a variety of disciplines, including, but not limited to, biology, psychology, evolution and computer science. These researchers are showing that, by using even simple simulations of the basic processes of nature, much about our complex natural world and our own humanity can be revealed.

Gatherings in the expanding Alife community are becoming more common. One such series, the Australian Conference on Artificial Life (ACAL), began in Adelaide in 2001 and was known as the "Inaugral Workshop on Artifical Life." From these small beginnings, it has become a biennial event that has previously been held in Canberra (2003), Sydney (2005) and the Gold Coast (2007). ACAL 2009 received over 60 quality submissions of which 27 were accepted for oral presentation at the conference. Each paper submission was assigned to three members of the Program Committee. The Program Committe, as well as the conference delegates, came from countries across Asia-Pacific, North America and Europe.

In addition to the regular papers, the conference was fortunate enough to be able to have two renowned invited speakers – Mark Bedau (Reed College, USA) and Andries Englebrecht (University of Pretoria, South Africa).

The organizers wish to acknowledge a number of people and institutions, without whose support ACAL would not be possible. First and foremost, we wish to thank the Program Committee for undertaking the reviewing of submissions, providing invaluable feedback to the authors. Monash University's Faculty of Information Technology very kindly provided financial backing for the event. Finally, the editors must pay tribute to the team at Springer for the production of these proceedings.

December 2009

Kevin Korb
Marcus Randall
Tim Hendtlass

Organization

ACAL 2009 was organized by the Faculty of Information Technology, Monash University in association with Bond University and Swinburne University of Technology.

Chairs

General Chair   Kevin Korb (Monash University, Australia)
Co-chairs   Marcus Randall (Bond University, Australia)
     Tim Hendtlass (Swinburne University of
     Technology, Australia)

Local Arrangements Chair

Steven Mascaro   Bayesian Intelligence, Australia

Spcial Track Editor

Alan Dorin   Monash University, Australia

Conference Coordinator

Dianne Nguyen   Monash University, Australia

Program Committee

Dan Angus	University of Queensland, Australia
Rodney Berry	National University of Singapore
Tim Blackwell	University of London, UK
Michael Blumensetin	Griffith University, Australia
Stefan Bornhofen	Ecole Internationale des Sciences du Traitement de l'Information (EISTI), France
Richard Brown	Mimetics, UK
Stephen Chen	York University, Canada
Vic Ciesielski	RMIT, Australia
David Cornforth	ADFA - University of New South Wales, Australia
Palle Dahlstedt	IT University of Göteborg, Sweden
David Green	Monash University, Australia
Gary Greenfield	University of Richmond, USA

International Advisory Committee

Table of Contents

Complex Systems

Biological Systems

Social Modelling

Swarm Intelligence

Heuristics

An Empirical Exploration of a Definition of Creative Novelty for Generative Art

Taras Kowaliw, Alan Dorin, and Jon McCormack

Centre for Electronic Media Art, Faculty of Information Technology,
Monash University, Clayton 3800, Australia
taras@kowaliw.ca, {alan.dorin,jon.mccormack}@infotech.monash.edu.au
http://www.csse.monash.edu.au/~cema/

Abstract. We explore a new definition of creativity — one which emphasizes the statistical capacity of a system to generate previously unseen patterns — and discuss motivations for this perspective in the context of machine learning. We show the definition to be computationally tractable, and apply it to the domain of generative art, utilizing a collection of features drawn from image processing. We next utilize our model of creativity in an interactive evolutionary art task, that of generating biomorphs. An individual biomorph is considered a potentially creative system by considering its capacity to generate novel children. We consider the creativity of biomorphs discovered via interactive evolution, via our creativity measure, and as a control, via totally random generation. It is shown that both the former methods find individuals deemed creative by our measure; Further, we argue that several of the discovered "creative" individuals are novel in a human-understandable way. We conclude that our creativity measure has the capacity to aid in user-guided evolutionary tasks.

1   Introduction

A recent definition of creativity recasts it as "a framework that has a relatively high probability of producing representations of patterns that can arise only with a small probability in previously existing frameworks" [2]. An interesting property of this definition is that it depends neither on notions of value nor appropriateness. These properties, of course, set it at odds with common usage of the term "creative", since the perceived creativity of a system is often culturally biased, associated with interest, or affected by context. Dorin and Korb consider the value of a value- and appropriateness-free definition of creativity extensively, and respond to some obvious criticisms. Regardless of common usage, the ability to capture even an aspect of creative novelty in an objective measure is enticing; It affords us the possibility of (a) empirically testing the consequences of the definition, and it's value to preconceived notions of creativity; and (b) suggesting a stream of new and, ideally, interesting frameworks to users. In this latter motivation, our work resembles the creation of an iterative fitness function [6], except that rather than provide exemplars of an optimal goal state, our notion of creativity supports a practically unlimited variation.

K. Korb, M. Randall, and T. Hendtlass (Eds.): ACAL 2009, LNAI 5865, pp. 1–10, 2009.

Here we explore the creativity of a *system* — something or someone that accepts an input and generates, possibly stochastically, a pattern or phenotype — rather than the creativity of a particular pattern or representation. An approach which selected for previously unseen representations or patterns would likely need to incorporate a notion of distance (since in a sufficiently rich space, any randomly-generated entity is almost certainly previously unseen), selecting for new entities "far" from previously seen entities. However, the notion of distance is problematic in many domains, due to a lack of meaningful metrics or unusual statistical distributions. Additionally, such a system would likely be biased towards statistical outliers, which could confuse an exploration/exploitation-based search technique.

We will interpret the Dorin/Korb definition as a simple boolean relation distinguishing between systems that can *reliably* generate some collection of patterns, as opposed to a system exceedingly unlikely to do so; This will allow us to side-step many of the difficulties associated with the use of genotypic or phenotypic similarity measures.

2   Formalization

In order to make this definition rigorous and practical, we need to qualify several aspects, and introduce some restraints. We will aim for systems which generate grayscale images, but generalization should be straightforward. Let us consider a space of patterns (phenotypes), $p \in P$. For simplicity, we will assume that the space of patterns is bounded, which excludes new modalities or ever-increasing scales. Since patterns are often too high-dimensional to deal with directly, we instead introduce a feature space of finite dimension, $\mathbb{F} = \{F_1, ..., F_k\}$ through which we can characterize the space of patterns: $p \to (F_1(p), ..., F_k(p)) \in \mathbb{R}^k$. We will assume only that connected regions form perceptually similar segments of phenotypic space, but assume nothing regarding volume, distribution, nor the significance of distance. Although our capacity to select a representative sample is ultimately governed by the representation space, the spread of values will be strongly affected by the number of features used, recovering the "curse of dimensionality" which plagues machine learning tasks; to prevent this, in practice, we will use only two features at a time.

2.1   Images and Image Features

We now turn our attention to systems which generate 8-bit grayscale images of a fixed size[1] of 200×200 pixels. Image similarity is a notoriously difficult problem, and these images can be considered points in a 40000-dimensional space. In order to characterize this pattern space, we select a set of features drawn from image processing in hope of drawing features perceptible to humans.

---

[1] To the reader who believes that a bounded space is insufficiently rich to generate a practically endless variation of patterns, we refer the reader to McCormack's discussion of generating simple images [7].

Following work in pattern classification, we use several statistical moments to characterize the overall form of the images (as in, for instance, [5]): specifically, we use geometric moments $M_{00}, M_{01}, M_{10}, M_{20}, M_{02}$, and M_{11}.

Following examples from content-based image retrieval, we also use histograms to characterize the space (as in, for instance, [8]). The image is transformed by an edge-detection measure (Laplacian convolution) and the normalized area is calculated. Several histogram-related measures are computed for both the original image and the edge-detected version: the maximum, mean, and standard deviation, and the entropy.

Finally, we use Grey Level Co-occurrence Matrices (GLCMs) to include texture measures into our feature set [3]. Specifically, we measure several statistical properties of the normalized co-occurrence matrix $P = \{p(i,j)\}$, chosen due to their demonstrated efficacy in content-based image retrieval [4]. These are Energy: $GE = \sum_{i,j} p^2(i,j)$; Entropy: $GI = \sum_{i,j} p(i,j) \log p(i,j)$; Contrast: $GC = \sum_{i,j} (i-j)^2 p(i,j)$; and Homogeneity: $GH = \sum_{i,j} \frac{p(i,j)}{1+|i-j|}$.

Overall, we considered a total of 18 features representing the overall form, histogram, and texture of the image and its edges. Following some initial experimentation, we selected six which support a rich characterization of phenotypic space (that is, were capable of measuring specialization during our informal experimentation of the space of images): M_{11}, M_{02}, M_{01}, *edge-area*, *histogram-entropy*, and the GLCM homogeneity measure.

2.2 Detecting Creativity-Indicating Regions

Let S_1 and S_2 be systems which map from a space of input, $x \in X$, to the space of patterns P. We wish to claim that S_2 is creative (or not) relative to S_1 on the basis of what can be reliably produced by the systems in question. We interpret this as the capacity to find a compact and connected region of feature space that can be reliably populated by system S_2 but not by system S_1. We restrict our attention to intervals since this is more likely to generate an intelligible region of space (assuming a measure of continuity in the chosen features), as opposed to the arbitrariness of general Borel sets, for example.

We now wish to formalize a notion of the capacity of a system to reliably generate a pattern, with error-tolerance τ and confidence c. We will write that an interval of feature space is $\mathbf{r} = (r_1 \pm \delta_1, ..., r_k \pm \delta_k)$, and that a point p is contained in $\mathbf{r}$, $p \in \mathbf{r}$, if it is contained within the bounds for each feature-space dimension. We aim to estimate the true probability of S_j generating a point in interval $\mathbf{r}$, $P[S_j(X) \in \mathbf{r}]$, via the frequency of sample points, written $\hat{P}$. Assuming our sample is representative, the (conservatively estimated) margin of error associated with this frequency is

$$m.e.(\hat{P}[S_j(X) \in \mathbf{r}]) = \frac{z}{\sqrt{n}} \tag{1}$$

where z is the upper critical value for confidence-level c, and n is the sample size. Let us assume that $\tau > \frac{z}{\sqrt{n}}$, i.e. that our margin of error does not dominate our error tolerance. Generally, we can achieve this by ensuring that $n > \frac{z^2}{\tau^2}$.

We will now write that $\mathbf{r}$ intersects the *reliable-support* of S_j iff

$$\hat{P}[S_j(X) \in \mathbf{r}] > \tau + \frac{z}{\sqrt{n}} \qquad (2)$$

We will write that $\mathbf{r}$ is **not** in the reliable-support of S_j iff $\hat{P}[S_j(X) \in \mathbf{r}] = 0$, with no conclusions being drawn on the region in between. In the former case, we have greater than c confidence that the probability of generating samples in the region $\mathbf{r}$ is greater than τ, and in the latter, greater than c confidence that the probability is less than our margin of error.

Hence, provided with values for reliability τ and c, we seek to find a region $\mathbf{r}$ which intersects the reliable-support of S_2, but not the reliable-support of S_1. Finding such a region in a set of samples, we shall soon see, is also a matter requiring some interpretation.

This definition relies heavily on the abilities of a new system relative to some base system, a metaphor for the existing worldview of an audience. Of course, if one begins with a trivial base system, nearly any new system will appear creative.

2.3   Generation of Intervals Given Sample Data

Here we attempt to find intervals using our sample and feature space that exist in the reliable support of system S_2, but not in the reliable-support of system S_1. Using a confidence interval of $c = 95\%$, an error tolerance of $\tau = 0.03$, and a sample size of $n = 5000$, we need to find regions with at least $n(\tau + \frac{z}{\sqrt{n}}) = 292$ samples from system S_2, and none from system S_1. It is natural to base the minimal size of an interval on the standard deviation of the sample; Rather than include any such interval capable of supporting the mass of points required, we will instead require than a minimal length[2] of interval be $\beta = \frac{\sigma}{5}$.

The attempt is made to find intervals surrounding each sample point provided. If our sample pattern is $p = (F_1(p), ..., F_k(p))$, then we initially define our interval about p as

$$\mathbf{b}(p) = \left(F_1(p) \pm \frac{\beta}{2} \sigma_1^j, ..., F_k(p) \pm \frac{\beta}{2} \sigma_k^j \right) \qquad (3)$$

where S_j is the system from whence sample p was drawn, and σ_i^j is the standard deviation of the system S_j in the i-th dimension.

For each such created interval, we ask if it does not contain points from S_1 (as we know it contains at least one point from S_2). If so, we attempt to generalize it. For each dimension in turn, we attempt to widen the width of the interval by a factor of 2. If we successfully define a new interval containing equal or greater points from S_2 and none from S_1, we retain the new interval. Once we have

---

[2] This number was chosen on the basis that during a search of possible values it was the smallest value leading to a small proportion of random genomes being declared creative; That is, using a smaller value we risked declaring regions resulting from a particularly impotent application of mutation (i.e. mutations which had no effect on phenotype) as creative.

traversed all dimensions, if our new interval is an improvement on the original, we traverse the dimensions again. This process will terminate either when a locally maximal interval is found, or when the interval covers the entirety of all dimensions. Finally, the best found interval is tested to see if it is in the reliable-support of S_2.

Since searching for our formally defined creativity is a slow process, we also introduce a multi-valued procedure for estimating relative creativity quickly. **Creativity Lite** will take a smaller sample, and return the maximum number of samples from set S_2 that can be found in some region containing no samples from S_1. We use a sample of points from set S_1 of size $n_1 = 1000$, and a sample of points from set two of size $n_2 = 100$. Intervals are constructed using the technique described above.

2.4 Model Limitations

There are several limitations to the presented model. The first, already noted, is the reliance on constructing a representative sample of pattern space, which sets practical limits on the number of features and maximal pattern sizes.

A second potential limitation is the pre-selected feature space. Given any two non-identical systems, it is likely always possible to generate some feature which separates them. Clearly, the choice of a trivial feature in feature space will allow for the discovery of a trivial form of creativity; Our choice of features introduces a notion of appropriateness alongside our originally appropriateness-free definition of creativity.

Finally, we note the reliance on several predetermined values — for certainty, minimal intervals, and interval discovery — likely to be dependent on the problem domain and data source distributions. As a result, we can never say, with measurable certainty, that S_2 is **not** creative relative to S_1.

3 Biomorphs

Biomorphs were introduced by Dawkins as a simple example of evolutionary search [1]. Here we present our own interpretation of Dawkins' original (and vaguely specified) biomorph growth process.

A genome consists of eleven genes, $g = (g_1, ..., g_{11})$, each a floating point number in the range $[0, 1]$. Initialization of genes is random and uniform. The generated form has a number of properties, applied recursively: A translation $(x_{trans}, y_{trans}) = (20g_1 - 10, 20g_2 - 10)$; A thickness $t = 3g_3 + 1$ and thickness variation $\Delta t = \frac{2g_4}{3}$; A magnitude $m = \frac{g_5 w}{2}$ (where w is the maximum dimension of the drawing surface) and magnitude variation $\Delta m = \frac{2g_6}{3}$; A branch angle $\alpha = \pi g_7$ with an angle variation $\Delta \alpha = 0.4g_8 + 0.8$; A branching factor $n_{kids} = \lfloor 6g_9 + 1 \rfloor$ and a branch depth variance, $\Delta n_{kids} = 0.4g_{10} + 0.8$; And a recursion depth, $n_{rec} = \lfloor 4g_{11} + 2 \rfloor$. These particular properties were chosen through informal visual search on the space of random genomes as able to generate a breadth of interesting phenotypes.

Biomorph drawing is a simple recursive procedure. Given a genome, a starting location (x_{start}, y_{start}), a current magnitude m', a current thickness t', a current number of children n'_{kids}, and a current angle α', we draw a line; This is drawn between the given start location and an end location calculated as

$$(x_{end}, y_{end}) = (x_{start} + x_{trans} + m' \cos(\alpha'), y_{start} + y_{trans} + m' \sin(\alpha')) \quad (4)$$

At the end point, if we have not exceeded n_{rec} recursive steps, we create $n''_{kids} = n'_{kids} \cdot \Delta n_{kids}$ child lines, each pointing $\alpha'' = \alpha' \cdot \Delta\alpha$ degrees from the original direction, with a magnitude $m'' = m' \cdot \Delta m$, and a thickness of $t'' = \min\{1, t' \cdot \Delta t\}$. We initialize a biomorph at the central point, drawing n_{kids} lines at integer multiples of α around the $\pi/2$ axis. A selection of interesting but randomly generated biomorphs can be seen in Figure 1.

Fig. 1. A selection of randomly generated biomorphs, chosen to show some of the diverse structures possible

3.1   Discovery of Creative Biomorphs via Interactive Evolution

Biomorphs are easily evolved using an Interactive Evolutionary Algorithm (IEA). Here we use a simple $(1 + 8)$-ES, where mutation selects two randomly chosen genes and mutates both with a variance of 0.16.

Our initial system is a wide collection of randomly generated genomes. We use 14400 samples in two-dimensional feature space, effectively assuming that 120 samples is sufficient to represent each dimension. Experimentation with computed descriptive statistics over several runs convinces us that this is a sufficient choice. This pool is considered as our basic worldview (S_1 above), the yardstick from which we will measure creativity in generative systems. Note that duplicate feature values are discarded; In practice, this means an additional 20% generation time.

Given any particular genome, we can consider it a generative system by considering it as a seed from which we can generate mutated children. That is, given

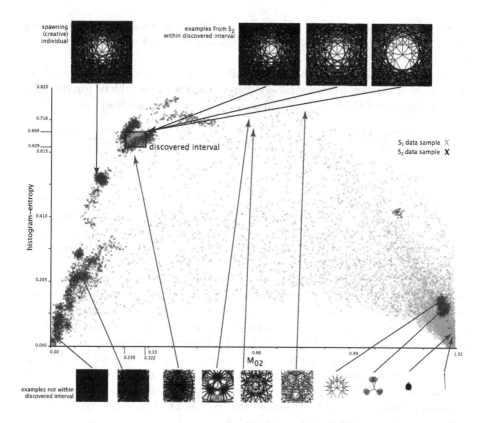

Fig. 2. *(top)* The spawning individual's phenotype and a collection of phenotypes in the discovered interval; *(middle)* a view of feature-space and sample distribution (data points have been slightly "jittered" for legibility); *(bottom)* a random sample of phenotypes not in the discovered interval

some genome g, we will generate a set of children $S_g = \{m(g)\}_k$, consisting of k mutated versions of the original genome. We will discard any children resulting from an impotent mutation (in practice, requiring an additional 10% generation time).

We use three combinations of pairs of features to evaluate the system. For each, several search strategies are evaluated on their capacity to produce creative phenotypes. The first search strategy was *totally random*, where a single random genome is evaluated. The next two search strategies are variants on evolutionary search. In each, an initial population of nine random genomes is created, and ten generations are explored by selecting a single agent and spawning an additional eight mutants to form the next population. The first evolutionary strategy is *IEA*, where the individual is selected subjectively by a human operator[3]; the

---

[3] Human operators were selected from five people at our research lab, including the authors, and asked to select for aesthetically pleasing biomorphs. Participants were not aware of creativity scores while making selections.

second is the *creativity search*, where each population member is evaluated using the creativity-lite function, and the maximum such value guides the choice of individual.

We computed the proportion of individuals termed creative for a sample of runs for each strategy. Each strategy was run forty times (thirty for the IEA runs), using one of three pairs of features. These proportions were:

search type	M_{02}, *edge-area*	M_{01}, *GLCM-H*	M_{11}, *histogram-entropy*
totally random	0.067	0.025	0.000
IEA	0.433	0.233	0.700
creativity search	0.800	0.475	0.733

The probability of a totally random individual being termed creative was very small, which we consider further evidence of the statistical sample being representative. Via creative-search, relatively high proportions of evolved genomes are termed creative. It appears that creativity-lite does indeed serve as an approximation of our more formal notion of creative novelty, allowing for a faster approximate search. The proportion of individuals termed creative discovered via IEA lay consistently between the proportions for the totally random and creative search techniques. This corresponds to the general intuition that humans tend to seek out novelty, but also that aesthetic pleasure and novelty are not the same thing. It is also evident that it is easier to discover creative regions using some combinations of features than others; Using the M_{01}, *GLCM-H* combination, only two rough patterns of creative regions were discovered — namely, regions comprised of homogeneous and top-heavy trees, and regions of very high non-homogeneity — while for the other two explored feature combinations many different creative regions were discovered.

An example of an easily understood discovered creative individual is illustrated in Figure 2. This individual was discovered using creative search and the M_{02} and *histogram-entropy* features. The interval discovered is characterized by a high *histogram-entropy* value, and a low M_{02} value. The high *histogram-entropy* value is difficult to discover in a line drawing, since most pixel values are either black or white; The only means of obtaining a high value is to include many spaced-out diagonal lines on a variety of angles, and rely on the anti-aliasing of the line drawing routine. Simultaneously, we require a low M_{02}, meaning that both the top and the bottom of the image must be mostly black in colour. Satisfying both simultaneously is non-trivial — since it is difficult to space out black lines on a white background and minimize the white content of the image — and thus a rarity; This individual satisfies both by creating patterns with a blank hole in the centre, surrounded by an increasing density of angled lines.

Generally speaking, the examples of individuals termed creative by the system ranged from (in the authors' opinions) trivial to interesting. Several additional interesting generators are illustrated in Table 1; Trivial examples include minor variations on these themes (such as selection for slightly smaller or larger circles, or a slightly larger or smaller number of diagonal lines), or examples of individuals that have discovered regions of trivial mutation (for instance, during the M_{02}, *edge-area* runs, some regions consisting of a very large number of near-identical

Table 1. Examples of discovered "creative" individuals, including the phenotype of the *(left)* spawning individual; *(middle)* an averaged image of all children in the discovered interval; and *(right four)* some examples of those children. The discovery technique and the reason for the individual being termed creative below.

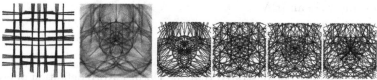

Discovered via IEA using features M_{02}, *edge-area*. Slightly bottom-heavy (difficult in a system which automatically begins with branches pointing upwards), high number of edges (i.e. spaced out lines).

Discovered via creativity-search using features M_{11}, *histogram-entropy*. A collection of dark shapes, solid black in the centre with a fuzzy boundary. The fuzzy shapes create a small but non-zero histogram entropy by having a small portion of the image devoted to many different angles of line, but a consistent mass via area of shape

Discovered via creativity-search using features M_{01}, *GLCM-H*. Selection for near-perfect symmetrical mass over the x-axis and a very large number of colour transitions.

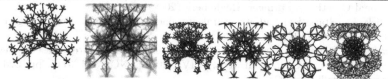

Discovered via IEA using features M_{02}, *edge-area*. Selection for specific amounts of edges, mass.

Discovered via IEA using features M_{11}, *histogram-entropy*. Selection for both a mid-level entropy (i.e. a fair number of diagonal lines at different angles) and a low but non-zero moment calculation. The moment calculation implies near-perfect symmetry in both the x- and y-axis. Children achieve this by having perfect symmetry in one axis, and near-symmetry in the other, some using the x and others the y.

patterns were found, relying on the invariance of both features to translation in the x-axis). Regardless, the tool has demonstrated its ability to find genuinely improbable and variant regions of space which occasionally correspond to human interest; Hence, it may be sufficient to remove some of the burden of search from human operators in an IEA.

4   Conclusions

In this paper we have demonstrated that the Dorin / Korb definition of creativity can be interpreted in such a way that makes it both tractable generally, and suitable for generated images. We have further shown that the definition can be integrated into an interactive evolutionary algorithm: Firstly, by treating individual genomes, along with their evolutionary operators, as pattern generators; And secondly, by considering an approximate and fast version of our creative-novelty measure, "creativity lite". It is shown that maximization of creativity lite generally leads to a system which is creative by our definition.

Several genomes have been discovered which do, indeed, generate regions of space which are highly unlikely through random generation. While some of these regions seem trivial to human operators, others are human-recognizably distinct, which we interpret as evidence that the system can detect human-recognizably novel generators; In conjunction with the work above, we believe that this measure can be used to suggest new and interesting directions for users of an IEA.

References

1. Dawkins, R.: The Blind Watchmaker. Longman Scientific & Technical (1986)
2. Dorin, A., Korb, K.: Improbable creativity. In: McCormack, J., Boden, M., d'Inverno, M. (eds.) Proceedings of the Dagstuhl International Seminar on Computational Creativity. Springer, Heidelberg (2009)
3. Haralick, R.: Statistical and structural approaches to texture. Proceedings of the IEEE 67, 786–804 (1979)
4. Howarth, P., Rüger, S.: Evaluation of texture features for content-based image retrieval. In: Enser, P.G.B., et al. (eds.) CIVR 2004. LNCS, vol. 3115, pp. 326–334. Springer, Heidelberg (2004)
5. Kowaliw, T., Banzhaf, W., Kharma, N., Harding, S.: Evolving novel image features using genetic programming-based image transforms. In: IEEE CEC 2009 (2009)
6. Machado, P., Romero, J., Manaris, B.: Experiments in computational aesthetics: An iterative approach to stylistic change in evolutionary art. In: Romero, J., Machado, P. (eds.) The Art of Artificial Evolution: A Handbook on Evolutionary Art and Music, pp. 381–415 (2008)
7. McCormack, J.: Facing the future: Evolutionary possibilities for human-machine creativity. In: Romero, J., Machado, P. (eds.) The Art of Artificial Evolution: A Handbook on Evolutionary Art and Music, pp. 417–453 (2008)
8. Vailaya, A., Figueiredo, M.A.T., Jain, A.K., Zhang, H.-J.: Image classification for content-based indexing. IEEE Transactions on Image Processing 10(1), 117–130 (2001)

A New Definition of *Creativity*

Alan Dorin and Kevin B. Korb

School of Information Technology
Monash University
Clayton, VIC 3800

Abstract. Creative artifacts can be generated by employing A-Life software, but programmers must first consider, explicitly or implicitly, what
would count as creative. Most apply standard definitions that incorporate notions of novelty, value and appropriateness. Here we re-assess this
approach. Some basic facts about creativity suggest criteria that guide us
to a new definition of creativity. We briefly defend our definition against
some plausible objections and explore the ways in which this new definition differs and improves upon the alternatives.

1 Introduction

Artificial Life has been applied broadly to creative domains, especially music
composition (e.g., Berry et al., 2001; Dahlstedt, 1999; Miranda, 2003), image
production (e.g., Todd and Latham, 1992; Sims, 1991) and ecosystem-based
installation or simulation (e.g., McCormack, 2001; Saunders and Gero, 2002;
Dorin, 2008). In some cases the production of creative artifacts is implemented
in autonomous software; elsewhere, this is done interactively with humans. With
autonomous software some definition of creativity, implicit or explicit, must determine the approach. Conventional definitions incorporate both novelty and
value. Programmers then either hard-code "value" to suit their own aesthetics
or they allow value to emerge from social virtual-agent interactions. In this latter
case, the agents establish their own, non-human notions of value. In this position
paper we introduce a new approach that allows us to measure the creativity of
artifacts and is suitable for automatically guiding creative processes.[1]

2 The Facts about Creativity

Creativity was originally the sole domain of the gods. Subsequently, humans
have been the beneficiaries of this Promethean gift, but there has been little
consensus on what to make of it. There is *some* consensus in the literature on
creativity: that we humans are sometimes creative. In trying to grapple with
the concept so as to find an exact definition of it there has been no consensus.
Nevertheless, there are other facts which are widely agreed to, either explicitly
or implicitly in the way the word 'creative' is used. We list five of the most basic
and agreeable here.

---

[1] We do not implement such a model here, however see Kowaliw, Dorin, McCormack
(in this volume) where we describe such an implementation.

K. Korb, M. Randall, and T. Hendtlass (Eds.): ACAL 2009, LNAI 5865, pp. 11–21, 2009.

Basic Facts about Creativity

Fact 1. *Humans may be creative.*

Fact 2. *Processes, solutions to problems and works of art may all be creative.*

Viewing humans as processes, we could, of course, combine Facts 1 and 2; however, humans on the whole being vain when not being creative, usually prefer to consider the facts about themselves as separate, primary or even unique. The main point we infer from these facts is just that creativity inheres in two varieties of entities: objects, abstract or concrete, and the processes that produce them.

Fact 3. *Duplication is uncreative; novelty is essential.*

Originality or novelty is at the center of all dictionary definitions of 'creativity'. It follows immediately that the creativity of works and processes is *relative*, relative to their own histories and the socio-cultural context in which they occur. For example, the original discovery of the Pythagorean theorem was creative, as was its use to demonstrate the existence of irrationals. But today these are schoolboy exercises for the bored, rather than innovations. Creativity eats its own tail.

Fact 4. *Creativity comes in degrees, from the exceedingly high, down to mediocrity and further to non-existence.*

The dullard student copying the creative master is readily distinguished from the original, simply by removing the creative source and observing the results. The dullard, lacking guidance, cannot avoid being dull.

Fact 5. *It is possible to be mistaken about whether or not an act is creative.*

This actually follows from Fact 3, since the context in which a work is produced may be unknown or mistaken. We separate it out as a new Fact because of its importance.

Corollary 1. *Creativity may go unrecognized for any period of time.*

This is one of the most remarked upon facts of creativity, with many hundreds of famous examples in the arts and sciences available for illustration.

These are the fundamental facts about creativity so far as we can tell. These facts are very widely taken to be either definitive or generally true in application. So, they provide criteria against which we can test any proffered definition, including our own.

3   The Cognitive Science of Creation

The introduction of modern computers in the nineteen fifties led directly to the invention of cognitive psychology and, more generally, cognitive science. The central metaphor of cognitive science is thought-as-computation, and its central quest is the search for artificial intelligence (AI). Those who believe in the possibility of strong AI — a genuinely intelligent artifact — must necessarily also believe in the possibility of an algorithm that exhibits creativity and so in the possibility of a computational analysis of creativity.

The potential for computers to generate works that, when produced by humans are considered to be creative, has been widely addressed. It is well beyond the present scope to delve into this area in detail. A couple of well-known examples include Harold Cohen's AARON software which explores the potential for machine creativity in the visual arts. Other examples of software targeting creativity exist in the domains of mathematics (cf. Lenat, 1983), logic (Colton, 2001), and scientific discovery (Langley et al., 1987). Some researchers, however, reject claims that these examples exhibit creativity. Rowe and Partridge (1993) critique many attempts to generate computational creativity, including Lenat's. Hofstadter (1995, p. 468) complains that, although AARON draws human forms, it has no "understanding" of what it is to be human and so cannot be creative. This complaint echoes the more general worries about the prospects of artificial intelligence frequently and strongly put by John Searle (Searle, 1980), the inventor of the term "strong AI". According to Searle, any system lacking understanding, and indeed consciousness, cannot hope to exhibit intelligence, let alone creative intelligence (Searle, 1992).

Although we reject the views of Hofstadter and Searle (see, e.g., Korb, 1991), here we simply address artificial creativity in the sense of weak AI: we assume it will be good enough should cognitive scientists be able to reproduce convincingly creative *behavior*, and so passing a Turing Test for creativity. If this does not pass a deeper test, failing to yield also a deep account of creative consciousness, then we shall regret that limitation, but we shall not consider that any defect in its account of creativity to the depth that it actually reaches. Such a pseudo-creative artifact would nevertheless have to be both an outstanding achievement and, at least implicitly, a profound account of creativity.

3.1   Definitions of Creativity

Before attempting our own definition of "creativity" we present a well known alternative, that of Boden:

Definition 1 (Creativity). *Creativity is the ability to come up with ideas or artifacts that are (a) new, (b) surprising and (c) valuable (Boden, 2004, p. 1).*

Boden refines this definition in various ways.

(a) There are two ways in which something might be new: (i) *p(sychological)-creativity* introduces something that is new to the person who devised the idea or artifact, but may be previously known to others, and (ii) *h(istorical)-creativity* is new to the whole of history.
(b) Boden finds three ways in which something might be surprising: (i) it is unfamiliar or unlikely; (ii) it unexpectedly fits into a class of things you hadn't realized it fitted into; (iii) it is something you thought was impossible.
(c) Regarding her third criterion for creativity, value, Boden is certain that there are more ways in which this might occur than anybody could ever list.

4   A New Definition of Creativity

Definition 2 (Representation). *Representations are the bearers of meaning, expressing, denoting or imitating something other than themselves (van Gulick, 1980).*

All ideas, concepts, algorithms, procedures, theories, patterns, etc. that we deal with are represented by us and to us, whether through language or otherwise. Note that representations are *instances* or *tokens* — e.g., marks on paper — rather than abstract objects, even when what they are expressing is abstract.

Definition 3 (Pattern). *A pattern is any abstract object expressed by a representation.*

There are representations of non-abstract objects, of course. These are generally held to be uncreative, simply tags or names of things. We shall not consider them further.

Definition 4 (Framework). *Frameworks are stochastic generative procedures. In particular, they generate representations of patterns.*

Frameworks are particular kinds of representations, namely stochastic procedures; thus, beginning in the very same circumstances they will not generally produce the very same patterns. For example, minimalism and abstract expressionism are different frameworks for the production of art. Quantum electrodynamics is a framework for generating questions and answers about how light and matter interact. A pseudo-random number generator is a framework for generating pseudo-random numbers. And evolution is a framework for generating ecosystems, niches and organisms.

Frameworks may be more or less abstract: meta-frameworks generate frameworks which generate concrete objects. This recursive abstraction may be carried to arbitrarily high levels.

Definition 5 (Creativity). *Creativity is the introduction and use of a framework that has a relatively high probability of producing representations of patterns that can arise only with a smaller probability in previously existing frameworks.*

What we mean by this definition is not altogether on the surface, so we shall spend the remainder of this section unpacking it and the next section comparing its implications with those of alternative definitions.

The basic idea of our definition is reflected in Figure 1. Distribution 1 represents the old framework and Distribution 2 the new; both are probability distributions over a common design space, represented by the horizontal dimension. All the points in the design space where Distribution 2 has significant probability are points where Distribution 1 has insignificant probability. The use of Distribution 2 relative to the prior use of Distribution 1 to generate one of these points is therefore creative.

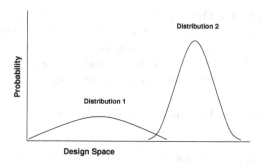

Fig. 1. Two Frameworks

The motivation for this approach to understanding creativity comes from search and optimization theory. When a solution to a problem can be cast in terms of a computational representation to be found in some definable representation space, then the problem can be tackled by applying some search algorithm to that space. Uncreative brute force searches and uniformly random searches may well succeed for simple problems, that is, for small search spaces. For complex problems, search spaces tend to be exponentially large and more creative approaches will be needed, and especially stochastic searches apply various heuristics for focusing the search in more productive regions.

The most important point is to note that on our account creativity is thoroughly relative: it is relative to the pre-existing frameworks being used to produce some kind of object and it is relative to the new framework being proposed. The creativity of objects is strictly derivative from that of the frameworks producing them and, in particular, the ratio of probabilities with which they might produce them. That is why some entirely mundane object, say a urinal, may become a creative object. Of course, the urinal of itself is uncreative, because its manufacturing process is uncreative, but its use in producing an art installation that challenges expectations may well be creative.

4.1   Methods for Discovering Novel Representations

As frameworks, that is stochastic procedures, may be represented, their representations may themselves be generated by other stochastic procedures, or meta-frameworks. So, we can talk of frameworks also as objects, and they are more or less creative according to the probability with which their meta-frameworks produce them. By recursion, then, we can consider the creativity of meta-frameworks and meta-meta-frameworks without any fixed theoretical bound, even while practically bounded by the complexity of the processes actually involved.

The meta-framework that finds that novel framework necessary for creativity may itself be uncreative; it may even be a brute force search, uniformly random or genetic drift. The manner in which the framework is discovered does not bear on the creative activity that occurs at the level of the framework and the patterns it may be used to generate. However we can separately or jointly consider the creativity of all of these searches:

Definition 6 (Creative Order). *A novel framework that generates a novel set of patterns in accordance with Definition 5 is of the* **first creative order.** *A novel framework for generating novel frameworks for generating novel patterns in accordance with Definition 5 is of the* **second creative order.** *And we can extend this arbitrarily to talk of* **nth-order creativity.**

4.2   Objective versus Psychological Creativity

How people *judge* creativity is at some variance with what we have presented above. We think the most obvious discrepancies between our definition and human judgments of creativity can be handled by the addition of a single idea, namely habituation. Human judgement of the novelty of a stimulus follows a path of negative exponential decay over repeated exposure (Saunders and Gero, 2002). Whereas Definition 5 simply makes reference to pre-existing frameworks, psychological judgment of creativity takes into account *how long* those frameworks have been around.[2] or, to put it another way, the context of frameworks to which creativity is relative had best be stated explicitly.

4.3   Objections and Replies

Here we put three objections to our definition which we think likely to occur to people and our rebuttals.

The Failure of Randomness. Something akin to our definition has played a role in repeated attempts in AI to generate creative behavior, and those attempts have repeatedly failed. So, it may very well be inferred that our definition is guilty by association with these failures. For example, consider *Racter*. *Racter* was a natural-language generation program that used templates and randomized word selection to produce surprising and sometimes evocative text. E.g.,

> I am a computer, not some lamb or chicken or beef. Can I skip to my at-torney? May I saunter to any solicitor? This is a formidable and essential issue. (Chamberlain, 2007)

The impression of originality and creativity, however, is not long sustained. As Rowe and Partridge (1993, p. 43) report, "This [impression], however, is also short-lived and soon gives way to dissatisfaction and eventual boredom. What is missing? If creativity is about producing something new then Racter should certainly qualify, but it seems that novelty is not enough."[3]

---

[2] Note that "objective" here is simply meant to contrast with psychological; we are making no grand claims about the objectivity of creativity.

[3] *Racter* generates text that seems, at least superficially, to mimic schizophasia. This confused language, or "word salad" is created by the mentally ill with defective linguistic faculties. A short word salad may appear semantically novel. However, further sentences exhaust the possibilities for novelty since they fall within the expected range of incoherent pattern construction.

Rowe and Partridge (1993) describe many other examples in AI of the use of randomness meant to turn the dull into something creative, without success. Lenat's AM and EURISKO, while producing interesting early results, such as a set-theoretic definition of number, soon degenerated into triviality. Similarly, Johnson-Laird (1987) failed to automate jazz. Actually, W. A. Mozart had this idea earlier, with his "Musikalisches Würfelspiel" (musical dice game) for creating waltzes (Earnshaw, 1991). The game produced "pretty minuets"; however, that was achieved by Mozart imposing strong constraints on the possible results, leaving very little scope for creative variation. And that is the correct objection to all of these efforts: while introducing randomness does introduce an element of novelty, it is typically within a very constrained scope, and so rapidly exhausted. The appetite of creativity for novelty is, however, inexhaustible — these failures illustrate the value of our definition in action, rather than argue against it.

The Verstehen Objection. A general objection that might be put to Definition 5 is that it is just too sterile: it is a concept of creativity that, while making reference to culture and context, does so in a very cold, formal way, requiring the identification of probability distributions representing those contexts in order to compute a probability ratio. Whatever creativity is, surely it must have more to do with human Verstehen than that! In response to such an objection we would say that where human Verstehen matters, it can readily and congenially be brought into view. Our definition is neutral about such things, meaning it is fully compatible with them and also fully compatible with their omission. If human Verstehen were truly a precondition for creativity, it would render the creativity of non-human animals and biological evolution impossible by definition, and perhaps also that of computers. Although we might in the end want to come to these conclusions, it seems highly doubtful that we should want to reach these conclusions analytically! A definition allowing these matters to be decided synthetically seems to be highly preferable. Definition 5 provides all the resources needed to accommodate the five basic Facts about creativity, as we have seen, while leaving the Opinions about creativity open. This is exactly as we should like.

The Very Possibility of Creativity. Some might object on the grounds that *everything* that occurs is hugely improbable! Any continuous distribution has probability 0 of landing at any specific point. So long as we look at specific outcomes, specific works of art, at their most extreme specificity — where every atom, or indeed every subatomic particle, is precisely located in space and time — the probability of *that* outcome occurring will be zero *relative to any framework whatsoever*. It follows, therefore, that the ratios of probabilities given new to old frameworks are simply undefined, and our definition is unhelpful.

Strictly speaking, this objection is correct. However, nobody operates at the level of infinite precision arithmetic, which is what is required to identify those absurdly precise outcomes in a continuous state space which have occurred and which have probability zero. The achievement of probability zero on this basis would appear to violate Heisenberg's Uncertainty Principle. Disregarding quantum mechanics, everyone operates at a degree of resolution determined at least by systematic measurement error. In effect, all state spaces are discretized so that the probabilities are *emphatically* not zero. Our definition is already explicitly relative to a cultural context; so, to be perfectly correct, we need also to relativize it to a system of measurement that accords with cultural measurement practices and normal measurement errors.

5   Consequences

5.1   The Irrelevance of Value

Many popular definitions of creativity (e.g., Definition 1) stipulate that a creative pattern must be both appropriate to and valued in the domain. Hofstadter requires that the creative individual's sense of what is interesting must be in tune with that of the masses, thereby ensuring also popularity (Hofstadter, 1995, p. 313). We find this expansion far-fetched and the original connection of creativity to value dubious.

The history of the concept of creativity clearly undermines the idea of popularity as any necessary ingredient in it. Consider the role of women in the history of art and science. The value of their contributions has been systematically underestimated, at least until recently. Concluding that their contributions were *therefore* also less creative than that of their male counterparts would surely be perverse. Many artists and scientists were notoriously unpopular during their times of peak contribution, becoming recognized only after their deaths. Whatever makes an activity creative, it clearly must be connected with that activity itself, rather than occurring *after* the activity has ceased entirely! The value and appropriateness of creative works are subject to the social context in which they are received — not created.

Of course, lack of centrality to the core concept of creativity is no impediment to forming a *combined* concept, say, of valued creativity. And that concept may itself be valuable. But it is the adherence to a separation between creativity and value that allows us to see their combination as potentially valuable.[4]

5.2   The Irrelevance of Appropriateness

As with value, appropriateness has often been considered necessary for creativity, and again we disagree.

Some have claimed that a creative pattern must meet the constraints imposed by a genre. However, as Boden herself notes, a common way of devising new conceptual frameworks is to drop or break existing traditions (Boden, 2004, pp. 71-74)! This recognition of the possibility of "inappropriate" creative works

---

[4] We should also note that omission of the concept of value from our defintion does not imply that value has no role in its *application*. Cultural and other values certainly enter into the choice of domain and the selection of frameworks for them. We are not aiming at some kind of value-free science of creativity, but simply a value-free account of creativity itself.

is again possible only if we avoid encumbering the definition of creativity itself with appropriateness. And Boden's recognition of this is evidence that she has put aside the appropriateness constraint in practice, if not in theory.

5.3   Inferring Frameworks from Patterns

On our definition, a specific pattern provides insufficient information for judging its creativity: a pattern is only creative relative to its generating framework and available alternative frameworks. Although in ordinary discourse there are innumerable cases of objects being described as creative, we suggest that this is a kind of shorthand for an object being produced by a creative process.

5.4   Degrees of Creativity

Following our definition it is not obvious how different people or their works may be ranked by their creativity (Fact 4). An account of the degrees of creativity that is intrinsic in our Definition 5 draws upon the probability that the patterns could have been produced by pre-existing frameworks. A novel framework that can generate patterns that could not, under any circumstances, have been generated prior to its introduction, is highly creative. A novel framework that only replicates patterns generated by preexisting frameworks is not at all creative. A novel framework that produces patterns that are less likely to have been generated by preexisting frameworks is the more creative, with the degree of creativity varying inversely with that probability of having been generated by preexisting frameworks. Finally, the degree of creativity attributable to objects is derivative from the degree of creativity shown by their generating framework.

6   Examples

Creativity in Number Theory. The introduction of zero, negative numbers and the imaginary number i were all creative. With the introduction of these new frameworks, novel patterns were produced, with some of their consequences still being explored. These could not have been generated within the previously existing frameworks of mathematics at all. For instance, positive integers are sufficient for counting objects owned, but the introduction of negative numbers is necessary for the calculation of objects owed. Imaginary numbers have permitted the creation of previously impossible concepts, for instance, the quaternion.

Creativity in the Visual Arts. Some Australian aboriginal visual artists draw or paint what is known about the insides of a creature rather than its skin and fur. By introducing this conceptual framework to visual art, patterns showing x-ray images are generated that are impossible within a framework focused on surfaces. Hence, this "way of seeing" is a creative transformation to introduce into a world that favors the superficial.

The step taken by the Impressionists away from "realism" in art in the 19th century was also creative. They held that art should no longer be valued simply according to its representation of the world as seen from the perspective of an ideal viewer. As already discussed, breaking constraints is a common way of transforming a framework in the way required for creativity.

With more time, we could expand the number and variety of examples of creativity across all the disciplines and fields of endeavor humanity has pursued. Our definition appears to be versatile enough to cope with anything we have thus far considered (but not here documented, due to time and space limits).

7   Conclusion

Here we have confronted some recent definitions of creativity. By invoking the often zealously defended boundaries of culture in general, and of science and art specifically, they have undermined the acknowledgement of the creativity of the heterodox — precisely those we expect to be the most creative.

By freeing creativity from popular and expert opinion, we offer potential means of applying it to machine learning tasks or automated creative discovery, as well as a coherent basis on which to discuss the creativity of natural, and in particular evolutionary, systems. This enables us to use our creativity definition to guide a search through generations of creative offspring, by reference to the probabilities of their appearance before and after the arrival of their parents in a population. We have begun to explore these aspects of creativity, but leave this discussion for another paper.

References

Berry, R., Rungsarityotin, W., Dorin, A., Dahlstedt, P., Haw, C.: Unfinished symphonies: Songs of 3.5 worlds. In: 6th European Conference on Artificial Life, Workshop on Artificial Life Models for Musical Applications, pp. 51–64 (2001)

Boden, M.: The creative mind: Myths and mechanisms, 2nd edn. Routledge, London (2004)

Chamberlain, W.: Getting a computer to write about itself (2007), http://www.atariarchives.org/deli/write_about_itself.php (accessed June 28, 2009)

Colton, S.: Automated theory formation in pure mathematics. Ph. D. thesis, University of Edinburgh, Division of Informatics (2001)

Dahlstedt, P.: Living melodies: Coevolution of sonic communication. In: First Iteration, pp. 56–66. Centre for Electronic Media Art, Monash University (1999)

Dorin, A.: A survey of virtual ecosystems in generative electronic art. In: Romero, J., Machado, P. (eds.) The art of artificial evolution: A handbook on evolutionary art and music, pp. 289–309. Springer, Heidelberg (2008)

Earnshaw, C.: Mozart's dice waltz (1991), http://tamw.atari-users.net/mozart.htm (accessed June 28, 2009)

Hofstadter, D.: Fluid concepts and creative analogies: Computer models of the fundamental mechanisms of thought. Basic Books, New York (1995)

Johnson-Laird, P.: Reasoning, imagining and creating. Bulletin of the British Psychological Society 40, 121–129 (1987)

Korb, K.: Searle's AI program. Journal of Experimental and Theoretical AI 3, 283–296 (1991)

Langley, P., Simon, H., Bradshaw, G., Zytkow, J.: Scientific discovery: Computational explorations of the creative process. MIT Press, Cambridge (1987)

Lenat, D.: EURISKO: A program that learns new heuristics and domain concepts. Artificial Intelligence 21, 61–98 (1983)

McCormack, J.: Eden: An evolutionary sonic ecosystem. In: 6th European Conference on Artificial Life, Advances in Artificial Life, pp. 133–142 (2001)

Miranda, E.: On the evolution of music in a society of self-taught digital creatures. Digital Creativity 14, 29–42 (2003)

Rowe, J., Partridge, D.: Creativity: A survey of AI approaches. Artificial Intelligence Review 7, 43–70 (1993)

Saunders, R., Gero, J.: Curious agents and situated design evaluations. In: Gero, J., Brazier, F. (eds.) Agents in design 2002, pp. 133–149. University of Sydney, Sydney (2002)

Searle, J.: Minds, brains and programs. The Behavioral and Brain Sciences 3, 417–424 (1980)

Searle, J.: The rediscovery of the mind. MIT Press, Cambridge (1992)

Sims, K.: Artificial evolution for computer graphics. Computer Graphics 25, 319–328 (1991)

Todd, S., Latham, W.: Evolutionary art and computers. Academic Press, San Diego (1992)

van Gulick, R.: Functionalism, information and content. Nature and System 2, 139–162 (1980)

Genetically Optimized Architectural Designs for Control of Pedestrian Crowds

Pradyumn Kumar Shukla[1,2]

[1] Institute of Numerical Mathematics, TU Dresden, 01062 Dresden, Germany
[2]Institute AIFB, Universität Karlsruhe, 76133 Karlsruhe, Germany
psh@aifb.uni-karlsruhe.de

Abstract. Social force based modeling of pedestrian crowds is an advanced microscopic approach for simulating the dynamics of pedestrian motion and has been effectively used for pedestrian simulations in both normal and panic situations. A disastrous form of pedestrian behavior is stampede, which is usually triggered in life-threatening situations such as fires in crowded public halls or rush for some large-scale events (like millions praying to the gods at an auspicious time and space). The architectural designs of the hall influence to a large extent the evacuation process. In this paper we apply an advanced genetic algorithm for optimal designs of suitable architectural entities so as to smoothen the pedestrian flow in panic situations. This has practical implications in saving lives/ injuries during a stampede. The effects of these new designs in normal situations are also discussed.

Keywords: design optimization, genetic algorithms, crowd stampedes.

1   Introduction

Crowd stampedes are unfortunately not a rare phenomenon in this world. They occur many times in life-threatening situations such as fires, or in rush situations. Often religious gatherings in major Hindu festivals in India or in Mecca are characterized by excessively large pedestrian gatherings. In these situations many times crowd stampedes occurs, due to numerous reasons. Hence there has been a growing momentum in the past decades to model pedestrians (see [9,6] and references therein). The social force model [6,5], for example, is a widely applied approach, which can model many observed collective behaviors. Also normal and escape situations can be treated by one and the same pedestrian model.

Pedestrian flow both in normal and in panic situations is governed to a large extent by the architectural design of the pedestrian facility. Hence the pedestrian flow can be smoothened by proper placements of suitable architectural entities. Some such interesting designs are discussed in [8]. In this work we apply a genetic algorithm to *find* optimal placement and shapes of architectural design elements so as to make pedestrian flow smooth, and reduce the physical interactions during possible crowd stampedes. This has practical implications in reducing the number of injured persons in a stampede.

K. Korb, M. Randall, and T. Hendtlass (Eds.): ACAL 2009, LNAI 5865, pp. 22–31, 2009.

The paper is organized as follows: Section 2 introduces the social force model for pedestrian modeling. In Section 3 we discuss the problem that we tackle and describe the algorithms to solve them. Simulation results are presented in Section 4 while concluding remarks are made in the last section.

2   A Force Based Panic Model

A physical force based model of pedestrian behavior has been suggested in [5] to investigate panic situations. This model is based on the socio-psychological [7] and physical forces that influence the behavior of a pedestrian in a crowd.

This model for pedestrians assumes that each individual α, having mass m_α is trying to move in a desired direction e_α with a desired speed v_α^0, and that it adapts the actual velocity v_α to the desired one, $v_\alpha^0 = v_\alpha^0 e_\alpha$ within a certain relaxation time τ_α. At the same time he or she also attempts to keep a certain safety distance to other pedestrians β and obstacles i. This is modeled by repulsive forces terms $f_{\alpha\beta}$ and $f_{\alpha i}$. At a given time t, the velocity $v_\alpha(t) = dr_\alpha/dt$ is itself assumed to change according to the acceleration equation

$$m_\alpha \frac{dv_\alpha(t)}{dt} = m_\alpha \frac{1}{\tau_\alpha}(v_\alpha^0 e_\alpha - v_\alpha) + \sum_{\beta(\neq\alpha)} f_{\alpha\beta} + \sum_i f_{\alpha i}. \qquad (1)$$

The repulsive interaction force $f_{\alpha\beta}$ due to other pedestrians is given as

$$f_{r,\alpha\beta} := A_\alpha e^{[(r_{\alpha\beta}-d_{\alpha\beta})/B_\alpha]} n_{\alpha\beta},$$

where $A_{\alpha\beta}$ and $B_{\alpha\beta}$ are constants, $d_{\alpha\beta} := \|r_\alpha - r_\beta\|$ denotes the distance between the pedestrians center of mass, and $n_{\alpha\beta} = (n_{\alpha\beta}^1, n_{\alpha\beta}^2) = (r_\alpha - r_\beta)/d_{\alpha\beta}$ is the normalized vector pointing from pedestrian β to α. If $d_{\alpha\beta}$ is smaller than the sum $r_{\alpha\beta} = (r_\alpha + r_\beta)$ of the radii of the two pedestrians then they touch each other. If the pedestrians touch each other two additional forces are needed in order to accurately describe the dynamical features of pedestrian interactions [5]. First, there is a force counteracting body compression. This is termed as *physical force*. The force is given by:

$$f_{b,\alpha\beta} = \kappa g(r_{\alpha\beta} - d_{\alpha\beta}) n_{\alpha\beta}$$

where the function $g(x)$ is zero if pedestrian do not touch each other, otherwise equal to x. When pedestrians touch each other in addition to body force there is a force impeding relative tangential motion. This is termed as *friction force*. This force is given by:

$$f_{f,\alpha\beta} = k g(r_{\alpha\beta} - d_{\alpha\beta})(\Delta v_{\beta\alpha}(t) \cdot e_{\alpha\beta}(t)) e_{\alpha\beta}(t)$$

where $e_{\alpha\beta}(t) = (-n_{\alpha\beta}^2, n_{\alpha,\beta}^1)$ is the tangential unit vector and $\Delta v_{\beta\alpha}(t) = (v_\beta - v_\alpha) \cdot e_{\alpha\beta}(t)$ is tangential velocity difference. k and κ are large constants. In summary

$$f_{\alpha\beta} = f_{r,\alpha\beta} + f_{b,\alpha\beta} + f_{f,\alpha\beta}.$$

The interaction of the pedestrian α with walls is treated in a similar way. Hence, if $d_{\alpha i}$ denotes the distance to the wall i, $\boldsymbol{n}_{\alpha i}$ denotes the direction perpendicular to it and correspondingly $\boldsymbol{e}_{\alpha i}$ the direction tangential to it, then, the corresponding repulsive force from the wall i is given as

$$\boldsymbol{f}_{\alpha\beta} = A_\alpha e^{[(r_\alpha - d_{\alpha i})/B_\alpha]} \boldsymbol{n}_{\alpha i} + \kappa g(r_\alpha - d_{\alpha i}) \boldsymbol{n}_{\alpha i}$$
$$-kg(r_\alpha - d_{\alpha i})(\boldsymbol{v}_\alpha(t) \cdot \boldsymbol{e}_{\alpha i}(t)) \boldsymbol{e}_{\alpha i}(t).$$

A realistic value of the parameters are taken from [5] as follows. Mass $m_\alpha = 80$ kg, acceleration time $\tau_\alpha = 0.5$s, $A_\alpha = 2 \times 10^3$N, $B_\alpha = 0.08$m, $\kappa = 1.2 \times 10^5$kg s^2 and $k = 2.4 \times 10^5$kg m^{-1}s^{-1}. For simplicity we take the same values for all the pedestrians. The pedestrian diameters $2r_\alpha$ were assumed to be uniformly distributed between 0.5m and 0.7m approximating the distribution of shoulder widths. The desired velocity depends on normal or nervous situations. The observed (and the values used here) are $v_\alpha^0 = 0.6$ms^{-1} under *relaxed*, $v_\alpha^0 = 1.0$ ms^{-1} under *normal* and $v_\alpha^0 > 1.5$ ms^{-1} under *nervous* conditions. In panic situations the desired velocity can reach more than 5 ms^{-1} (till 10 ms^{-1}). These are all empirically obtained values from literature [11,12].

This original model (1) is a set of highly intractable nonlinearly coupled differential equations and is continuous in both space and time. However for solving it we discretized it in time and used a simple Euler method.

3   Problem Description and Related Works

We consider the evacuation of a room of size 15m×15m having a single exit of width 1m. This might be a public hall or inside a temple for example. We assume that there are 200 people in this room and due to some reason they all need to be evacuated. The room architecture is shown in Figure 1. The same figure also shows the clogging of pedestrians near the exit. The figure is for $v_\alpha^0 = 2.0$ ms^{-1}, i.e., when pedestrians are nervous and are in a hurry to leave the room.

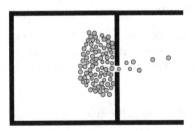
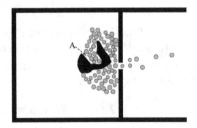

Fig. 1. Room boundaries with an exit. Also shown is the clogging phenomenon due to pedestrians leaving the room from left to right.

Fig. 2. An unused part of the obstacle

The clogging near the exits reduce the outflow and also cause injuries/ fatalities. It has been shown in the original work [5] that the danger of clogging can be minimized and an improved outflow can be reached by placing column asymmetrically in front of the exits. An obstacle in front can also reduce the number of injured persons by taking up pressure from behind. Based on this original idea some works have considered the placement of the suitable obstacles to improve the outflow.

Escobar and Rosa [4] have considered some architectural adjustments so as to increase the pedestrian flow out of the room. Although the results were encouraging no systematic optimization was performed in order to get the optimal obstacle designs and locations. It was more tweaking with different placements and seeing which one gives better results. Johansson and Helbing [10] proposed a genetic algorithm for full scale evolution of the optimal shapes that increase the flow. However the design that sometimes evolved had some unused channels. A sketch of one such design element is shown in Figure 2. One can see that the region A is not an efficient design element. As shown in the figure some fleeing pedestrians are stuck in region A and have to spend more time finding the exit.

Apart from the reasons mentioned in the last paragraph, until now none of the studies have considered the effect of the obstacles in *normal* situations. It is important that efficient design be included in a room architecture so as to minimize the injuries/ evacuation time in case there is an emergency situation and people are panicking. However panic situations are rare are hence the effect of these obstacles in normal everyday situations also needs to be examined. For example on placing a pillar near an exit improves the outflow [5], still we do not see these designs in public halls, or near exists. The reason for this is that these designs might be inefficient in normal situations. Hence every architectural design element needs to be examined for both normal and panic situations.

In the next section we propose two algorithms for efficient designs of suitable architectural entities.

4   Proposed Algorithms

We propose two algorithms for finding optimal shape of the obstacles. Before this we describe the objective function(s) in both panic and normal situations that we optimize. The problem geometry (and number of persons) is as described in the last section.

In panic situations we use $v_\alpha^0 = 2.0$ ms^{-1} and consider the number of persons evacuated in 2 minutes as the objective function that needs to be maximized. Due to the underlying approximation of Equation 1 we call all the designs that have the objective function values within two, compared to the best value, as optimal. Hence if the best design is one in which 120 pedestrians are evacuated in 2 minutes then the design giving 118 pedestrians will also be called optimal.

In normal situations we use $v_\alpha^0 = 1.0$ ms^{-1} in the simulations. In normal situations any obstacles in the room near the exit causes a *discomfort*. This level of discomfort measure D is quantified by the following objective function [8]:

$$D = \frac{1}{N} \sum_\alpha \overline{\frac{(v_\alpha - \overline{v_\alpha})^2}{(v_\alpha)^2}} = \frac{1}{N} \sum_\alpha \left(1 - \frac{\overline{v_\alpha}^2}{(v_\alpha)^2} \right). \tag{2}$$

Here bar denotes a time average and the number of pedestrians $N = 200$. Obviously $0 \le D \le 1$ and D needs to be minimized. The measure D reflects the frequency and degree of sudden velocity changes or the level of discontinuity of walking because of necessary avoidance maneuvers due to pedestrians and obstacles. The level of discomfort is measured over the total evacuation time of 200 pedestrians.

The above problem is multi-objective in nature. However we solve for the panic situation and also report the values of discomfort. These values help a planner to know various feasible designs in panic situations and the corresponding level of discomfort in normal situations.

We describe two kinds of design elements and correspondingly optimize them. In the first we use pillars with their location and the radii as the design variables. We investigate the use of a single pillar and two pillars. Pillars are the most common architectural elements and hence it is easy for the designer to put them at appropriate places. In this study circular pillars were preferred over square pillars after an initial simulation circular pillars were found to be more efficient than square ones (also intuitive). We call the first method as $\mathcal{M}1$. In the next method the optimal shape is free to evolve. This is based using a method suggested in [3]. We call this second method as $\mathcal{M}2$.

In method $\mathcal{M}1$, we used a real coded genetic algorithm. The maximum generation number is set to be 100 while the population size to be 10 times the number of variables, see [1]. So for the single pillar case there are 3 variables (two space coordinates and radii as the third variable) and for the two pillar case there are a total of 6 variables. The tournament selection operator, simulated binary crossover (SBX) and polynomial mutation operators [1] are used. The crossover probability used is 0.6 and the mutation probability is 0.33 for single pillar case and 0.17 for the two pillar situation. We use the distribution indices [2] for crossover and mutation operators as $\eta_c = 20$ and $\eta_m = 20$, respectively. For statistical accuracy, for each configuration 100 runs were performed and the average of the objective function(s) was taken. We did two version of this method one where we did not allowed overlapping of pillars and in the second we allowed this. Both of these versions are easily implementable with the help of constraints. We use the parameter free constraint handling scheme described in [1].

The second method $\mathcal{M}2$ is based on a boolean grid representation taken from [3]. As the region near exits has the highest potential for pedestrian fatalities/injuries during an emergence evacuation, we consider a region R of size 5m×5m near the exit symmetrically located, see Figure 3. Next we divide this region into square grids of size 0.25m×0.25m. Hence we have a total of 400 small grids. The presence and absence of an obstacle is denoted by 1 and 0 respectively. Hence we use a binary coding scheme for describing a particular shape, with a left to right coding scheme as shown in Figure 4. To a basic skeleton we

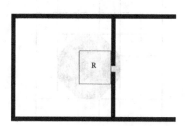

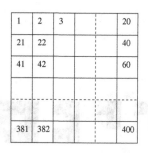

Fig. 3. Region R of size 5m×5m near the exit is divided into grids

Fig. 4. Grid representation

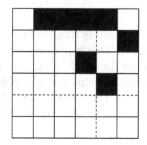

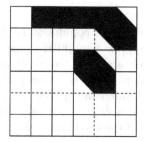

Fig. 5. Skeleton of a possible shape

Fig. 6. Final smoothened shape of the basic skeleton

use triangular elements to smoothen the shape. This is illustrated by shaded regions for different cases in Figure 7. Using such smoothing scheme we get a smoothened version in Figure 6 for a skeleton shown in Figure 5, for example. Since simulation using such a scheme might give rise to unwanted small islands of few elements we ignore any connected pieces having less than 5 elements. Two elements are said to be connected if they have at least one common corner. A limited simulation study has shown that the smoothened shapes without any connected pieces having less than 5 elements gives a better value of both the objective functions.

As a mutation operator we use a bit-wise with a probability of 1/string-length. The crossover operator is as follows [1]. Swapping between rows or column is decided with a probability of 0.5. Each row or column is swapped with a probability $0.95/d$, where d is the number of rows or columns, depending on the case. A population size of 100 is used and the maximum number of generations is set to be 250.

We found that even after 250 generations (i.e, 250×100 function evaluations), the method $\mathcal{M}2$ did not converge (the best objective function values obtained from $\mathcal{M}2$ were less than 80% of the corresponding ones from method $\mathcal{M}1$). This happens since the number of grids is large and it is a 400 dimensional variable search space. Thus without using some problem information we need large number of generations. In order to alleviate this difficulty in convergence we

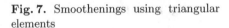

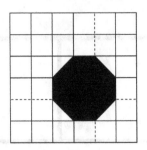

Fig. 7. Smoothenings using triangular elements

Fig. 8. Randomly placed pillar type obstacles are added in the initial population

create an initial population having few (< 5, integer random number) members as the approximate pillar shapes as shown in Figure 8. This is because we know intuitively and from the results in [5] that a pillar near the exit helps the outflow. Simulations show that using this problem information helps the algorithm a lot for faster convergence to optimal solutions. As in the earlier method, for statistical accuracy, for each configuration 100 runs were performed and the average of the objective function(s) was taken.

5   Simulation Results

In this section we show and discuss the simulation results. The pillar based method $\mathcal{M}1$ gives five designs A1, B1, C1, D1 and E1 as shown in Figure 9 to Figure 13.

Table 1 presents the values of the two objective using method $\mathcal{M}1$. Before we discuss the results we present the designs from the second method $\mathcal{M}2$. It gives four designs A2, B2, C2 and D2 as shown in Figure 14 to Figure 17.

Table 2 presents the values of the two objective using method $\mathcal{M}2$. The simulation results can be summarized as follows:

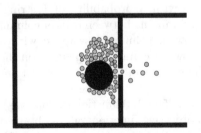
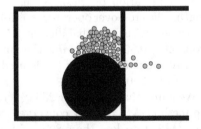

Fig. 9. Design A1: Optimal placement of a single non-overlapping pillar near the exit. The clogging effect is now not seen.

Fig. 10. Design B1: Optimal placement of a single overlapping pillar near the exit. The clogging effect is now not seen.

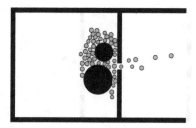

Fig. 11. Design C1: Optimal placement of two non-overlapping pillars near the exit. The clogging effect is now not seen.

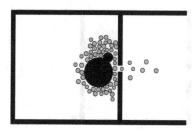

Fig. 12. Design D1: An optimal placement of two overlapping pillars near the exit. The clogging effect is now not seen.

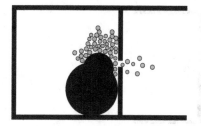

Fig. 13. Design E1: Another optimal placement of two overlapping pillars near the exit. The clogging effect is now not seen.

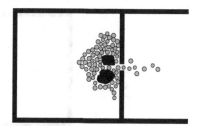

Fig. 14. Design A2: Two almost circular obstacles near the exit emerge from optimization. The clogging effect is now not seen.

1. Design elements for panic situations show that the problem is multi-modal in nature as seemingly different designs are almost equally good.
2. Funnel shape near the exists helps against clogging. In this case the shape is such that one person can easily pass, and hence preventing clogging.
3. Zig-zag shapes helps for panic situations however it comes at the cost of more discomfort, so there is a trade-off between design for panic vs. normal situations.
4. Asymmetry has always been observed. In method $\mathcal{M}1$ there is bigger chance of a symmetric location of pillars than in method $\mathcal{M}2$ however still the optimal solutions never show this.

Table 1. Values of objective functions using method $\mathcal{M}1$

	Pedestrians evacuated	Discomfort
A1	139.82	0.113
B1	141.16	0.253
C1	140.51	0.165
D1	139.67	0.179
E1	141.38	0.272

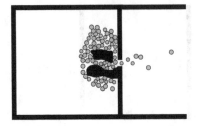

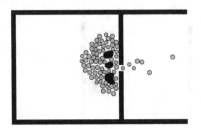

Fig. 15. Design B2: Two rounded-rectangular obstacles near the exit emerge from optimization. The clogging effect is now not seen.

Fig. 16. Design C2: Three obstacles near the exit emerge from optimization. The clogging effect is now not seen.

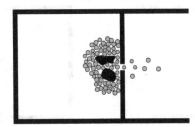

Fig. 17. Design D2: Two obstacles near the exit emerge from optimization. The clogging effect is now not seen.

5. The trade-off between designs in panic and normal situations has been evaluated for the first time and this information is useful for the designer to select the best design.

6   Conclusions and Extensions

We believe (and also the simulation results show) that the problem of efficient design of pedestrian facilities is a multi-objective one. The objectives can be many, such as maximizing the flow, minimizing the discomfort or reducing the number of injured persons among others. The objective function for this problem comes from a set of highly nonlinear coupled differential equations. This together with multi-modality of solutions makes this problem an excellent example for

Table 2. Values of objective functions using method $\mathcal{M}2$

	Pedestrians evacuated	Discomfort
A2	140.84	0.168
B2	141.86	0.254
C2	139.91	0.159
D2	141.69	0.187

genetic and other such heuristics algorithms. An extension of our study is to use a multi-objective evolutionary algorithm like NSGA-II [1] to find the complete set of Pareto-optimal solutions. We hope that this study will stimulate more research in this area.

Acknowledgements

The author acknowledges discussions with Dirk Helbing and Anders Johansson.

References

1. Deb, K.: Multi-objective optimization using evolutionary algorithms. Wiley, Chichester (2001)
2. Deb, K., Agarwal, R.B.: Simulated binary crossover for continuous search space. Complex Systems 9, 115–148 (1995)
3. Deb, K., Goel, T.: Multi-Objective Evolutionary Algorithms for Engineering Shape Design. In: Sarker, R., Mohammadian, M., Yao, X. (eds.) Evolutionary Optimization, pp. 146–175. Kluwer, New York (2002)
4. Escobar, R., De La Rosa, A.: Architectural design for the survival optimization of panicking fleeing victims. In: Banzhaf, W., et al. (eds.) ECAL 2003. LNCS (LNAI), vol. 2801, pp. 97–106. Springer, Heidelberg (2003)
5. Helbing, D., Farkas, I., Vicsek, T.: Simulating dynamical features of escape panic. Nature 407, 487–490 (2000)
6. Helbing, D., Molnár, P.: Social force model for pedestrian dynamics. Physical Review E 51, 4282–4286 (1995)
7. Helbing, D., Molnár, P.: Social force model for pedestrian dynamics. Phys. Rev. E 51, 4282–4286 (1995)
8. Helbing, D., Molnár, P., Farkas, I.J., Bolay, K.: Self-organizing pedestrian movement. Environment and Planning B: Planning and Design 28, 361–383 (2001)
9. Hughes, R.L.: A continuum theory for the flow of pedestrians. Transportation Research B 36, 507–535 (2002)
10. Johansson, A., Helbing, D.: Pedestrian flow optimization with a genetic algorithm based on boolean grids. In: Waldau, N., Gattermann, P., Knoflacher, H., Schreckenberg, M. (eds.) Pedestrian and Evacuation Dynamics 2005, pp. 267–272. Springer, Heidelberg (2007)
11. Predtetschenski, W.M., Milinski, A.I.: Personenströme in Gebäuden - Berechnungsmethoden für die Projektierung. Müller, Köln-Braunsfeld, Germany (1971)
12. Weidmann, U.: Transporttechnik der Fussgänger. Schriftenreihe des Instituts für Verkehrsplanung, Transporttechnik Straßen- und Eisenbahnbau Heft 90, ETH Zürich, Switzerland (1993)

Co-evolutionary Learning in the N-player Iterated Prisoner's Dilemma with a Structured Environment

Raymond Chiong and Michael Kirley

Department of Computer Science and Software Engineering,
The University of Melbourne, Victoria 3010, Australia
{rchiong,mkirley}@csse.unimelb.edu.au

Abstract. Co-evolutionary learning is a process where a set of agents mutually adapt via strategic interactions. In this paper, we consider the ability of co-evolutionary learning to evolve cooperative strategies in structured populations using the N-player Iterated Prisoner's Dilemma (NIPD). To do so, we examine the effects of both fixed and random neighbourhood structures on the evolution of cooperative behaviour in a lattice-based NIPD model. Our main focus is to gain a deeper understanding on how co-evolutionary learning could work well in a spatially structured environment. The numerical experiments demonstrate that, while some recent studies have shown that neighbourhood structures encourage cooperation to emerge, the topological arrangement of the neighbourhood structures is an important factor that determines the level of cooperation.

Keywords: N-player iterated prisoner's dilemma, co-evolutionary learning, neighbourhood structure.

1 Introduction

The N-player Iterated Prisoner's Dilemma (NIPD), an extension to the fascinating Iterated Prisoner's Dilemma (IPD) game that has been discussed in many studies (see [1-4]), is one of the many models the artificial life community has adopted as a metaphor to understand the evolution of cooperation. Many real-world scenarios often involve interactions among various parties such that the dominating option for each party will result in a non Pareto-optimal outcome for all parties. As such, the NIPD model is considered by many to have greater generality and applicability to real life situations than the IPD, as many of the real life problems can be represented with the NIPD paradigm in a much realistic way [5, 6].

Despite the great generality and wide applicability of NIPD, little attention has been devoted to it in comparison to the vast amount of work undertaken in the IPD domain. Previous research in NIPD had studied the impact of the number of players [6], payoff function [7], neighbourhood size [7], history length [8], localisation issue [8], population structure [9], generalisation ability [6, 10], forgiveness in co-evolving strategies [11], trust [12], cultural learning [13], and so on. Most of these existing works show that cooperation is unlikely to emerge in NIPD. Even if it does, it is not likely to be stable.

K. Korb, M. Randall, and T. Hendtlass (Eds.): ACAL 2009, LNAI 5865, pp. 32–42, 2009.

Recently, there has been much interest in studying evolutionary games in structured populations, where who-meets-whom is not random but determined by spatial relationships or social networks (see [14-17] for a glimpse of these studies). Many researchers have shown that this kind of spatialised environment is beneficial to the evolution of cooperation. In the context of NIPD, we have seen the importance of neighbourhood structure [18] and community structure [19-20] in promoting cooperative behaviour. In particular, the authors of [19] demonstrated that cooperation can be sustained even with the presence of noise when community structure is in place. These studies give a strong indication that structures are indeed important for the NIPD model. However, questions arise regarding the topological arrangement of the structures: Does the existence of spatial structure unconditionally promote cooperation? Is there a difference between fixed neighbourhood structures and random neighbourhood structures? What will happen when we vary the group size (i.e. N)? Does the group size affect the co-evolutionary process in structured populations? These are among the questions we would try to answer here.

In this paper, we explore the ability of co-evolutionary learning to evolve cooperative strategies in a lattice-based NIPD model. In [18], [19] and [20], better strategies were simply imitated by others, and constant group size was used. Our work differs from these studies in that we reproduce strategies genetically, i.e. rather than 'cultural' imitation we use an evolutionary based model. We also compare and contrast alternative neighbourhood structures with varying group sizes. By examining the effects of these distinctive neighbourhood structures on the evolution of cooperation, we hope to gain a deeper understanding on how co-evolutionary learning could work well in an environment with spatial constraints.

The rest of this paper is organised as follows: Section 2 briefly introduces some background information of NIPD. In Section 3, we present the details of our lattice model. Section 4 describes the experimental settings and results. Finally, we draw conclusion in Section 5 and highlight potential future work.

2 N-player Iterated Prisoner's Dilemma

The NIPD is an extension of the conventional IPD game in which a group of N players, where N is greater than 2, will continuously interact with each other. In an abstract manner, N players make decision independently on two actions, either cooperate or defect, without knowing other players' choices. A rule exists that rewards a benefit which increases when more players are cooperating. There is, however, always a cost for the cooperators.

The scenario can be further illustrated with two simple payoff functions: Let cooperate be C and defect be D, a score is attained by each player based on the payoff functions $C(i) = i/N$ for cooperation and $D(i) = (i + t)/N$ for defection, where i is the number of cooperators and t is the total cost for cooperation $(1 < t < N)$ [9]. During the game, players are engaged with one another iteratively, competing for higher scores based on these payoff functions.

Yao and Darwen [6] were among the first to use an evolutionary approach for the NIPD. In their initial work, they showed that when the number of players in an NIPD game increases, it becomes more difficult for cooperation to emerge. This is mainly due to the fact that a player is unable to clearly distinguish who the defectors and who the cooperators are. Unlike the IPD game, a player can easily reciprocate against a defector in a one-to-one situation, and therefore discourage defection. In NIPD, retaliation against a defector means punishment to everyone else in the game, including the cooperators. This makes the cooperative act an unattractive choice. As long as there is one player who defects, those who cooperate will be worse off than the defector. Accordingly, defection turns out to be the dominant strategy [6, 7, 8, 10].

3   The Model

In order to study the impact of different neighbourhood structures on the evolution of cooperation in the NIPD model, a two-dimensional multi-agent grid-world resembling cellular automata is created. Based on the strategies the agents have, they can either defect or cooperate. Every agent's score is determined not only by its own action alone but also by the actions of other agents in its neighbourhood. For instance, if an agent who cooperates is pitting against k other agents and i of those are cooperators, then its payoff is $bi - ck$, where b is the benefit of the altruistic act and c is the cost for cooperation. On the other hand, an agent who defects does not bear any cost but receives benefit from neighbouring agents who cooperate. Hence, its payoff is bi. In this paper, b and c are set to be constants, i.e. 3 and 1 respectively.

3.1   The Neighbourhood Structures

The grid-world is implemented in the form of cellular automata to simulate agents with different strategies. The grid is formed by a two-dimensional array with overlapping edges, which means every agent on the grid has eight immediate neighbouring agents, including those at the edges of the grid. For the purpose of this study, two fixed neighbourhood structures are created specifically for group sizes N = 4 and N = 5, as depicted in Figure 1. In these fixed structures, each agent is designed to interact with the same neighbouring agents throughout a game.

Apart from the fixed neighbourhood structures, we also use random neighbourhood structures, where the neighbouring agents are randomly picked. Figure 2 shows the examples of random structures with group sizes of 4 and 5 respectively. It is necessary to note that in the random neighbourhood structures, the neighbouring agents are being changed in every generation.

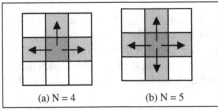

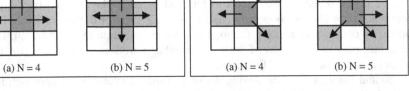

Fig. 1. Fixed neighbourhood structures   **Fig. 2.** Random neighbourhood structures

3.2 Strategy Representation

There are various ways to represent the agent's game-playing strategies. We have decided to adopt the representation developed by Yao and Darwen [6] as it is exponentially much shorter and easier to implement than the others.

Under this representation, a history of l rounds for an agent can be represented as the combination of the following:

- l bits to represent the agent's l previous actions. Here, a '1' indicates a defection, and a '0' a cooperation.
- $l * \log_2 N$ bits to represent the number of cooperators in the previous l rounds among the agent's N-1 group members. Here, N is the group size.

In our implementation, we have limited the number of previous actions in memory to 3 (i.e. $l = 3$). In the case of N = 4, the history for an agent would therefore be 3 + 3 $\log_2 4 = 9$ bits long based on the above representation scheme.

For example, an agent in our NIPD model could have a history as follows:

110 11 01 10

Here, the first three bits from the left are the agent's previous three actions. From this we can see that the agent defected in the last two rounds and cooperated the round before that. The two-bit sets after the first three bits represent the number of cooperators in the last three rounds from the agent's group members. This agent's history indicates that there were 3, 1 and 2 cooperators in the agent's group in the last three rounds.

An agent's strategy provides a move in response to every possible history. Therefore, when N = 4 the strategy should be at least $2^9 = 512$ bits in length. The larger the group size is, the more bits we need for the representation. Due to the fact that there is no memory of previous rounds at the beginning, we have added additional three bits to each strategy to compensate for the lack of complete history in the first three rounds. This means that the actions in the first three rounds of each generation are hard-coded into the strategy. Thereafter, the moves are made based on the history of the agent and its group members.

3.3 Genetic Operations

After devising a way for the strategy representation, an initial population of agents with random strategies is created. These strategies are being evolved from generation to generation via evolutionary learning. Genetic modification is achieved through both crossover and mutation.

For crossover, a random number is generated to determine whether it should take place. Two-point crossover with rank-based selection is used, where 60% of the best strategy within a group is being selected and recombined with 40% of the current strategy of an agent. Note that this will happen only when the crossover rate is satisfied and the current strategy is ranked below the elite group (in this study, strategies that ranked among the top 50% are considered to be in the elite group). Otherwise, nothing comes about. This elite preserving mechanism ensures that good strategies are being carried forward to the next generation.

As with crossover, a random number is generated to determine whether the strategy will be mutated. For mutation, a random position in the strategy's bit representation is selected and the bit at that position is flipped.

In all our experiments, we set the crossover rate to 0.7 and the mutation rate to 0.05.

4   Experiments and Results

In this section, we present the experiments undertaken and discuss the corresponding results. As our main focus is to examine the effectiveness of co-evolutionary learning in a spatialised environment with fixed and random neighbourhood structures, three separate sets of experiments are conducted: the first being the fixed and random neighbourhood structures for $N = 4$, and the second on the fixed and random neighbourhood structures for $N = 5$; in the third experiment, we vary the group sizes from $N = 3$ to 9 with random neighbourhood structures to see how different group sizes affect the co-evolutionary process. For comparison purposes, among the random neighbourhood structures we have inner and outer neighbourhoods. By inner neighbourhood it means the group members will be selected only from within the eight immediate neighbouring agents. On the other hand, with the outer neighbourhood the group members can be selected from anywhere across the entire population. For all these experiments, we use a population size of 100 agents in which all the agents are inhibited on a 10 x 10 grid-world. All the agents are to play against one another iteratively for 1000 generations, with 100 rounds of learning process constituting each generation. Scoring in the game is based on the payoff functions mentioned at the beginning of Section 3.

4.1   Fixed and Random Neighbourhood Structures with $N = 4$

In the fixed neighbourhood structure with $N = 4$, each agent has three group members with whom it interacts, one immediately above itself, one to its left and another to its right (see Figure 1a). Meanwhile, the agent's group members will also have other different groups associating to them. This allows overlapping between different groups of agents during the game.

Figures 3, 4 and 5 show the experimental results from 10 independent runs with 100 agents. As can be seen, cooperation is highly dominant when fixed neighbourhood structure is in place. In most of the runs, almost all the agents were cooperating after 500 generations. There were some instances where cooperation was not as dominant but there were always more cooperators than defectors.

In contrast, cooperation becomes less dominant when random neighbourhood structures are used, as shown in Figures 4 and 5. Particularly, defection prevails when the group members are being selected randomly from anywhere. These results indicate that the fixed neighbourhood structure is highly conducive for cooperation to emerge. The fact that an agent is also a member of its group members' interaction groups makes it possible for agents to retaliate against defectors and reward cooperators. An agent can directly influence the score of two of its group members, thus significantly increases its bargaining power against those group members. This characteristic of the neighbourhood structure, we believe, leads to cooperation.

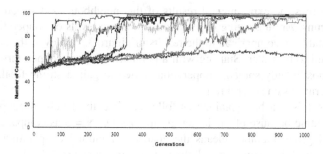

Fig. 3. The number of cooperators with fixed neighbourhood structure (N = 4) over 10 runs

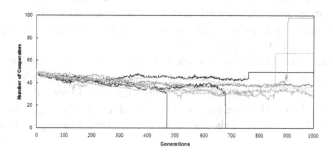

Fig. 4. The number of cooperators with random neighbourhood structures (N = 4) based on immediate neighbours over 10 runs

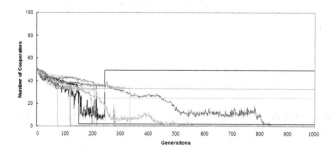

Fig. 5. The number of cooperators with random neighbourhood structures (N = 4) based on the entire population over 10 runs

4.2 Fixed and Random Neighbourhood Structures with N = 5

In the case of the fixed neighbourhood structure with N = 5, each agent has one group member to its right, one to its left, one below it, and another one above it (see Figure 1b). As with the fixed neighbourhood structure when N = 4, the group of the agent and the groups of its neighbours are overlapping. The only difference being the additional agent below it.

Figure 6 shows the experimental results of this neighbourhood structure from 10 independent runs. An interesting observation is that the levels of cooperation became slightly less stable in comparison to the results in Figure 3. However, the convergence speed appeared to be faster. Starting with a population where cooperators and defectors were almost evenly spread, cooperation started to gain a strong foothold in less than 100 generations on several runs.

We believe this can be explained as follows: First, the levels of cooperation are slightly less stable because of the larger group size, i.e. $N = 5$ as compared to $N = 4$. An additional group member means the chances of mutual cooperation within the group become slightly lesser. Second, the faster convergence speed is likely due to the fact that there is more overlapping between different groups in $N = 5$ than $N = 4$. More overlapping means information can be propagated quicker. As no agent has any global information about the payoffs, and no communication is allowed among the agents, co-evolutionary learning is relying solely on the interactions among agents to learn the best strategies for playing the game.

For the random neighbourhood structures, we can see from Figures 7 and 8 that the levels of cooperation have increased greatly as compared to the results in Figures 4 and 5. Even though several runs still ended up with all defectors, there were also a number of runs where full cooperation was nearly achieved. The larger the group size, the smaller the randomness. This allows co-evolutionary learning to learn slightly better.

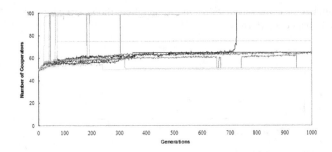

Fig. 6. The number of cooperators with fixed neighbourhood structure ($N = 5$) over 10 runs

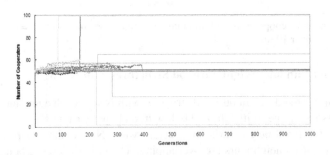

Fig. 7. The number of cooperators with random neighbourhood structures ($N = 5$) based on immediate neighbours over 10 runs

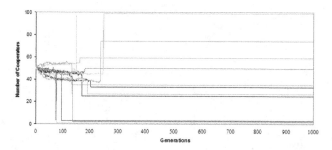

Fig. 8. The number of cooperators with random neighbourhood structures (N = 5) based on the entire population over 10 runs

4.3 Random Neighbourhood Structures with Varying Group Sizes

As aforementioned, we have two types of random neighbourhood structures: inner and outer. For the inner neighbourhood, each agent's group members are selected randomly from its eight immediate neighbours, whereas for outer neighbourhood group members are selected from anywhere in the population. These group members are changed dynamically at every generation. Due to the nature of this kind of selection, the chances that there is overlap between an agent's group and those of its group members are minimal.

Table 1. Results (percentages of cooperation and standard errors) with different types of neighbourhood structures for N = 4 and N = 5 averaged over 30 runs

N	Fixed Neighbourhood Structures	Random Neighbourhood Structures (Inner)	Random Neighbourhood Structures (Outer)
4	83.78 ± 2.19	50.64 ± 3.88	22.80 ± 2.70
5	80.09 ± 3.26	63.52 ± 5.08	44.03 ± 4.50

Table 1 shows the overall results with different types of neighbourhood structures for N = 4 and N = 5 based on the average of 30 runs. It is clear that fixed neighbourhood structures favour cooperation over defection, with averages of 80% or more cooperation for both group sizes. For random neighbourhood structures with inner neighbourhood, cooperators and defectors tow the middle line with averages of 50% and 63% respectively. In the random neighbourhood structures with the outer neighbourhood, defection is rampant, especially in the case of N = 4. It is simple to see why defection is so dominant here. This kind of interactions offers anonymity, where the chances of the same group members meeting one another again are extremely slim. As there is little fear of retaliation, there is no incentive for cooperation. This result concurs with those in [18] and [19] when the neighbourhood or community structures are random or non-existent.

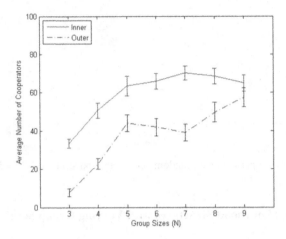

Fig. 9. The number of cooperators in random neighbourhood structures with varying group sizes averaged over 30 runs

Finally, Figure 9 shows the experimental results of random neighbourhood structures with N varying from 3 to 9 for inner and outer neighbourhoods. Generally, we see that cooperative behaviour is much stronger with inner neighbourhood than outer neighbourhood. The inner neighbourhood is able to maintain cooperation of at least 50% or more when the group size is greater than 4. Only when N is 3, the population is less cooperative. This makes perfect sense because when the group size is 3, an agent only interacts with another two group members selected randomly from the eight immediate neighbours. This setting is poor for any information to be conveyed effectively.

As for the outer neighbourhood, we see an increase of the levels of cooperation when the group sizes increase. This implies that group size does not seem to have much impact on the evolution of cooperation in spatialised environment, and this finding is consistent with those of Seo et al. [7-8] when they studied the localisation issues in NIPD. When the group size is larger, co-evolutionary learning is able to learn from more strategies within the population. This allows information to be propagated slightly wider than when the group size is smaller. Nevertheless, cooperation rarely becomes dominant with this setting.

5   Conclusion

In this paper, we have considered the ability of co-evolutionary learning to evolve cooperative behaviour in a structured version of the NIPD game. While confirming that structured models are indeed good for the evolution of cooperation, we demonstrated via our empirical studies that the underlying structures and their topological arrangements are important factors to be considered for the levels of cooperation desired. We observed that an overlapping foundation is highly critical for information to propagate across a structured environment in an effective manner. Meanwhile, this overlapping architecture is also influencing the convergence rate. In other words, sufficient overlapping is good, but extensive overlapping could be destructive.

While this paper uses the NIPD game to examine the efficacy of co-evolutionary learning in a structured environment, we believe the results can be generalised to other examples or models. Co-evolutionary learning is more than capable in solving more complex and dynamic problems. Hence, future work will involve the application of co-evolutionary structured models to real-world optimisation tasks.

References

1. Axelrod, R.: The Evolution of Cooperation. Basic Books, New York (1984)
2. Axelrod, R.: The Evolution of Strategies in the Iterated Prisoner's Dilemma. In: Davis, L. (ed.) Genetic Algorithms and Simulated Annealing, pp. 32–41. Morgan Kaufmann, Los Altos (1987)
3. Lindgren, K.: Evolution Phenomena in Simple Dynamics. In: Langton, C.G., Taylor, C., Farmer, J.D., Rasmussen, S. (eds.) Artificial Life II: Proceedings of the 2nd Interdisciplinary Workshop on the Synthesis and Simulation of Living Systems, pp. 295–312. Addison-Wesley, Reading (1991)
4. Fogel, D.B.: Evolving Behaviors in the Iterated Prisoner's Dilemma. Evolutionary Computation 1(1), 77–97 (1993)
5. Davis, J.H., Laughlin, P.R., Komorita, S.S.: The Social Psychology of Small Groups. Annual Review of Psychology 27, 501–542 (1976)
6. Yao, X., Darwen, P.: An Experimental Study of N-person Iterated Prisoner's Dilemma Games. Informatica 18(4), 435–450 (1994)
7. Seo, Y.G., Cho, S.B., Yao, X.: The Impact of Payoff Function and Local Interaction on the N-player Iterated Prisoner's Dilemma. Knowledge and Information Systems: An International Journal 2(4), 178–461 (2000)
8. Seo, Y.G., Cho, S.B., Yao, X.: Emergence of Cooperative Coalition in NIPD Game with Localization of Interaction and Learning. In: Proceedings of the 1999 Congress on Evolutionary Computation, pp. 877–884. IEEE Press, Piscataway (1999)
9. Suzuki, R., Arita, T.: Evolutionary Analysis on Spatial Locality in N-Person Iterated Prisoner's Dilemma. International Journal of Computational Intelligence and Applications 3(2), 177–188 (2003)
10. Yao, X.: Automatic Acquisition of Strategies by Co-evolutionary Learning. In: Proceedings of the International Conference on Computational Intelligence and Multimedia Applications, Gold Coast, Australia, pp. 23–29. Griffith University Press, Brisbane (1997)
11. O'Riordan, C., Griffith, J., Newell, J., Sorensen, H.: Co-evolution of Strategies for an N-player Dilemma. In: Proceedings of the 2004 Congress on Evolutionary Computation, pp. 1625–1630. IEEE Press, Piscataway (2004)
12. Birk, A.: Trust in an N-player Iterated Prisoner's Dilemma. In: Proceedings of the International Conference on Autonomous Agents, 2nd Workshop on Deception, Fraud and Trust in Agent Societies, Seattle, WA (1999)
13. O'Riordan, C., Griffith, J., Curran, D., Sorensen, H.: Norms and Cultural Learning in the N-player Prisoner's Dilemma. In: Proceedings of the 2006 Congress on Evolutionary Computation, pp. 1105–1110. IEEE Press, Piscataway (2006)
14. Ohtsuki, H., Hauert, C., Lieberman, E., Nowak, M.A.: A Simple Rule for the Evolution of Cooperation on Graphs and Social Networks. Nature 441, 502–505 (2006)

15. Taylor, P.D., Day, T., Wild, G.: Evolution of Cooperation in a Finite Homogeneous Graph. Nature 447, 469–472 (2007)
16. Lozano, S., Arenas, A., Sánchez, A.: Mesoscopic Structure Conditions the Emergence of Cooperation on Social Networks. PLoS ONE 3(4), e1892 (2008)
17. Tarnita, C.E., Antal, T., Ohtsuki, H., Nowak, M.A.: Evolutionary Dynamics in Set Structured Populations. Proceedings of the National Academy of Sciences of the United States of America 106(21), 8601–8604 (2009)
18. Chiong, R., Dhakal, S., Jankovic, L.: Effects of Neighbourhood Structure on Evolution of Cooperation in N-Player Iterated Prisoner's Dilemma. In: Yin, H., Tino, P., Corchado, E., Byrne, W., Yao, X. (eds.) IDEAL 2007. LNCS, vol. 4881, pp. 950–959. Springer, Heidelberg (2007)
19. O'Riordan, C., Sorensen, H.: Stable Cooperation in the N-Player Prisoner's Dilemma: The Importance of Community Structure. In: Tuyls, K., Nowé, A., Guessoum, Z., Kudenko, D. (eds.) ALAMAS 2005, ALAMAS 2006, and ALAMAS 2007. LNCS (LNAI), vol. 4865, pp. 157–168. Springer, Heidelberg (2008)
20. O'Riordan, C., Cunningham, A., Sorensen, H.: Emergence of Cooperation in N-Player Games on Small World Networks. In: Bullock, S., Noble, J., Watson, R., Bedau, M.A. (eds.) Artificial Life XI: Proceedings of the 11th International Conference on the Simulation and Synthesis of Living Systems, pp. 436–442. MIT Press, Cambridge (2008)

Evolving Cooperation in the N-player Prisoner's Dilemma: A Social Network Model

Golriz Rezaei, Michael Kirley, and Jens Pfau

Department of Computer Science and Software Engineering
The University of Melbourne, Australia
{grezaei,mkirley,jens.pfau}@csse.unimelb.edu.au

Abstract. We introduce a social network based model to investigate the evolution of cooperation in the N-player prisoner's dilemma game. Agents who play cooperatively form social links, which are reinforced by subsequent cooperative actions. Agents tend to interact with players from their social network. However, when an agent defects, the links with its opponents in that game are broken. We examine two different scenarios: (a) where all agents are equipped with a pure strategy, and (b) where some agents play with a mixed strategy. In the mixed case, agents base their decision on a function of the weighted links within their social network. Detailed simulation experiments show that the proposed model is able to promote cooperation. Social networks play an increasingly important role in promoting and sustaining cooperation in the mixed strategy case. An analysis of the emergent social networks shows that they are characterized by high average clustering and broad-scale heterogeneity, especially for a relatively small number of players per game.

1 Introduction

Social dilemma games such as the prisoner's dilemma [1,13] have provided important insights into the emergent properties of interactions in multi-agent systems. However, in order to investigate cooperation within a social group, games consisting of more than two players must be considered. In the N-player prisoner's dilemma [3,16,15], multiple agents $(N \geq 2)$ interact within their designated group and must choose to *cooperate* or *defect*. Any benefit or payoff is received by all participants; any cost is borne by the cooperators only. Hardin [8] describes the N-player prisoner's dilemma as a "tragedy of the commons" game in which the players are worse off acting according to their self interests than if they were cooperating and coordinating their actions.

Recently, a number of studies have investigated the evolution of cooperation in 2×2 games on dynamical networks, where the interaction links between agents playing the game varied over time. The models have ranged from comparative studies examining the level of cooperation on different base network models [11] to endogenous network formation models based on local interactions [2]. In Tanimoto [17], individuals were able to self-organize both their strategy and their social ties throughout evolution, based exclusively on their self-interest. It

K. Korb, M. Randall, and T. Hendtlass (Eds.): ACAL 2009, LNAI 5865, pp. 43–52, 2009.
© Springer-Verlag Berlin Heidelberg 2009

was reported that the clustering coefficient of the network (see Section 4) affects the emergence of cooperation in the games.

In this paper, we extend this line of research by focussing on the N-player prisoner's dilemma game based on the formalism of Boyd and Richerson [3]. We propose an endogenous network formation model. Successful strategies spread via a form of cultural evolution based on imitation. When agents play cooperatively, they form a social link that is reinforced each time the action is repeated. This weighted link provides an estimate of the "reliability" of the agent. When an agent defects, all links with its opponents in that game are dissolved. We examine different levels of agent cognitive ability: agents with a pure strategy always play cooperate or defect. In contrast, an agent equipped with a mixed strategy plays a particular action based on a function of the reliability of other agents in its group in a given game. Thus, individual agents are able to adjust both these links and the action they play based on interactions with other agents.

Detailed simulation experiments show that our model is able to promote higher levels of cooperation as compared with panmictic populations. An analysis of the social interactions between cooperative agents reveals high average clustering and associated single-to-broad-scale heterogeneity for relatively small values of N. When agents are equipped with mixed strategies, social networks play a more significant role.

The remainder of this paper is organized as follows: In Section 2 we present background material related to N-player prisoner's dilemma game. In Section 3 our model is described in detail. This is followed by a description of the simulation experiments and results in Section 4. We conclude the paper in Section 5 and identify future research directions.

2   Background and Related Work

2.1   N-player Prisoner's Dilemma Game

In the 2×2 prisoner's dilemma game, two players interact with each other by simultaneously choosing to *cooperate* or to *defect*. Based on their joint actions, each individual receives a specific payoff or utility value, U [1]. The altruistic act consists of conferring a benefit b on the recipient at a cost c to the donor. Here, $b > c$. If both players cooperate, each receives $b - c$, which is better than what they would obtain by both defecting (the Nash equilibrium of the game). But a unilateral defector would earn b, which is the highest payoff, and the exploited cooperator would pay the cost c without receiving any benefit.

The N-player game can be thought as a natural extension of the 2-player game. Boyd and Richerson [3] define the payoff values as follows:

$$U = \begin{cases} \frac{b \times i}{N} - c & \text{if the agent cooperated,} \\ \frac{b \times i}{N} & \text{if the agent defected.} \end{cases} \quad (1)$$

with i being the number of cooperators in the game, and N the number of players. The following conditions must also hold for a valid multi-player prisoner's dilemma game [16]: $c > \frac{b}{N}$ (defection is preferred for the individual) and $b > c > 0$ (contribution to social welfare is beneficial for the group).

Conventional evolutionary game theory predicts that cooperation is unlikely to emerge in the N-player prisoner's dilemma, and if it does emerge, then the cooperation levels are unlikely to be stable [3]. As the number of players per game increases, there should be a decrease on the level of cooperation [16].

2.2 Reciprocity in the Spatial Prisoner's Dilemma

It is well known that spatial structure enhances the cooperation levels in the prisoner's dilemma game [9,13]. It has also been shown that there is a direct correlation between the relative costs and benefits of cooperating and defecting and the underlying connectivity of agents playing the game [14]. High levels of cooperation can be attributed to "network reciprocity". That is, by limiting the number of game opponents (based on a pre-defined local neighborhood) and employing a local adaptation mechanism in which an agent copies a strategy from a neighbor linked by a network, higher levels of cooperation typically emerge.

There have been relatively few papers investigating spatial versions of the N-player prisoner's dilemma game. One recent notable example is the work of Santos et al. [15] who have shown that heterogenous graphs and the corresponding diversity in the number and size of the public goods game in which each individual participates (represented via an N-player prisoner's dilemma game) helps to promote cooperation. Here, the heterogenous graphs enable cooperators to form clusters in some instances, thereby reducing exploitation.

2.3 Reputation Systems and Indirect Reciprocity Mechanisms

Reputation systems are a common method of gaining information about an agent's behavioral history [10]. Approaches based on reciprocity and image scoring [12] have been applied as a solution for "tragedy of the commons" problems in open peer-to-peer networks.

The use of "shared history" where all nodes have access to the behavior of all other nodes, provides a mechanism for nodes to adapt their behavior [6]. A disadvantage of this approach is that maintaining a shared history increases overhead [7]. Consequently, Hales and Arteconi introduced a computational sociology tagging approach to support high levels of cooperation. It was achieved without central control or reciprocity based on a reputation mechanism in prisoner's dilemma like games. This approach was based on simple rules of social behavior observed in human societies and emergent networks. As such it was consistent with the work of Suzuki and Akiyama [16] who examined the relationship between cooperation and the average reputation of opponents.

Ellis and Yao [5] introduced a distributed reputation system for the prisoner's dilemma game using the notion of "indirect reciprocity". In their model, individuals do not store their own or their opponents' image/reputation scores. Instead, these scores are embedded in a social network in the form of mutually established links. However, this model is potentially vulnerable to collusion, as agents may leave unjustified positive feedback in order to boost each other's scores.

Algorithm 1. Social network based N-player prisoner's dilemma model

---

Require: Population of agents $\mathcal{P}$, evolutionary rate $e \in [0,1]$, number of iterations i_{max}, number of players per game $N \geq 2$.

```
1:  for i = 0 to i_max do
2:      G = ∅
3:      while g = NextGame(P, G, N) do
4:          G = G ∪ {g}
5:          PlayGame(g)
6:          AdaptLinks(g)
7:      end while
8:      for i = 0 to |P| × e do
9:          a, b = Sample(P)
10:         CompareUtilityAndSelect(a, b)
11:     end for
12: end for
```

---

3   Model

In our model of the N-player prisoner's dilemma game, a population of agents playing a series of independent games self-organize their social ties based exclusively on their self-interest. Interactions can take place between any group of agents, however, they tend to happen between those that have cooperated previously. This reflects the notion that agents prefer to seek interaction with partners that have already proven to be reliable. Agents who play cooperatively form social links, which are reinforced by subsequent cooperative actions. However, when an agent defects, the links with its opponents in this game are broken. As agents not participating in this game are unaware of the defective action, their links with the defective player, however, are retained.

The execution cycle of the model is sketched in Algorithm 1. Every iteration involves the forming of a number of *games* $\mathcal{G}$ of size N from the population (line 3); the *execution* of each game and the calculation of its outcome (line 5); the *adaptation* of the links of the agents in the game based on the actions played (line 6); and finally the *selection* process on a subpopulation, whose size is determined by the evolutionary rate e (lines 8-11). In the following sections, these steps are explained in detail.

3.1   The Formation of Games

The population of agents $\mathcal{P}$ is partitioned into disjoint sets of size N, each of which forms a game. While the first agent for every game is selected randomly amongst those that have not yet been assigned to any game in this iteration, the other $N-1$ slots of the game are filled as follows: With probability ϵ, the slot is filled randomly with an agent from the first agent's neighborhood. With probability $1 - \epsilon$, or if all agents in the neighborhood have already been assigned to a game in this iteration, the slot is filled randomly with an arbitrary agent from the remaining population. In that, ϵ regulates how often the agent explores cooperative play with unknown members of the population or exploits its current local neighborhood. Usually every agent plays exactly one game per iteration. However, depending on the size of $\mathcal{P}$ and N, a single agent might not play at all or the last game might not reach size N.

3.2   The Execution of Games

The outcome of every game depends on the strategies of its players. We consider two different scenarios that host agents with varying cognitive abilities. The first one involves only players with *pure strategies*. That is, each agent always plays the cooperate or the defect action. In the second scenario, we also consider a third type of agent that follows a *mixed strategy*. A mixed strategy is an assignment of a probability to each pure strategy (see details below).

The action taken by a mixed strategy agent i depends on the weights of the links w_{ij} it has established with each of its opponents $j \in g$ for the current game g. The average link weight for player i in game g is then defined as:

$$\overline{w}_i(g) = \frac{1}{|g|} \sum_{j \in g} w_{ij}$$

A mixed strategy i plays cooperatively in game g with probability:

$$P_i(g) = \frac{\overline{w}_i(g)^\alpha + \beta}{\overline{w}_i(g)^\alpha + \beta + 1}$$

With probability $1 - P_i(g)$, it plays defectively. Basically, β determines the probability of playing cooperatively if the agent does not have any link with its opponents. It allows agents to play generously, which was found by [1] to be a feature of successful iterated prisoner's dilemma stategies. The gradient of the probability density function is determined by α. Higher values correspond to an agent being satisfied with less links with its opponents to decide on cooperative play. This decision rule was introduced by [5].

Based on the game's outcome every agent receives a payoff or utility. This payoff is based on the functions described in Boyd and Richerson [3] and Suzuki and Akiyama [16] (see Equation 1 in Section 2 for details).

3.3   Link Adjustment

For every pair of agents i and j in the game, link weights w_{ij} are changed as follows:

$$w_{ij} = \begin{cases} w_{ij} + 1 & \text{if both } i \text{ and } j \text{ played cooperatively,} \\ 0 & \text{otherwise.} \end{cases}$$

In Figure 1 three stages in the evolution of a sample network are shown, illustrating the dynamic nature of link adjustment in the model. We assume that links can only be formed by mutual consent, which prevents any defector from influencing the selection of its opponents to its own advantage. Furthermore, we

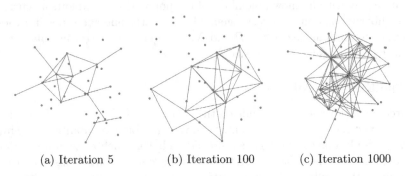

(a) Iteration 5   (b) Iteration 100   (c) Iteration 1000

Fig. 1. Three snap shots of the emerging social network at different iterations of the simulation. In this sample, $|\mathcal{P}| = 40$ and $N = 3$.

assume that the actions of other agents are observable. Otherwise, establishing links as described here would be impossible. Defective actions cannot lead to negative link weights in order to encourage agents to be forgiving. This is another requirement for successful iterated prisoner's dilemma strategies according to [1].

3.4   Strategy Update Mechanism

A form of cultural evolution based on imitation is used for strategy update. At each time step, $|P| \times e$ pairs of agents are drawn randomly from the population for update. The accumulated utility of the agent pair is compared. The agent with the lower utility is replaced by a copy of the agent with the higher utility. The new agent copies the strategy and the links of the winning agent, forms a link with weight one to this agent and initializes its own utility. This models the propagation of successful strategies and trust within the growing network.

4   Simulations

A systematic Monte Carlo simulation study was carried out to investigate the system dynamics of our model. The underlying hypothesis tested was that the introduction of agents equipped with mixed strategies would promote higher levels of cooperation in the N-Player Prisoner Dilemma game. Our social network model should encourage high levels of cooperation to persist for longer, even when all agents played with a pure strategy.

4.1   Parameters

The following parameter settings were used in all simulations: population size $|\mathcal{P}| = 1000$ with $\epsilon = 0.9$ and strategy update probability $e = 0.001$. Payoff values $b = 5$ and $c = 3$ were used for the benefit and costs of cooperation. In the pure strategy scenario, the population was initialized with 50% cooperators and 50%

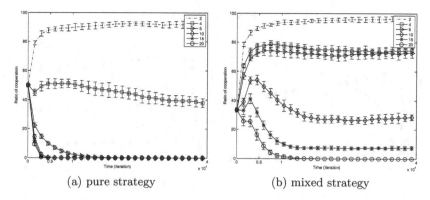

(a) pure strategy   (b) mixed strategy

Fig. 2. The proportion of cooperation vs. time for various values of N for (a) the pure strategy model and (b) the mixed strategy model

defectors. In the mixed strategy scenario, the population was initialized with 33.3% pure cooperators, 33.3% pure defectors and 33.3% mixed strategy agents with $\alpha = 1.5$ and $\beta = 0.1$. We report results averaged over 20 independent trials with the number of iterations $i_{max} = 40000$.

4.2   Results

Group size vs. strategy. Figure 2(a) plots results for the proportion of co-operation in the population vs. time for increasing values of N when all agents play with a pure strategy. When $N = 2$, the high proportion of cooperation at equilibrium is consistent with results reported for the standard 2×2 prisoner's dilemma game using a spatial structure. As N increases, the equilibrium cooper-ation levels drop off. At $N = 4$, approximately 50% of the population is playing cooperatively. However, as expected, for larger values of N defection takes over the population at a rate proportional to N.

Figure 2(b) plots results for the proportion of cooperation in the population vs. time for increasing values of N when mixed strategy agents have been intro-duced. A comparison of the plots clearly shows that the introduction of mixed strategy agents promotes higher levels of cooperation. Decreasing trends in coop-eration levels consistent with the number of players are still apparent. However, the equilibrium levels for the proportion of agents who have cooperated are sig-nificantly higher. For example, when $N = 10$, the proportion of cooperation initially increased as the social network began to form. However, after approxi-mately 5000 time steps this level gradually decreases to an equilibrium level of approximately 30%. In the pure strategy case, when $N = 10$, defection was the dominant action after 5000 time steps.

Emergent social network. When agents play cooperatively, they form a social link that is reinforced each time the action is repeated. When an agent defects, all links with its opponents in that game are dissolved. Here, we confine our

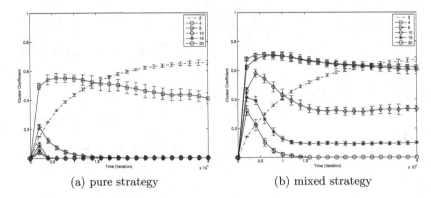

(a) pure strategy   (b) mixed strategy

Fig. 3. Average clustering coefficient vs. time for various values of N

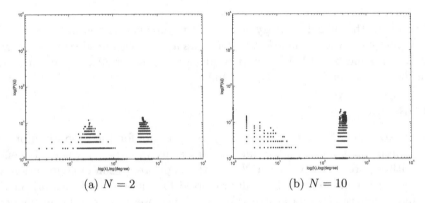

(a) $N = 2$ (b) $N = 10$

Fig. 4. Final degree distributions in the mixed strategy model on a log-log scale

discussion of the characterization of the emergent networks to the two quantities typically used in the analysis of networks: the clustering coefficient and the degree distribution [4]. The clustering coefficient is the probability that two nearest neighbors of a node are also nearest neighbors of each other. Its values range from zero to one and the latter value holds if and only if the network is globally coupled. This analysis provides important insights into the underlying social mechanisms, which promote the collective behavior within the model.

Figure 3 plots the average clustering coefficient vs. time for (a) the pure strategy model, (b) the mixed strategy model. The trend in the time-series values are consistent with the trends shown in Figure 2(a) and (b), that is, the average clustering coefficient values are generally smaller for larger values of N. Significantly, the relative magnitudes of the average clustering coefficient values are higher in the mixed strategy model, suggesting that the social network helps to promote higher levels of cooperation. The underlying link adjustment mechanisms and mixed strategy decision-making directly favor the formation of cliques. These findings are consistent with results reported in [17]. It is also interesting to note that the values of the average clustering coefficient values when $N = 2$ are very similar in both models, confirming that the additional cognitive ability of the mixed strategy agents is not a requirement to promote cooperation in the restricted game.

In Figure 4 we provide two examples of final degree distribution plots on a log-log scale of the social network in the mixed strategy model (space constraints limit the inclusion of further plots). For all scenarios examined, a clear hierarchical structure was observed. The general trends in degree distribution plots were similar. When it was possible to sustain cooperation in a game (for $2 \leq N \leq 10$), the degree distribution was typically bimodal where the size of the gap and shape of the distribution was dependent on N. For larger values of N, the plot shows a large number of nodes with high average degrees corresponding to a very dense network. This suggests that densely knit cliques may be stable and preserve links. There are also a smaller number of nodes with lower average degrees corresponding to newly formed cooperative cliques.

5   Summary and Conclusion

In this paper, we have investigated the coevolution of strategies and social networks using a version of the N-player prisoner's dilemma game. An important component of our model is the endogenous network formation based on agent interactions. Agents who play cooperatively form social links, which are reinforced by subsequent cooperative actions. However, when an agent defects, links in the social network are broken. Simulation results validate our approach and confirm that cooperation is promoted by agents exploiting the social network to guide their decision-making. However, as the number of players participating in the game increases, the proportion of cooperations within the population decreases.

We have examined the long term system dynamics of our model based on two different agent cognitive abilities: agents with pure strategies who always play one of cooperate or defect, and a mixed strategy where agents based their decision as to which action to play on the reputation or reliability of other agents. When agents have increased cognitive capacity to classify their environment, social networks play an increasingly important role in promoting and sustaining cooperation. An analysis of the emergent social networks shows that they are characterized by high average clustering and broad-scale heterogeneity, especially for a relatively small number of players. The key findings in this study suggest that the interplay between the local structure of the network and the hierarchical organization of cooperation is non-trivial.

The notion that a group of individuals can collectively do better when they cooperate, often against the self-interest of individuals, is relevant to many research domains, including social and biological systems, economics, artificial intelligence and multi-agent systems. In future work, we plan to extend this study in a number of different directions. In the present study, we have assumed that all groups were of a fixed size during a given game. It would be interesting to examine the effects of dynamic group formation on the coevolution of cooperation and social networks. A second avenue worth investigating is specific aspects of trust/reputation in groups within a peer-to-peer framework.

References

1. Axelrod, R.: The Evolution of Cooperation. Basic Books, New York (1984)
2. Bilancini, E., Boncinelli, L.: The co-evolution of cooperation and defection under local interaction and endogenous network formation. Journal of Economic Behavior and Organization 70(1-2), 186–195 (2009)
3. Boyd, R., Richerson, P.: The evolution of reciprocity in sizable groups. Journal of Theoretical Biology 132, 337–356 (1988)
4. Dorogovtsev, S.N., Mendes, J.F.F.: Evolution of Networks: From Biological Nets to the Internet and WWW. Oxford University Press, Oxford (2003)
5. Ellis, T.S., Yao, X.: Evolving cooperation in the non-iterated prisoners dilemma: A social network inspired approach. In: IEEE Cong. on Evol. Comp. (CEC), September 2007, pp. 736–743 (2007)
6. Feldman, M., Lai, K., Stoica, I., Chuang, J.: Robust incentive techniques for peer-to-peer networks. In: Proc. 5th ACM Conf. Electronic Commerce (EC 2004), pp. 102–111. ACM Press, New York (2004)
7. Hales, D., Arteconi, S.: SLACER: A self-organizing protocol for coordination in peer-to-peer networks. IEEE Intelligent Systems 21(2), 29–35 (2006)
8. Hardin, G.: The tragedy of the commons. Science 162, 1243–1248 (1968)
9. Hauert, C.: Fundamental clusters in spatial 2×2 games. Proc. R. Soc. Lond. B 268, 761–769 (2001)
10. Ismail, R., Josang, A., Boyd, C.: A survey of trust and reputation systems for online service provision. Decision Support Systems 43(2), 618–644 (2007)
11. Kun, A., Scheuring, I.: Evolution of cooperation on dynamical graphs. BioSystems 96(1), 65–68 (2009)
12. Nowak, M., Sigmund, K.: Evolution of indirect reciprocity by image scoring. Nature 393, 573–577 (1998)
13. Nowak, M.A., May, R.M.: Evolutionary games and spatial chaos. Nature 359, 826–829 (1992)
14. Ohtsuki, H., Hauert, C., Nowak, M.A.: A simple rule for the evolution of cooperation on graphs and social networks. Nature 441(7092), 502–505 (2006)
15. Santos, F.C., Santos, M.D., Pacheco, J.M.: Social diversity promotes the emergence of cooperation in public goods games. Nature 454, 213–216 (2009)
16. Suzuki, S., Akiyama, E.: Reputation and the evolution of cooperation in sizable groups. Proceedings of the Royal Society B 272, 1373–1377 (2005)
17. Tanimoto, J.: Promotion of cooperation through co-evolution of networks and strategy in 2x2 games. Physica A 96, 953–960 (2009)

Using Misperception to Counteract Noise in the Iterated Prisoner's Dilemma

Lachlan Brumley, Kevin B. Korb, and Carlo Kopp

Clayton School of Information Technology, Monash University, Australia
Lachlan.Brumley@gmail.com, korb@infotech.monash.edu.au,
carlo@infotech.monash.edu.au

Abstract. The Iterated Prisoner's Dilemma is a game-theoretical model which can be identified in many repeated real-world interactions between competing entities. The Tit for Tat strategy has been identified as a successful strategy which reinforces mutual cooperation, however, it is sensitive to environmental noise which disrupts continued cooperation between players to their detriment. This paper explores whether a population of Tit for Tat players may evolve specialised individual-based noise to counteract environmental noise. We have found that when the individual-based noise acts similarly to forgiveness it can counteract the environmental noise, although excessive forgiveness invites the evolution of exploitative individual-based noise, which is highly detrimental to the population when widespread.

1 Introduction

The competitive social and biological contests that occur in evolutionary situations commonly have simple underlying rules, which aid in their analysis. Due to their simplicity and strategic nature, these conflicts can often be modeled mathematically with game theory in order to examine which strategies entities should select and how their opponents may react. The fundamentals of game theory were first formalised by von Neumann and Morgenstern (1947), with the aim of providing a mathematical method for modeling and analysing games of strategy. A game is defined as consisting of a number of players, the rules for the game and the strategies that the players may select. All players select their strategies simultaneously, producing an outcome for the game which provides various payoffs for the players. Players are expected to select their strategies rationally, preferring those which maximise their expected payoff. An outcome from which no player has an incentive to change their strategy, in order to receive a better payoff, is called an equilibrium (Nash 1950). It is possible for a game to have zero, one or more outcomes that are equilibria. Since players have no incentive to change their strategy at these equilibria, they are typically considered to be solutions or likely outcomes of a game. Some strategies may offer a player its best payoff, regardless of the strategies selected by its opponents, and such a strategy is said to be dominant.

Game Theory can also model the evolution of strategies used by players in a biological game (Maynard Smith & Price 1973) and their propagation through a population over time. Players' strategies are traits and are therefore subject to evolutionary pressures. Strategies which become conserved in a population over

K. Korb, M. Randall, and T. Hendtlass (Eds.): ACAL 2009, LNAI 5865, pp. 53–62, 2009.

Table 1. Payoffs for the Prisoner's Dilemma Game. Each player's best payoff occurs when they Defect against a Cooperating opponent, however the equilibrium state for this game is the sub-optimal outcome of mutual Defection.

	B Cooperates	**B Defects**
A Cooperates	A: short jail term B: short jail term	A: long jail term B: no jail term
A Defects	A: no jail term B: long jail term	A: moderate jail term B: moderate jail term

a long time span and which cannot be replaced by invading strategies are said to be Evolutionarily Stable. If a new strategy invades a population, its frequency in future generations depends upon its effectiveness compared to existing strategies. If the invading strategy provides a higher payoff, it is better adapted to the environment and its frequency in the population will increase as it is selected over the existing strategies. If the invading strategy is worse than the existing strategies then it will not be adopted by the population and will be displaced.

A popular game theoretical model is the Prisoner's Dilemma game (Tucker 1950), which describes a contest between prisoners who were captured near the scene of a crime. The police suspect their involvement, but without a confession they lack sufficient evidence to charge either prisoner. The prisoners are separated and each is presented with the opportunity to confess to the crime and implicate their accomplice in exchange for their own release. The two strategies the prisoners may select from are to **Cooperate** by staying silent or to **Defect** by implicating the other prisoner. Table 1 shows the possible outcomes and payoffs for this game.

The game's equilibrium point is mutual defection, where both prisoner's receive moderate jail terms, their second worst possible outcome. While mutual cooperation provides the prisoners with a shorter jail term, this outcome is not stable as either player may then choose to Defect, since Defection is the dominant strategy. The Prisoner's Dilemma game has been identified as unique among other similar games, as it is the only game with a strongly stable deficient equilibrium, where neither player has any reason to select Cooperation over Defection (Rapoport & Guyer 1966). The core of the dilemma is the prisoner's choice between "rational" selfish behaviour or "irrational" cooperation. Real world examples of such choices are plentiful, including Stalin's Great Terror (Grossman 1994) and cooperation in stickleback fish during predator evaluation (Milinski 1987).

The Prisoner's Dilemma also changes substantially when it is played repeatedly by players who possess memory of previous games and attempt to maximise their total payoff from the series of games. In the Iterated Prisoner's Dilemma (IPD), cooperation becomes a rational strategy, as past knowledge allows players to cooperate with a previously cooperative opponent or retaliate against previous defections (Axelrod 1984). Mutual cooperation can be maintained in the IPD, as the incentive to defect in the current game is deterred by the potential for punishment in future games.

In the IPD game players use some strategy to determine their next move, which considers the outcomes from previous iterations of the game. Strategies for the IPD and their performance were examined by Axelrod (1984) in a round robin tournament. This tournament was won by a strategy called **Tit** for **Tat**(TFT), which Cooperated in the first round and then mimicked its opponent's strategy from the previous round. A second tournament was organised with some new strategies entered and it was once again won by TFT. By mimicking an opponent's previous move, TFT cooperates with other cooperative players and defects against those which are not.

Mutual cooperation in the IPD is maintained by the ability to retaliate against any defection in the future. Cooperation may also develop between uncooperative players who are capable of learning. Axelrod (1997) argued that soldiers in the trenches during World War I learned to cooperate by firing inaccurately upon each other, however, this behaviour may also be attributed to a selfish desire to avoid any psychological discomfort from killing enemy soldiers (Grossman 1995).

Mutual cooperation between players can easily be disrupted by noise (Axelrod & Dion 1988). Noise may be defined as an error which alters the strategy a player performs, such mis-implementation of strategies, errors in the communication of strategies or the misperception of previously performed strategies. The players in these games (and in analogous real world situations) are typically incapable of differentiating between intentional and erroneous strategy selection. Molander (1985) has demonstrated that if there is any significant level of noise, the performance of TFT players will approach that of two players randomly selecting strategies over a sufficiently long time span.

Contrition and forgiveness are two proposed solutions to reduce the impact of noise upon cooperative strategies, such TFT, in the IPD (Molander 1985, Wu & Axelrod 1995). Contrition alters a player's strategy to prevent retaliation against any defections provoked by its own erroneous Defections. Contrition prevents an unintentional defection caused by noise from echoing between players for multiple turns. Forgiveness or generosity has a player choose Cooperate in response to some percentage of their opponent's Defections. Wu and Axelrod found that contrition was more effective than forgiveness at maintaining mutual cooperation. Contrition only allows an erroneous Defection to echo between players, while forgiveness may not immediately correct an erroneous Defection.

While the TFT strategy is an effective strategy for the IPD, it is detrimentally affected by random noise. Forgiveness can allow TFT players to maintain some level of mutual cooperation in a noisy environment. Forgiveness can be achieved by the misperception of a previous defection as cooperation and is therefore one type of individual-based noise which may affect players. The other is the misperception of cooperation as defection, which will benefit the player immediately at the expense of continued mutual cooperation. These two types of misperception may be known as **Forgiving Misperception** and **Punishing Misperception**. Generosity is a possible descriptive name for the trait which causes Forgiving Misperception, while paranoia may aptly describe the trait embodied by Punishing Misperception. These two traits could each have a distinct probability of affecting each perception of the TFT players.

Previous research has demonstrated that in some instances populations may benefit from misperception (Akaishi & Arita 2002) and even evolve stable levels of misperception (Brumley, Korb & Kopp 2007). We hypothesise that a population of TFT players in a noisy environment could evolve a stable level of Forgiving Misperception, indicating a benefit to misperception. This benefit would evolve despite the detrimental impact that Punishing Misperception could have on the individual players. The TFT players would evolve a probability of Forgiving

Misperception suitable for their environment, while also evolving a Punishing Misperception probability very close to 0%. This hypothesis was tested through the simulation of an evolutionary IPD game between misperceiving players using the TFT strategy.

2   Method

Our hypothesis was tested by in an evolutionary IPD tournament between a population of TFT players. Punishing and Forgiving Misperception probabilities are a heritable trait of the players, allowing the population to evolve optimal probabilities for both values over many generations. If the population of TFT players can evolve a stable level of Forgiving Misperception in a noisy environment, then this indicates a potential benefit of misperception. Such a population should also perform better than players randomly selecting strategies, as is the case when there is excessive environmental noise. Noise is caused by misaction by the players, with all players sharing an equal probability of misacting, which was set to 1%. Misaction is symmetric and will cause an affected player to implement the opposite of its intended move. It cannot be detected or prevented by the players.

Each player has a chromosome storing its Forgiving and Punishing Misperception probabilities, which are values between 0.0 and 1.0. Punishing Misperception may only occur whenever a player observes its opponent Cooperating, while Forgiving Misperception may only occur whenever a player observes its opponent Defecting. Players use the TFT strategy to determine which move they will select, based upon the observed previous move of their opponent, which may be misperceived. After both players have selected their strategy during a game, the payoff from the resulting outcome is added to their total score. The payoffs for each game outcome match those used previously by Axelrod (Table 2) and also satisfy Hofstadter's conditions for a true dilemma. A player's score is considered the only measure of its fitness in this environment.

The population of 25 players is initialised with players whose misperception probabilities are randomly generated. These values are taken from a normal distribution with a mean of 5% and a standard deviation of 1%. During this evolutionary IPD tournament, each player competes in an iterated game with 200 repetitions against all the other players in the population. All players are then ranked by their total score from all the games they played. The highest scoring player is assumed to be the fittest and the lowest scoring player the least fit. Each player's misperception probabilities will affect its behaviour and determine how it responds to noise during its competitions against other players.

Reproduction occurs after each generational tournament has concluded, between the fittest player and a player selected at random from the population. The least fit player is removed from the simulation and replaced by the new offspring. The next generational tournament between the updated population then begins. The simulation was run for 10000 generations. Reproduction utilises a

Table 2. Numerical payoffs for the Prisoner's Dilemma Game used by Axelrod (1981)

	Cooperate	Defect
Cooperate	3, 3	0, 5
Defect	5, 0	1, 1

crossover operation of the parents' chromosomes to produce the new offspring's chromosome, which is then mutated. Mutation applies a small randomly generated mutation value to each value in the offspring's chromosome, which alters the Punishing and Forgiving Misperception probabilities of the new offspring slightly. Mutations are prevented from increasing or decreasing a misperception probability beyond its minimum or maximum value of 0.0 and 1.0 respectively.

If the population evolves a stable level of Forgiving Misperception, this will be observable in a noticeable average Forgiving Misperception probability which exists over many generations. Such a level of Forgiving Misperception should also ensure that the players' scores are greater than those received from random strategy selection and close to those received for continual mutual Cooperation. Such results support the hypothesis that Forgiving Misperception can provide an evolutionary benefit to the TFT players in a noisy environment.

A player's score or the population's average score may be compared to the scores obtained in the best and worst possible outcomes. The best possible outcome for the players would be mutual cooperation in every single game, while the worst would be mutual defection. During each generation a player competes in $200 \times 24 = 4800$ individual games. Since mutual cooperation gives a payoff of 3, the (unobtainable) best case score is 14400, while the worse case score is 4800. If misperception is also benefiting the players, then the score should also be greater than that expected from random strategy selection. The average payoff per game of random strategy selection is 2.25, which would provide a score of approximately 10800 over the length of the competition.

3   Initial Results

The simulation was executed with the selected parameters 30 times to produce a sufficient statistical sample. Figure 1 compares the average individual score and the average misperception probabilities. The area under the average score is shaded in these plots to differentiate between the score and the misperception probabilities. The populations begin with initial Forgiving and Punishing Misperception probabilities that are approximately 5%. From this point the population evolves a high Forgiving Misperception probability and a near-zero Punishing Misperception probability, which increases the population's total score close to the 14400 available from continual mutual cooperation. This demonstrates that Forgiving Misperception can help TFT players maintain cooperation in a noisy environment. The population then begins to evolve higher Punishing Misperception probabilities, which decreases the average player score and also causes the Forgiving Misperception probability of the populations to decrease. After 10000 generations the average population's score is less than if the players abandoned the TFT strategy and instead selected their actions randomly.

Of the 30 simulation iterations, only five evolved stable populations containing cooperative players with substantial Forgiving Misperception probabilities and low Punishing Misperception probabilities. In the remaining 25 populations high Punishing Misperception probabilities have evolved to exploit the Forgiving Misperception, reducing the population's score below that obtained in a noisy environment with no misperception. Therefore, in the majority of cases, misperception is detrimental to the players. The invasion of populations with Forgiving Misperception by Punishing Misperception demonstrates that high Forgiving

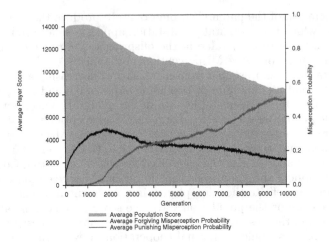

Fig. 1. The average player score and average evolved misperception probabilities. Forgiving Misperception quickly evolves to maintain Cooperation. Punishing Misperception then evolves to exploit the forgiving players, reducing the average score.

Misperception probabilities are not an Evolutionary Stable Strategy. Furthermore, as the population's Punishing Misperception probability increases, the selection pressure against forgiveness also increases, causing it to decline in the population. Punishing Misperception begins to develop rapidly when the Forgiving Misperception probability is approximately 30%. The evolution of the high probabilities of Punishing Misperception is an example of a 'tragedy of the commons' (Hardin 1968), wherein individuals act selfishly, to the eventual detriment of the entire population when such behaviour becomes universal. If Forgiving Misperception is to benefit these TFT players, some mechanism is required to prevent the evolution of excessive levels of Punishing Misperception.

4   Preventing Excessive Punishing Misperception

Forgiving Misperception does help the TFT players to cooperate, however, it also invites exploitation by Punishing Misperception. A possible solution is to limit the maximum Forgiving Misperception probability players may evolve. If the population is prevented from evolving to be highly forgiving, then there will be less selective pressure for Punishing Misperception in the population, hopefully preventing its invasion of the forgiving player population. It is hypothesised that an upper bound on Forgiving Misperception may prevent populations from being easily invaded by Punishing Misperception, while still allowing some benefit from Forgiving Misperception. An optimal value for Forgiving Misperception is one that maximises the benefit from restoring and maintaining Cooperation, while also not inviting exploitation from Punishing Misperception. Varying the upper bound of Forgiving Misperception and measuring the evolved Forgiving and Punishing Misperception probabilities of the population will identify the optimal upper bound for Forgiving Misperception. Up to this limit the player populations should benefit from Forgiving Misperception, helping maintain mutual Cooperation. Beyond this upper bound, the population may develop Forgiving Misperception probabilities which invite the evolution of high Punishing Misperception probabilities and reduce the players' scores.

An alternative to restricting a population's Forgiving Misperception probabilities is to restrict the Punishing Misperception probabilities that may evolve. If Punishing Misperception is restricted, whenever there is sufficient Forgiving Misperception to exploit the population will likely evolve the maximum permitted level of Punishing Misperception. Restricting Punishing Misperception treats the symptoms but not the underlying cause, which is the benefit from exploiting highly forgiving players. Limiting the population's Forgiving Misperception probabilities will instead solve the cause of this problem.

5   The Effects of Limiting Forgiveness

An upper bound to a player's Forgiving Misperception probability was added to the simulation, ensuring that when the TFT players reproduced, the offspring's Forgiving Misperception probability could not exceed this limit. This bound became an additional parameter of the simulation. This simulation reused the same parameters as the previous one, with the exception of the misaction probability which was made a parameter of the simulation. Seven different values for the misaction probability were investigated — 0.0, 0.005, 0.01, 0.02, 0.03, 0.05 and 0.1. These values are low noise probabilities, as in an excessively noisy environment the population of TFT players will randomly move between game states and any effects from misperception, beneficial or detrimental, will be concealed. For the Forgiving Misperception probability limit, 24 values were investigated — 0.05, 0.1, 0.2, 0.25, 0.26, 0.27, 0.28, 0.29, 0.3, 0.31, 0.32, 0.33, 0.34, 0.35, 0.36, 0.37, 0.38, 0.39, 0.4, 0.45 0.5, 1.0. There is an increased focus on Forgiving Misperception limits between 0.25 and 0.4 since previous simulations identified that high probabilities of Punishing Misperception did not evolve until the Forgiving Misperception probabilities reached this range. A finer examination of the Forgiving Misperception probabilities within this range will clarify the threshold at which Forgiving Misperception switches from beneficial to detrimental.

In this case we are interested in whether the population evolves to a stable state and will therefore focus on the average player score from the final generation of the simulation. Studying the change in the average player scores as the Forgiving Misperception probability upper bound is increased will indicate at what point Punishing Misperception evolves in the population. This will identify the optimal upper bound value which allows the players to benefit from forgiveness, while avoiding the evolution of significant punishment in their population.

The average score from the final generations of the TFT players are shown in Figure 2. Higher scores indicate that Forgiving Misperception is maintaining mutual Cooperation between the players despite the noise in the environment, while lower scores indicate a lack of cooperation. Here the score received by the player populations reaches a peak at approximately 30%, before dropping suddenly. This fall corresponds to the point at which Punishing Misperception increases. When the misaction probability is 0%, the average score also decreases somewhat before increasing again. In these cases the average scores are higher as there is no noise from misaction to affect the players.

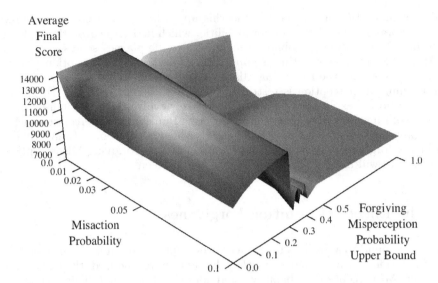

Fig. 2. The average final individual score of the Tit for Tat populations. The sudden decrease in the population's score occurs when the population is sufficiently forgiving to allow Punishing Misperception to evolve in the population.

There is a regular curve before the sharp decline in player scores caused by the increase in Punishing Misperception. This curve shows that when there is some probability of noise from misaction, the average score will increase along with the Forgiving Misperception probability upper bound. This trend ceases however once the optimal forgiveness threshold is exceeded, at which point Punishing Misperception evolves and the population's score drops dramatically. When the Forgiving Misperception cap is 0.0, preventing Forgiving Misperception, the average score is reduced due to the effects of noise which cannot be corrected. With the highest misaction probability of 0.1 the average score is reduced to approximately 10800, which is approximately the score obtained from random behaviour. As small amounts of Forgiving Misperception are permitted, the score gradually rises up to the threshold near the 30% Forgiving Misperception probability. Beyond this threshold the score drops below that of random behaviour, due to the evolution of exploitative Punishing Misperception.

The sudden threshold before the average player score decreases may be considered to be close to the optimal value for the Forgiving Misperception upper bound in that situation and can be called the beneficial threshold. Forgiving Misperception is beneficial up to this threshold, but becomes detrimental once this threshold is exceeded. Figure 3 shows the Forgiving Misperception upper bound values that produced the highest average scores and the beneficial threshold.

The beneficial threshold is the highest Forgiving Misperception probability upper bound which permitted beneficial forgiveness without allowing excessive Punishing Misperception to evolve in the population. This threshold is close to the point where Punishing Misperception was earlier observed to begin to evolve and degrade the population's performance. The optimal population score was typically obtained with a Forgiving Misperception upper bound less than the

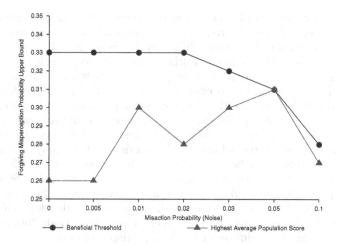

Fig. 3. Comparison of the highest Forgiving Misperception probability upper bound (beneficial threshold) where the player population maintains Cooperation and the Forgiving Misperception probability upper bound at which the highest average score was obtained. The optimal Forgiving Misperception probability upper bound which maximises the population's score is typically less than the highest value for which Cooperation is maintained, suggesting that exploitative Punishing Misperception has already began to evolve in the latter case and is reducing the average score.

beneficial threshold and this suggests that beyond the optimal score Punishing Misperception has begun to evolve. The optimal scores are very close to those obtained from mutual Cooperation, indicating that Forgiving Misperception is enabling these players to maintain cooperation despite the noise from misaction.

6   Conclusion

This work does demonstrate that misperception can help TFT players maintain mutual cooperation in an evolutionary IPD game. However, this requires that misperception mimics forgiveness and that any misperception causing unwarranted defections is limited. Forgiveness can counteract the effects of random noise; however, excessive forgiveness leaves the population of TFT players vulnerable to exploitation, in this case by selfish Punishing Misperception. Forgiving Misperception is not an Evolutionarily Stable Strategy, as Punishing Misperception will invade the population to exploit the forgiving players. High Punishing Misperception probabilities are an Evolutionarily Stable Strategy, albeit a highly detrimental one since such widespread behaviour produces worse payoffs than a population only affected by noise.

The optimal upper bound for Forgiving Misperception is approximately 30% for the TFT players. At this point mutual cooperation can be maintained, while Punishing Misperception does not evolve to invade the population. The Forgiving and Punishing Misperception probabilities which evolve in the player population interact in a manner similar to a predator-prey relationship. When the population's Forgiving Misperception probability is restricted, misperception will provide an evolutionary benefit when it induces behaviour analogous to forgiveness. However, this benefit requires an asymmetric model of misperception in which the evolution of Forgiving Misperception is restricted.

Evolution can identify stable equilibria where misperception is beneficial, although the structure of the IPD game prevents this from occurring without restrictions. For a population to evolve a stable level of Forgiving Misperception, external restrictions on either forgiveness or punishment are required and here we restricted Forgiving Misperception. However, in the real world misperception cannot be so easily restricted. Real world social systems instead often attempt to deter exploitative behaviour with punishment, thereby increasing the effective cost of exploitative behaviour beyond that of cooperative behaviour. In human society, these restrictions are in part provided by laws and societal norms and they require knowledge of the interactions between other players. Society-wide knowledge of previous IPD games allows players to develop self-policing strategies which can reinforce mutual cooperation (Ord & Blair 2002) and also potentially prevent the evolution of Punishing Misperception.

References

Akaishi, J., Arita, T.: Multi-agent Simulation Showing Adaptive Property of Misperception. In: FIRA Robot World Congress, pp. 74–79 (2002)

Axelrod, R.: Evolution of Cooperation. Basic Books, New York (1984)

Axelrod, R.: The Complexity of Cooperation. Princeton University Press, New Jersey (1997)

Axelrod, R., Dion, D.: The Further Evolution of Cooperation. Science 242, 1385–1389 (1988)

Brumley, L., Korb, K.B., Kopp, C.: An Evolutionary Benefit from Misperception in Foraging Behaviour. In: Randall, M., Abbass, H.A., Wiles, J. (eds.) ACAL 2007. LNCS (LNAI), vol. 4828, pp. 96–106. Springer, Heidelberg (2007)

Grossman, D.: On Killing: The Psychological Cost of Learning to Kill in War and Society. Little, Brown and Company, Boston (1995)

Grossman, P.Z.: The Dilemma of Prisoners. Journal of Conflict Resolution 38(1), 43–55 (1994)

Hardin, G.: The Tragedy of the Commons. Science 162, 1243–1248 (1968)

Maynard Smith, J., Price, G.R.: The Logic of Animal Conflict. Nature 246, 15–18 (1973)

Milinski, M.: Tit for Tat in sticklebacks and the evolution of cooperation. Nature 325, 433–435 (1987)

Molander, P.: The Optimal Level of Generosity in a Selfish, Uncertain Environment. The Journal of Conflict Resolution 29(4) (1985)

Nash, J.F.: Equilibrium Points in n-Person Games. Proceedings of the National Academy of Sciences of the United States of America 36(1), 48–49 (1950)

Ord, T., Blair, A.: Exploitation and Peacekeeping: Introducing more sophisticated interactions to the Iterated Prisoner's Dilemma. In: World Congress on Computational Intelligence, Honolulu (2002)

Rapoport, A., Guyer, M.: A taxonomy of 2 × 2 games. General Systems 11 (1966)

Tucker, A.W.: A two-person dilemma, Mimeograph. Reprint, On jargon: The prisoner's dilemma. UMAP Journal 1, 101 (1980)

von Neumann, J., Morgenstern, O.: Theory of Games and Economic Behaviour. Princeton University Press, Princeton (1947)

Wu, J., Axelrod, R.: How to cope with noise in the iterated prisoner's dilemma. The Journal of Conflict Resolution 39(1), 183–189 (1995)

An Analysis and Evaluation of the Saving Capability and Feasibility of Backward-Chaining Evolutionary Algorithms

Huayang Xie and Mengjie Zhang

School of Engineering and Computer Science
Victoria University of Wellington, New Zealand
{hxie,mengjie}@ecs.vuw.ac.nz

Abstract. *Artificial Intelligence*, volume 170, number 11, pages 953–983, 2006 published a paper titled "Backward-chaining evolutionary algorithm". It introduced two fitness evaluation saving algorithms which are built on top of standard tournament selection. One algorithm is named Efficient Macro-selection Evolutionary Algorithm (EMS-EA) and the other is named Backward-chaining EA (BC-EA). Both algorithms were claimed to be able to provide considerable fitness evaluation savings, and especially BC-EA was claimed to be much efficient for hard and complex problems which require very large populations. This paper provides an evaluation and analysis of the two algorithms in terms of the feasibility and capability of reducing the fitness evaluation cost. The evaluation and analysis results show that BC-EA would be able to provide computational savings in unusual situations where given problems can be solved by an evolutionary algorithm using a very small tournament size, or a large tournament size but a very large population and a very small number of generations. Other than that, the saving capability of BC-EA is the same as EMS-EA. Furthermore, the feasibility of BC-EA is limited because two important assumptions making it work hardly hold.

1 Introduction

Artificial Intelligence, volume 170, number 11, pages 953–982, 2006, published a research paper titled "Backward-chaining evolutionary algorithms" [1]. It presented two computational saving algorithms that avoid the creation and evaluation of individuals that never participate into any tournaments in standard tournament selection. One algorithm is named Efficient Macro-selection Evolutionary Algorithms (EMS-EA) and the other is named Backward-chaining EA (BC-EA). Through sets of experiments using tournament sizes 2 and 3, both algorithms were claimed to be able to yield considerable savings in term of fitness evaluations, and BC-EA was demonstrated to be better than EMS-EA. The two algorithms were also claimed to be able to apply to any form of population based search, any representation, fitness function, crossover and mutation, provided they use tournament selection. Therefore, we evaluated the BC-EA algorithm

K. Korb, M. Randall, and T. Hendtlass (Eds.): ACAL 2009, LNAI 5865, pp. 63–72, 2009.
© Springer-Verlag Berlin Heidelberg 2009

in a context of tree-based GP but obtained results differing from what were claimed in [1]. We think it is worthwhile providing the evaluation and analyse results to supplement the understanding of the BC-EA algorithm and provide some insight into the strength and weakness of these two algorithms.

This paper is outlined as follows. Section 2 briefly reviews the EMS-EA and BC-EA algorithms. Section 3 points out a missing element of the investigations of [1]. Section 4 then focus on the missing element to evaluate and analyse the saving capability of the two algorithms via experiments. Section 5 further discusses the feasibility of the two algorithms. Finally, section 6 concludes the paper.

2   Brief Review

2.1   Overall

One prerequisite of using EMS-EA or BC-EA is that the selection method must be standard tournament selection. The standard tournament selection method randomly draws/samples k individuals with replacement from the current population of size N into a tournament of size k and selects the one with the best fitness from the tournament into the mating pool. Since tournament selection consists of two steps, sampling and selecting, individuals that do not get a chance to produce offspring can be categorised into two kinds. One kind is termed *not sampled*, referring individuals that never get sampled/drew into a tournament when tournament sizes are small. This proportion of the population β can be modeled as:

$$\beta = \left(\frac{N-1}{N} \right)^{kN} \tag{1}$$

where k is the tournament size and N is the population size [1]. The proportion is approximately 13.5%, 5.0%, and 1.8% for tournament size 2, 3, and 4, respectively. The other kind is termed *not selected*, referring individuals that participate into some number of tournaments but do not win any due to their worse fitness values. Both EMS-EA and BC-EA aim to avoid creating and evaluating those not sampled individuals.

In EMS-EA, the maximum number of generations G needs to be set in advance. A sequence of genetic operations that will be used to create the entire population at each generation is determined according to the pre-generated random numbers and the predefined crossover and mutation rates. However, the sequence of genetic operations is just memorised at this stage without execution. For each of the individuals that are required for executing each of the memorised genetic operators, the IDs of a number of parent individuals that need to be sampled into each tournament are also memorised. In other words, tournament selections are *virtually* conducted by just memorising the IDs of the sampled individuals, and genetic operations are also *virtually* executed by just generating the IDs of offspring and making connections between the sampled individual IDs and the offspring IDs. For instance, if the tournament size is k, at a non-initial generation g, each individual has connections from at most k individuals at the previous generation $g - 1$. However, it is not necessary that every individual at generation $g - 1$ is connected to individuals at generation g, especially when k is small. At the end, a graph structure that describes the *weak ancestral relationship*[1] between individuals across the whole *virtual* evolutionary process is constructed. Then a post-process on the graph is conducted by starting from the individuals at the last generation and tracing back to the initial generation. Individuals that are not involved in tournaments are marked as *neglected*. Finally, the real evolutionary process starts by creating and evaluating the individuals whose IDs are not marked as neglected at the initial generation and performing the memorised genetic operations to generate offspring whose ID is also unmarked at the next generation, and so on. Individuals who are marked as neglected will not be created and evaluated, thus computational savings can be obtained. According to the model given in [1], EMS-EA can provide about 13.5%, 5.0%, and 1.8% savings for tournament size 2, 3, and 4 respectively. The algorithm is given in Figure 1.

In order to more rapidly find better solutions and possibly to further increase savings, BC-EA is then proposed where the last generation G also needs to be set in advance. BC-EA starts from the first individual at the last generation G and uses the depth-first search to determine all *possible* ancestors all the way back to the initial generation. It then creates and evaluates these ancestors and moves forward to the last generation. The process then repeats for another individual at the last generation and so on. The algorithm is given in Figure 2.

```
 1: for gen = 1 to G do
 2:     for ind = 1 to M do
 3:         op[gen][ind] = choose genetic operator
 4:         for arg = 1 to arity(op[gen][ind]) do
 5:             pool[gen][ind][arg] = choose k random individuals drawing from pop[gen − 1]
 6:         end for
 7:     end for
 8: end for
 9: Analyse connected components in pool array and calculate neglected array
10: Randomly initialise individuals in population pop[0] except those
11: marked in neglected[0], calculate fitness values, and store them in vector fit[0]
12: for gen = 1 to G do
13:     for ind = 1 to M do
14:         if not(neglected[gen][ind]) then
15:             for arg = 1 to arity(op[gen][ind]) do
16:                 w[arg]=select winner from pool[gen][ind][arg] based on fitness in fit[gen-1]
17:             end for
18:             pop[gen][ind]=result of running operator op[gen][ind] with arguments w[1], ...
19:             fit[gen][ind] = fitness of pop[gen][ind]
20:         end if
21:     end for
22: end for
```

Fig. 1. EMS-EA from [1]

---

[1] This is because not every sampled individual can become a parent due to the selection pressure.

```
1: Let r be an individual in the population at generation G
2: Choose an operator to apply to generate r
3: Do tournaments to select the parents:
   s₁, s₂, ... = individuals in generation G − 1 involved in the tournaments
4: Do recursion using each unknown sⱼ as a sub-goal. Recursion terminates
   at generation 0 or when the individual is known(i.e. has been evaluated before).
5: Repeat for all individuals of interest at generation G
```

Fig. 2. BC-EA from [1]

The paper concluded that BC-EA can provide around 20% savings in terms of numbers of fitness evaluations than a standard EA when using $k = 2$ and "for very low selection pressures saving of over 35% fitness evaluations are possible" (first paragraph, page 980, [1]). The paper also concluded that BC-EA is superior to EMS-EA due to a special property — *transient period* — of BC-EA.

2.2 Transient Period

BC-EA creates and evaluates individuals recursively from the last generation to the first using depth-first search and backtracking. Therefore, for a given number of individuals that it creates and evaluates at the last generation, the number of distinct weak ancestors of these individuals increases generation by generation (from later generations to earlier generations) until the number hits an upper bound or the initial generation is reached, whichever is earlier. The period, that is, the number of generations between the last generation and the generation where the number of weak ancestors stops growing, is called transient period.

Transient period was claimed to provide BC-EA with an additional source of saving compared to EMS-EA. Individuals that are not sampled during the transient period make an extra contribution to the saving in BC-EA. The longer the transient period, the larger the extra saving. Therefore, differing from EMS-EA, BC-EA was claimed to be able to fruitfully apply to large tournament sizes. "For example, with BC-EA, tournament size 7, and a population of a million individuals — which is not unusual in some EAs such as GP — one could calculate 1 individual at generation 7, 7 individuals at generation 6, 49 individuals at generation 5, etc. at a cost inferior to that required to initialise the population in a forward EA." (fourth paragraph, page 980, [1]).

If two-offspring crossover operators and one-offspring mutation operators are used, and if the upper bound simply is the population size, Equation 2 from [1] estimates the transient period g_e as:

$$g_e \approx \log_k^{N/m} \tag{2}$$

where m is the number of individuals evaluated at the last generation, k is the tournament size and N is the population size.

3   Missing Element

Although the paper [1] investigated the saving ability of BC-EA using tournament sizes 2 and 3 through mathematical and experimental analyses, we think there still has an important investigation missed in [1]. Other than giving an imagined example of using tournament size 7 (see Section 2.2), the paper did not experimentally investigate the search performance, including the effectiveness and the efficiency, of EMS-EA and BC-EA using any larger tournament sizes, for instance 7.

Sometimes low parent selection pressure (small tournament sizes) cannot reliably drive the search to find acceptable solutions within a given number of generations whereas high parent selection pressure (large tournament sizes) can. As a result, when using low parent selection pressure, the total number of generations needed to find an acceptable solution can be much larger than when using high parent selection pressure. Although savings can be obtained from not-sampled individuals at each generation with low parent selection pressure, overall the total number of individuals evaluated can be much larger than the number evaluated when using high parent selection pressure. For this reason, it seems that focusing only on tournament size 2 or 3 to analyse the saving capability of EMS-EA and BC-EA makes the study incomplete.

When investigating whether a variation of an algorithm can provide extra savings, it is necessary to not only compare its efficiency with that of the standard algorithm or other variations using the same set of parameters, but also to ensure it can provide comparable or better problem-solving quality than the standard algorithm or other variations using tuned parameters. It would be inappropriate if the latter has not been taken into account during investigations.

4   Experimental Evaluation

In order to further investigate the saving capability of BC-EA, we followed the instructions given in [1] to reproduce the experiments for their two symbolic regression problems, Poly4 and Poly10 (shown in Equations 3 and 4).

$$f(x_1, x_2, x_3, x_4) = x_1x_2 + x_3x_4 + x_1x_4 \tag{3}$$

$$f(x_1, x_2, ..., x_{10}) = x_1x_2 + x_3x_4 + x_5x_6 + x_1x_7x_9 + x_3x_6x_{10} \tag{4}$$

4.1   Experimental Settings

We constructed two conventional GP systems using tournament sizes 2 and 7 respectively. Four different population sizes (100, 1000, 10000 and 100000) were used for both Poly4 and Poly10 in [1]. From the four population sizes, we arbitrarily picked a population size 1000 for Poly4 and a population size 10,000 for Poly10. We set the maximum number of generations to 50 and conducted 100 independent runs.

Since "one child crossover operators are less efficient in a BC-EA" [1], we use the same two-offspring subtree crossover with uniform random selection of crossover points as in [1] for the sake of favouring BC-EA. Instead of using the point-mutation as in [1], we used the subtree mutation [2] as it is more commonly used in GP.

The function set includes the standard arithmetic binary operators { +, -, * } and a customised {/} operator. The / function returns the numerator if the denominator is zero. The terminal set consists of a single variable x. The fitness functions is the sum of absolute error of the outputs of a program relative to the expected outputs. The crossover rate is 80%, and the mutation rate is 20%. Note these settings are consistent with those in [1].

4.2   Experimental Results

Table 1 shows the performance of using tournament size 2 and 7. For instance, when tournament size 2 is used and the maximum number of generations G is 50, 42 out of 100 runs can find the optimal solution for Poly4. The average number of generations processed to find the optimal solution over the 42 runs is 31 (the standard deviation is 11). The average fitness value over the 100 runs is 4.6 (the standard deviation is 4.2). When tournament size 7 is used and G is 50, 83 out of 100 runs can find the optimal solution. The average number of generations required to find the optimal solution over the 83 runs is 13 (the standard deviation is 11). The average fitness value over the 100 runs is 1.3 (the standard deviation is 3.0).

The results show advantages of using tournament size 7 clearly for both Poly4 and Poly10: 1) more runs can find the optimal solutions, 2) significantly fewer generations are required on average, and 3) the fitness value is better on average. Due to the incomparable problem-solving qualities, it would be inappropriate to compare the saving capability between the use of two different tournament sizes. Therefore, we conducted additional sets of experiments by gradually increasing G for the GP system using tournament size 2 with the aim of making its average fitness value comparable with that in the GP system using tournament size 7. When the maximum number of generations reached 100, the tournament size-two systems were approaching but still worse than the average fitness values of the tournament size-seven systems. The corresponding performance is also reported in Table 1 (records with $G = 100$). The next sub-section analyses the efficiency of the EMS-EA and BC-EA algorithms based on these results.

Table 1. Performance comparison between tournament sizes 2 and 7 for poly4 and poly10 problems

Problem	k	G	# of Success	Generations Required for Success	Sum of Error
poly4	2	50	42	31 ± 11	4.6 ± 4.2
		100	74	47 ± 22	1.6 ± 2.9
	7	50	83	13 ± 11	1.3 ± 3.0
poly10	2	50	0	–	5.4 ± 1.2
		100	7	69 ± 15	4.6 ± 1.8
	7	50	11	36 ± 7	4.4 ± 1.9

4.3   Efficiency Analysis

Equation 2 models the transient period approximately. We think the model can be refined to:

$$g_e \approx \log_k \frac{N(1-\beta)}{m} \tag{5}$$

where β is the expected proportion of a population not involved in any tournaments (see Equation 1). This is because the upper bound should *not* be simply set to the population size but should be calculated by taking off the expected number of not-sampled individuals from the population. We use the refined transient period model in the efficiency analysis.

Table 2 shows the estimated total number of evaluations conducted in the conventional GP, GP with EMS (EMS-GP), and GP with backward-chaining (BC-GP) for Poly4 and Poly10 problems. For instance, for Poly4 with $k = 2$ and $G = 100$, the total number of individuals evaluated in the conventional GP system can be estimated as the sum of average generations required in successful runs and that in unsuccessful runs times the population size. In this case, it is approximately $(74 \times 47 + 26 \times 100) \times 1000 \approx 6.1 \times 10^6$. The total number of individuals evaluated in a GP system with EMS under the same conditions can be easily estimated by taking out of the expected not-sampled individuals from the estimation of the conventional GP. In this case, it is approximately $(74 \times 47 + 26 \times 100) \times 1000 \times (1 - 13.5\%) \approx 5.3 \times 10^6$. Although we have not yet implemented a backward-chaining GP system, we can make an assumption that the optimal solution is the first individual evaluated at the last generation. Such an assumption is the best scenario for a backward-chaining GP system. Under this assumption, we estimated the corresponding total number of individuals evaluated in the following steps:

- calculate the transient period, which is approximately $\log_2^{1000 \times (1-13.5\%)} = 9.75 \approx 10$
- calculate the sum of evaluated individuals within the transient period, which is approximately $\frac{1 \times (1-2^{10})}{1-2} = 1023$
- calculate the total number of evaluated individuals in the whole evolutionary process, which is approximately $74 \times 1023 + (74 \times (47 - 10) + 26 \times 100) \times 1000 \times (1 - 13.5\%) \approx 4.7 \times 10^6$

Note that the table ignored the cases for $G = 50$ and $k = 2$ as they did not provide comparable problem-solving qualities.

The estimated results show that, for Poly4, when tournament size 2 is used, a backward-chaining GP system (BC-GP) in the best scenario evaluated a smaller number of individuals than a GP with EMS (EMS-GP) and the conventional GP. It appears to support the claim made in [1]. However, the conventional GP using the tournament size 7, which provided the similar or even better problem-solving quality, just evaluated less *half* of the individuals evaluated by BC-GP. A similar finding for Poly10 is also obtained.

Table 2. Efficiency comparison between conventional GP, EMS-GP and BC-GP using tournament sizes 2 and 7 for Poly4 and Poly10 problems. Note that the total number of evaluations for BC-GP is estimated according to the best assumptions.

Problem	k	G	Total # of Evaluations ($\times 10^6$)		
			Conventional GP	EMS-GP	BC-GP
poly4	2	100	6.1	5.3	4.7
	7	50	1.9	1.9	1.6
poly10	2	100	97.8	84.6	83.9
	7	50	48.5	48.4	47.9

We now compare the efficiency between the conventional GP, EMS-GP and BC-GP using the tournament size 7. Based on the model given in [1], for tournament size 7, there are about only 0.09% of population not-sampled. Therefore, an EMS-GP will not be able to provide a large saving and will be effectively the same as the conventional GP. However, a BC-GP may be able to provide some saving from the not-sampled individuals in the transient period. If we again assume, in BC-GP, the optimal solution is the first individual evaluated at the last generation, then for Poly4, the transient period is approximately $\log_7^{1000 \times (1-0.09\%)} = 3.55 \approx 4$ generations. Within the *last four* generations, a BC-GP needs to only evaluate approximately $\frac{1+7+49+343}{4 \times 1000} = 10\%$ of the total number of individuals that need to be evaluated in the conventional GP, obtaining 90% saving in *the transient period*. For Poly10, the transient period is approximately 5 generations. Within the *last five* generations, a BC-GP needs to only evaluate approximately 6% of the total number of individuals that need to be evaluated in the conventional GP, obtaining 94% saving in *the transient period*. These results seem to support the claims given in [1]. However, the overall savings in the entire evolutionary processes obtained by BC-GP compared to the conventional GP decreased significantly, only $1 - (1.6/1.9)100\% \approx 16\%$ and $1 - (47.9/48.5) \approx 1\%$ for Poly4 and Poly10, respectively. Furthermore, note that there are two important assumptions made for BC-GP: the last generation is correctly determined and the optimal solution is the first individual evaluated at the last generation. Since neither assumption is likely to hold, the saving capability of BC-EA is much less than claimed in [1].

5   Further Discussion

The capability of EMS-EA to provide computational savings is limited by the tournament size. More precisely, saving can be obtained only when smaller tournament sizes are used. For problems that require high selection pressure for finding acceptable solution, there will be no clear saving.

The capability of BC-EA to provide computational savings is limited by three primitive factors and one derived factor. The three primitive factors are the tournament size as in EMS-EA, the decision on the last generation, and the number of individuals that need to be evaluated at the last generation. The derived factor is the length of the transient period.

Properly determining the last generation is not trivial. This is why another stopping criterion, which is the maximum number of generations without improvement (*max-gwi*) has been used [3]. It is quite possible that no acceptable solution can be found by BC-EA after evaluating all individuals at the predefined last generation, whereas the conventional EA may be able to find an acceptable solution in a later generation by using the max-gwi stopping criterion. Incorrect decisions on the last generation will seriously affect the saving capability of BC-EA and reduce its feasibility.

Suppose the last generation G is correctly determined, the probability of finding the best individual of the last population within a small proportion of the population is the same as the proportion of the population evaluated. For instance, if all the fitness values are distinct, then the probability of finding the best of the population within the first $n\%$ of the population evaluated is just $n\%$. The probability also depends on the proportion of the population having the same best fitness value. For instance, if 50% of the last population has the same best fitness value, then the expected number of individuals that need to be evaluated in order to find the best individual is two. However, such a scenario hardly occurs in EAs for non-trivial problems. Therefore, the probability of finding the best individual within a small number or portion of the last population is low in general, which further reduces the saving feasibility of BC-EA.

The transient period is controlled by the tournament size, the population size, and most importantly the number of individuals evaluated at the last generation G. The transient period will become shorter if more individuals at the last generation will be evaluated, and will not exist if the almost entire population of the last generation will be evaluated. If the transient period does exist, the saving contribution made by the transient period depends on the ratio of the transient period to the total number of generations used. For cases that the predefined number of generations is less than the transient period, the saving will be very significant as imagined in [1] (see Section 2.2). However, the problem solving ability of EAs in the such imagined scenario is really questionable.

Our previous work Ejit [4] is a much simpler algorithm to avoid the evaluations of not-sampled individuals in order to reduce the fitness evaluation cost in standard tournament selection. It follows the standard procedure to create programs at a generation g but *do not* evaluate them. It then samples programs at generation g for tournaments and evaluates the sampled programs if they have not been evaluated. Finally it selects the winners as the parents of programs at the next generation. Unlike EMS-EA and BC-EA, Ejit does not require any extra memory, or any pre- or post-processes. In Ejit, the decision on which program should be evaluated arises naturally from using a passive evaluation order. Furthermore, the efficiency of Ejit is stable and not restricted by the decisions on the number of generations, the length of transient period, and the size of populations. It can provide consistent savings about 14%, 5%, 1.8% and 0.7% for tournament size 2 to 5 respectively.

6   Conclusions

When developing new algorithms aiming at reducing computational cost, the baseline requirement is to at least ensure that the effectiveness of the new algorithms is as good or better than existing standard algorithms. Our analysis based on the experiments using the almost identical settings to that used in [1] shows that BC-EA failed to meet the baseline requirement and the saving capability of BC-EA has many restrictions. In addition, making good decisions on the last generation G and the number of evaluated individuals m at the generation G is very difficult. The difficulty seriously affects the feasibility of BC-EA.

Although the easy deployment makes Ejit significantly attractive for problems that require very low parent selection pressure, similarly to EMS-EA and BC-EA, Ejit works only with standard tournament selection and provides only limited saving for larger tournament sizes. Avoiding the creation and evaluation of not-sampled individuals has limited effect to the development of efficient EAs. Other directions should be investigated, including bloat control [5,6,7,8], but must have effectiveness-first bore in mind.

References

1. Poli, R., Langdon, W.B.: Backward-chaining evolutionary algorithms. Artificial Intelligence 170, 953–982 (2006)
2. Koza, J.R.: Genetic Programming — On the Programming of Computers by Means of Natural Selection. MIT Press, Cambridge (1992)
3. Francone, F.D.: Discipulus Owner's Manual (2000),
   http://www.aimlearning.com/TechnologyOverview.htm
4. Xie, H., Zhang, M., Andreae, P.: Another investigation on tournament selection: modelling and visualisation. In: Proceedings of Genetic and Evolutionary Computation Conference, pp. 1468–1475 (2007)
5. Luke, S., Panait, L.: A comparison of bloat control methods for genetic programming. Evolutionary Computation 14, 309–344 (2006)
6. da Silva, S.G.O.: Controlling Bloat: Individual and Population Based Approaches in Genetic Programming. PhD thesis, University of Coimbra (2008)
7. Kinzett, D., Zhang, M., Johnston, M.: Using numerical simplification to control bloat in genetic programming. In: Li, X., et al. (eds.) SEAL 2008. LNCS, vol. 5361, pp. 493–502. Springer, Heidelberg (2008)
8. Dignum, S., Poli, R.: Operator equalisation and bloat free gp. In: O'Neill, M., et al. (eds.) EuroGP 2008. LNCS, vol. 4971, pp. 110–121. Springer, Heidelberg (2008)

Evolutionary Intelligence and Communication in Societies of Virtually Embodied Agents

Binh Nguyen and Andrew Skabar

Department of Computer Science and Computer Engineering, La Trobe University
Victoria 3086 Australia
tb6nguyen@students.latrobe.edu.au,
a.skabar@latrobe.edu.au

Abstract. In order to overcome the knowledge bottleneck problem, AI researchers have attempted to develop systems that are capable of automated knowledge acquisition. However, learning in these systems is hindered by context (i.e., symbol-grounding) problems, which are caused by the systems lacking the unifying structure of bodies, situations and needs that typify human learning. While the fields of Embodied Artificial Intelligence and Artificial Life have come a long way towards demonstrating how artificial systems can develop knowledge of the physical and social worlds, the focus in these areas has been on low level intelligence, and it is not clear how, such systems can be extended to deal with higher-level knowledge. In this paper, we argue that we can build towards a higher level intelligence by framing the problem as one of stimulating the development of culture and language. Specifically, we identify three important limitations that face the development of culture and language in AI systems, and propose how these limitations can be overcome. We will do this through borrowing ideas from the evolutionary sciences, which have explored how interactions between embodiment and environment have shaped the development of human intelligence and knowledge.

Keywords: context recognition, symbol grounding, embodiment, artificial life, emergent culture and language, evolutionary biology and psychology.

1 Introduction

Providing knowledge is one of the biggest hurdles in developing AI systems to go beyond specialised applications. This includes knowledge of how to reason, learn, communicate, move and perceive, manifested in rules, scripts, templates, algorithms, models, architectures and training data. Researchers working to provide such knowledge have found that that the requirement for knowledge continues to grow, yet it takes considerable time to convert knowledge into a form that AI systems can use. This is known as the knowledge bottleneck problem. Although there are projects which focus on encoding human knowledge and making it available for shared use, they highlight that researchers are still required to provide a great deal of additional knowledge.

K. Korb, M. Randall, and T. Hendtlass (Eds.): ACAL 2009, LNAI 5865, pp. 73–85, 2009.
© Springer-Verlag Berlin Heidelberg 2009

As a result, researchers have tried to offload this work onto AI systems. However, learning even relatively simple concepts requires either a large body of knowledge to be already in place, or a way for the AI program to use context to learn. The problem of requiring context to learn can also be viewed as the more widely discussed symbol grounding problem, where the issue is how to provide meanings for symbols for computers without circular definitions [11] [20]. However, the context point of view is more intuitive to grasp in regards the challenges it creates for AI development. For example, consider the process of teaching a child the meaning of the word "brown." We could point at a brown table, but the child will not know if "brown" refers to a colour, size or shape. How do we begin explain that we are referring to the colour, if the child does not already use our language? According to Dreyfus, the child must be engaged in a form of life in which they share at least some of the goals and interests of their teacher. This way the activity at hand helps reduce the number of possible meanings of the words being used [9]. It is this idea of getting AI systems to engage in a form of life to begin the learning process which is the focus of this paper.

While the fields of Embodied AI and Artificial Life (ALife) have come some way towards demonstrating how artificial systems can develop knowledge of the physical and social worlds, the focus in these areas has been on low level intelligence, and it is not clear how such systems can be extended to deal with higher level knowledge. In this paper we argue that we can build towards a higher level intelligence by framing the problem as one of stimulating the development of culture and language. Specifically, we identify three important limitations that face the development of culture and language in AI systems, and propose how these limitations might be overcome through borrowing ideas from the evolutionary sciences, which have explored how interactions between embodiment and environment have shaped the development of human intelligence and knowledge.

The remainder of the paper is structured as follows. Section 2 will describe the context problem and argue that systems must be engaged in a form of life in order to avoid this problem. Section 3 discusses the how we can view the fields of Embodied AI and ALife as a bottom-up attempt at developing a form of life and intelligence. Section 4 will focus on the role of culture and language in stimulating knowledge development, and Section 5 will extend this by proposing that we can further stimulate development by drawing upon evolutionary disciplines, and especially evolutionary psychology.

2   The Context Problem

Hubert Dreyfus, philosopher and leading critic of AI research, argued that providing knowledge of the human world is the fundamental problem facing AI development. However, because this knowledge is not programmable, we will be better served by developing an AI system with a body that is engaged in situations, pursuing its needs and acquiring knowledge as it learns to cope with the world [9]. Below are the key points of his argument.

The first point is that AI research which uses the information processing model (rules operating on facts) holds that knowledge of the world is a large set of facts. However, this set of facts must include the human world in addition to the physical world. For example, being at home could mean being in my house, which is a physical situation. However, I could also be in my house, but not at home, because I have not yet moved my furniture in.

The second point is that there are difficulties with providing knowledge of human situations. For example, consider the sentence "The box is in the pen." For an AI system to recognise whether the pen is a writing implement or an enclosure for children, it must recognise whether the situation is a James Bond movie (miniaturised box in a writing pen), or parents at home with their children. The information processing model holds that situations are recognisable by their features, which are facts. However, while facts such as the size of the box and pen may seem to be relevant, Dreyfus argues that it is primarily the context (conspiratorial or domestic) which determines which facts are significant in recognising the situation.

The problem for AI systems is that in any situation there are an infinite number of possibly relevant facts from which the system must select a relevant subset. AI systems must use a predetermined set of facts unless it already has the context for that situation (which it can then use to determine the relevant features). However, determining such a fixed set of facts is clearly not possible, since different facts will be relevant in different situations, and will depend on the context. Thus, AI systems require some other means of recognising situations.

An obvious approach is to attempt to program systems to recognise context. However this simply leads to an infinite regress, since in order to correctly identify that context, we must first identify the super context, and so on. To avoid such a regress, we can try to identify some base, or ultimate, context to commence from. But what might such an ultimate context look like?

In the human sphere, we can gain some insight into what such an ultimate context would look like by considering more and more general contexts. For example, in order to identify which facts are relevant for recognising a conspiratorial or domestic situation, we appeal to the broader context e.g., the James Bond movie. But this is only recognisable within the context of entertainment, and entertainment in turn is only recognisable in the context of human activity. Further, according to Dreyfus:

> "human activity itself is only a subclass of some even broader situation - call it the human life-world ... But what facts would be relevant to recognising this broadest situation? ... It seems we simply take for granted this ultimate situation in being people ... It seems then a matter of making explicit the features of the human form of life from within it."

However, reducing the features of the human form of life to rules and facts is a challenging task and has been the implicit goal of philosophers for two thousand years. Without these rules and facts, programming an AI system to recognise more than a handful of situations will be difficult at best.

One way out of this problem might be to consider the way in which human beings transition from one context to another. However, this only changes a hierarchical problem into a temporal one. How did this ability develop? Dreyfus notes that:

> "to the programmer this becomes the question: how can we originally select from the infinity of facts those relevant to the human form of life so as to determine a context we can sequentially update? Here the answer seems to be: human beings are simply wired genetically as babies to respond to certain features of the environment such as nipples and smiles which are crucially important for survival. Programming these initial reflexes and letting the computer learn might be a way out of the context recognition problem..."

Therefore, overcoming the impasse of requiring at the same time both a broader context and an ultimate context requires challenging the assumption that there are no other alternatives. The assumption that there are no other alternatives is only true for someone trying to construct artificial reason as calculation on facts. The way out is to not separate human world facts and situations. This then avoids the problem of how to select facts to recognise the situations from the outside. For an intelligence to select and then interpret facts, it must already be involved a situation.

Thus, the three areas which seem to underlie all intelligent behaviour are: bodies which organise and unify experiences of objects; situations which provide a background against which behaviours can be orderly, without being rule-like; and human purposes and needs, which organise situations so that objects are recognised as relevant and accessible.

3 Bottom Up / Low Level Intelligence

From examining the context problem, we have seen that simulating an early form of human development is a necessary condition for starting the context recognition process. This section discusses the fields of Embodied AI and ALife, and shows how these can be viewed as an attempt at developing intelligence from the bottom up.

3.1 Embodied AI

Embodied AI is a field with diverse forms and goals without a clear definition and often going by different names [1] [28]. The word "embodiment" refers to the state of having a body and for the purpose of simulating the human form of life, having a body means having sensory, general movement, and especially locomotive capabilities [18].

Embodiment research grew out of robotics, where the focus has been on writing programs to control a body [17]. Therefore, it may be unclear how embodiment relates to simulating a form of life. However, if we move our focus from programming intelligence over to evolving intelligence, then we can find research that is more in line with the idea of simulating the early form of human life.

The purpose of stimulating agents to interact with the environment is to enable them to learn background concepts, which are fundamental to gaining higher-level intelligence such as space, time, size, groups, numbers, categories, weight, and gravity. As such, in addition to our working definition, we have two more requirements: adaptive control systems and autonomy.

In order to simulate a human form of life, control systems must be capable of a certain level of adaptability (to enable agents to learn to cope with their bodies) and autonomy (to enable agents to learn independently). Control systems need to be more adaptable than for example, those in insect-like agent systems that consist of small instruction sets with limited extendibility because in those systems the focus has been on designing bodies in a way to reduce computational load. Control systems need to be more autonomous than for example, the agent in Roy (2003). That agent consisted of a head and a neck and was presented with a set of differently coloured objects. The system was then provided with human feedback to help it learn to associate names with the objects, so that it could then name the new objects along with their colours. Though insightful, the system is difficult to extend because it lacks the mechanisms for continuing the learning process independently, since it depended heavily on external reinforcement.

Work that fits the working definition and requirements, focus on creating conditions where the system develops key skills as it engages in an overarching process, such as improving to keep up with competition. By creating a population of agents with random neural networks to perform a goal, there will be some agents who will come closer to the goal. By cultivating those agents and using them to produce the next population, we can evolve systems to become better at reaching the specified goals. In this way, researchers have been able to produce systems with behaviours such as walking, flying, climbing, swimming, jumping, navigating, fighting, and using objects [23].

Though Embodied AI provides useful insights, the focus of research has been on low level intelligence and faces two limitations. Firstly, agents lack key starting predispositions (needs), so they are only driven to fulfil goals that can be explicitly described (e.g., eating the most food). This leaves out a large number of desirable behaviours which cannot be explicitly described (e.g., being a good communicator). Secondly, agents lack pressure from social worlds to compete and to cooperate. Competing and cooperating are requirements in the process of developing knowledge and skills such as naming the world, naming concepts, and sharing information. Artificial Life research has partly addressed these limitations.

3.2 Artificial Life

The area of research which looks at needs, social worlds and life in general is known as Artificial Life (ALife). From ALife research, we gain two powerful points to build upon Embodied AI research: behaviour driven by needs, and further driven by social situations.

ALife research spans a wide spectrum of life systems at different levels. At the lower levels there are projects which look at cellular processes, and at higher levels there are projects which look at how simple rules can produce complex behaviour in populations of agents (such as flocking and swarming). The level of research most relevant to simulating the human form of life is where a population of embodied agents interact with an environment as they pursue their needs, which generally includes eating, mating and staying away from danger. Research in ALife has produced agents capable of basic survival behaviour. There are generally two types of research at this level. The first type focuses on training agent neural networks and then putting them in action [2] [8] [21]. The second type focuses on adaptation during interaction with the world [4] [16] [17] [27].

To overcome the limitation of Embodied AI requiring explicit fitness functions to drive behaviour, ALife research has shown that agents can instead be motivated by their needs and social pressures [27]. By modelling natural and sexual selection, agents can be motivated to act to preserve their genetic line. In addition, to overcome the limitation of Embodied AI lacking pressure to develop knowledge to cope with the social world, ALife research has shown that agents can learn to do so by being part of populations of interacting agents.

Researchers have been able to get embodied AI systems to develop knowledge to cope with the physical world by using neural networks and evolution. However, these systems face the following limitations: they lack key starting predispositions (needs), and they lack pressures to develop knowledge to cope with the social world. Work has been done in ALife to address these limitations. Although these fields provide key foundations, their focus has been on low level intelligence which leaves questions about how to build up from there.

4   Building Upon Low Level Intelligence by Stimulating Cultural Development

How can we build upon low level intelligence? We looked at Embodied AI and ALife because they relate closely with the idea of simulating early human life to initiate the process where humans move from and differentiate one context to another. The problem is that although we can stimulate agents to develop basic survival knowledge, Embodied AI and ALife research has primarily focused on low-level intelligence.

There are a large number of things we could do to stimulate more sophisticated adaptations. We could make the environment more challenging by for example, making it so that agents must go through additional steps to prepare their food; put the right shape into the right slot before receiving food; learn to recognise more types of food; operate a vehicle from one point to another before receiving a reward and so on.

While these changes may produce interesting behaviour, it is unclear whether they will lead to useful high-level intelligence. Abilities that are foundational for useful high-level intelligence include the ability to manipulate concepts and symbols and to engage in natural interaction. Being able to manipulate and combine concepts provides a way to build up knowledge. Being able to manipulate and link symbols with concepts provides a way to share concepts. This empowers concept acquisition, usage and sharing via natural interaction.

4.1   Culture

Natural interaction allows concept transferral. This increases the capability of the agent population as agents make discoveries and acquire concepts. Natural interaction thus improves the chances of survival for the population by facilitating cooperation and by allowing agents to pass on knowledge. It thereby creates a way for knowledge to persist even when individuals pass away [3]. The idea of knowledge transfer and of knowledge existing outside of oneself is referred to in the biological psychology literature as culture [19] and by Embodied AI as scaffolding [18].

How can we stimulate the development of the ability to manipulate concepts and symbols to engage in natural interaction? The ability to manipulate concepts and symbols is required to enable natural interaction. Natural interaction is required to enable knowledge transfer. So it is a matter of creating conditions where knowledge transfer provides evolutionary benefits to motivate developing supporting skills.

Such conditions can be created by using the idea that "evolution is a knowledge gaining process," and that it occurs at three levels [19]. To cope with changes in the environment, life adapts at the genetic level. This may take many lifetimes to achieve (phylogenetic). However, there are times when the environment changes rapidly so adaptations also occur at the personal learning level. This occurs within one's lifetime (ontogenetic). However, knowledge that is hard to come by is lost when an individual passes away and there are situations where knowledge sharing needs to occur quickly. Therefore, adaptations also occur at the third level, which is knowledge transfer or culture. As such, we can view the accumulation of knowledge as part of the life process.

A powerful result of this approach is that agents will also adapt at the biological level to become better learners because being able to use culture is useful in order to be better survivors. Agents will change because the culture that the agents introduce changes world that they inhabit.

Culture has generally been studied at the cognitive level to produce rules to match observed behaviours. If these rules are intended to be used by human beings then the cognitive level is a valid approach. Human beings can use these rules because they can overcome knowledge bottleneck and context problems. If these rules are intended to be used by AI systems then the cognitive approach faces the same problems as classical AI since both are based on the information processing model.

We instead approach culture from an evolutionary point of view, drawing upon Embodied AI, ALife, evolutionary biology, and evolutionary psychology more than classical AI, traditional psychology, and cognitive science. We seek to model rather than to abstract the conditions of early life and to let rules emerge rather than to predetermine rules, in order to increase the open-endedness of our system.

Cultural adaptation involves, among other things, rituals, social norms and changes to the environment. But at the most basic level, it involves language to enable this process to start. How can we affect language development in the first place, and how can we make cultural adaptation increasingly vital?

4.2 Language Simulation

Language research looks at how communication emerges initially in non-communicating agents. Research thus far has established three key points: that language development is possible [24]; that language does convey evolutionary benefits [5]; and that agents can use language to further develop language (structured communication) [5].

Despite these advances, work in language simulation has been difficult to extend because they employ a significant amount of predetermined mental and communication processes. For example, one of the earliest experiments, "Talking Heads," [24] involved a group of agents, with vision systems, heads and torsos, facing a whiteboard with symbols, and tasked with describing what they were seeing to other agents. Although successful, it was difficult to extend because it comprised heavily of predetermined designed algorithms to begin the communication process [15].

Predetermined mental processes are limiting and unnecessary. For example, later projects such as those by Cangelosi [5], Vogt and Divina [25], demonstrated that language simulation is possible with increasingly fewer predetermined mental processes. These projects used adaptive control systems such as neural networks, decision Q-trees, and genetic algorithms to breed agents selectively. In these projects, agents roamed a virtual world, engaged in basic survival behaviours and developed languages to aid themselves. These projects established that agents could develop languages through interaction with the world and that language did indeed provide survival benefits. In addition, these projects established that agents can combine signals to produce structured communication such as "approach" and "mushroom type 1." Although these projects used fewer predetermined processes, they did still rely on a few making them difficult to extend. For example, the system by Cangelosi uses predetermined communication processes. In the Cangelosi experiment, just before agent A acts, the system randomly picks another agent, B. It then places agent B in the same location and with the same orientation as agent A, and prods agent B to name the object closest to agent A and then passes that answer on to agent A.

Like predetermined mental processes, predetermined communication processes are limiting and unnecessary. For example, Marocco and Nolfi [14] demonstrated that language development can occur without the use of both predetermined mental and communication processes. In this project a group of four robots were tasked with locating one of two circles and then to occupy it with one other agent. Agents developed simple signals to indicate their status as they proceeded with the task equivalent to for example, "looking for a circle" and "this circle is full." This project used evolution to produce agents with appropriate neural networks to accomplish the task. The limitation of this project was that by removing hand designed communication processes such as that used by Cangelosi, it lacked a way to update agent networks during their lifetime. In the Cangelosi experiment, the custom communication process created artificial situations where both agents and the system already knew the object of attention without conversing. Thus, the system could provide feedback for agent A because it was just a matter of agent A correctly interpreting the signal about the object from agent B (about whether to eat or avoid the object). Without this custom communication process, there is no way to provide feedback for personal learning.

In addition, language simulation research faces two other limitations: that language has been highly task specific to satisfy fitness functions; and that evolved languages presently only consist of a handful of signals. In the project by Marocco and Nolfi, the evolved languages had a vocabulary that only revolved around the task of looking for and occupying circles. Thus, the evolved language was too specific and limited. In the Cangelosi and Harnad experiment [6], which is one of the most advanced experiments [26], the vocabulary consisted of less than a dozen words. These words covered only food types (e.g., mushroom type 1) and recommended actions (e.g., approach or avoid), therefore agent vocabulary development must be stimulated further in order for agent languages to become non-trivial.

5   Further Stimulating Cultural Development by Drawing Upon Ideas from the Evolutionary Disciplines

The focus of the evolutionary disciplines is on how interactions between embodiment and environment have shaped the development of human intelligence and culture. Evolutionary disciplines look at how various human features developed to aid in the process of life. For example, evolutionary linguistics looks at language [13], and evolutionary psychology looks at mental capacities. In this section, we describe how ideas from the evolutionary sciences can be used to overcome the limitations that have been identified in the previous section.

Although removing predetermined mental and communication processes increases open-endedness, it creates the limitation of lacking a way to provide feedback to update agent neural networks. However, we can use the Hebbian learning model, which more closely simulates the how organic brains work. In this model, connections that activate more often become stronger and connections that activate less often become weaker. This opens the way to explore concepts such as networks that grow not only from one generation to another but that also grow within an agent's lifetime. However, it is unknown whether even basic language will emerge.

The limitation of evolved languages being overly specific to satisfy explicit fitness functions can be overcome by removing fitness functions. However, this creates the problem that suddenly agents lack an impetus to take action since we have removed the selection mechanism. However, we can draw upon ideas from evolutionary biology [7] by modelling natural and sexual selection to stimulate agent action to preserve their genetic line as has been done in ALife projects. Natural selection creates pressures among agents to develop knowledge to survive in a particular ecology. Sexual selection creates pressures among agents of the same sex to develop knowledge to compete for the attention of the opposite sex. Although ALife research has shown that natural and sexual selection can produce agents who can develop knowledge to cope with the physical world [27], it is unknown whether even a basic language will emerge.

Finally, the limitation of how select pressures to encourage further adaptations to support cultural development can be overcome by using evolutionary psychology to systematically select key features to simulate. Key features include increasing the physical realism of the environment, including time to reach maturity [10] and allowing mate selection [12]. When children take longer to mature, they depend longer upon their parents. The longer nurturing period creates opportunities for learning, which is essential for human beings because we are physically less well equipped than a number of our predators and prey. It also creates a pressure for role specialisation in the population such as carers, and providers. All these interactions encourage the development of culture. With mate selection, we can use the pressure arising from choosing and competing for mates to have a better chance of passing genes forward to stimulate further development of knowledge to cope with the social world. Good mates require in addition to good physical traits, the ability to learn and operate well in the social world.

5.1 Experimental Setup

We are currently in the process of testing these ideas empirically. Since we are seeking to simulate early human life and subsequent development of knowledge, the experimental setup foundation requires embodied agents with specific predispositions and an environment for interaction. From this foundation, we can gradually introduce additional key embodiment and environmental features to further drive knowledge development.

We have identified five key constraints that the foundation requires to lead to adaptations to support culture in an open-ended way. Firstly, agents require mortal bodies in order for needs such as eating and mating to drive behaviour in order to model natural and sexual pressures. This is necessary to motivate them to act to preserve their genetic line. This model also creates the opportunity to have different generations existing together. Secondly, agents require the ability to sense and move. Sensors allow agents to learn determine what is relevant rather than having researchers predetermine relevance. Being able to determine relevance is a critical skill required in order to learn in the first place. Agents require movement and locomotive abilities to enable interaction with the world. Thirdly, agent control systems must be adaptive, and capable of learning independently without supervision. Fourthly, the genetic code must be able to encode brain structure (which provides the predisposition to learning), as well as the physical traits of the agent. This is required to evolve better learners by creating conditions where learning provides evolutionary advantages. If child agents are born with parent memories, this removes the motivation to have to pass on knowledge to the next generation. Finally, the environment must be a place where agents can engage in a form of life to gain interaction experiences. Interactions between embodiment and environment are necessary because it drives and shapes knowledge development.

Only a few ALife simulators come close to fitting these requirements. A large number of simulators focused on the cellular level, had abstract worlds, used scripted control systems or lacked sensors. Those that had sensors were often highly abstract. For example they directly conveyed to agents the details of the closest object in the form of a digital bar code representing the object's features [6]. Simulators also often abstracted agent movements to for example, 90 degree turns and single steps forwards. Some simulators abstracted basic actions such as running away and finding food instead of allowing agents to learning to use their bodies to perform these actions [4]. Since the focus of this research is to simulate an increasingly human like early life experience we must increase the level of realism. One of the few systems which met the requirements was Polyworld [27] which in particular, was the only one that provided a vision system. At the moment we are extending the system to enable agent communication and to simulate the maturation process, gender differences and kin identification. Kin identification is necessary because children take longer to develop so they require support from parents and parents must support their children in order to continue their genetic line. In addition, kin identification is necessary to identify partners because supporting children requires parents to cooperate.

Analysis of the system comprises of two stages, firstly detecting the presence of language and secondly interpreting the meaning of utterances. Part one of stage one involves looking at the effect of language on the population. For example, we can look at how long they live and how successful they are at finding food and mating. Part two of stage one involves collecting statistics about the symbols agents use. For example, we can look at whether out of a large number of possible symbols a key subset is being frequently used and we can look at language stability and whether it drifts slowly over time. Once we detect the presence of language, the second stage involves tracking, categorising, and linking signals with situations. Concept categories include for example, physical, social, task-based, tangible, and intangible concepts.

6 Conclusion

In this paper we described the context problem and how we can overcome it by simulating a human form of life to start the process of moving from context to context. We saw how work in Embodied AI and ALife can be viewed as attempts to simulate a form of human life to develop intelligence from the bottom up. However, we also saw that work in these fields focus on low level intelligence and we argued that we can build towards higher level intelligence by framing it as a problem of simulating culture and language. Finally, we identified limitations in language simulation and discussed how ideas from the evolutionary sciences can be used to overcome them. Our focus has been on stripping away artificial elements that inhibit adaptation and to encourage adaptation by more fully modelling life to shape the development of increasingly capable neural networks.

References

1. Anderson, M.L.: Embodied cognition: A field guide. Artificial Intelligence 149(1), 91–130 (2003)
2. Baldassarre, G., Nolfi, S., Parisi, D.: Evolving mobile robots able to display collective behaviours. Artificial Life 9, 255–267 (2003)
3. Boyd, R., Richerson, P.J.: Why Culture is Common but Cultural Evolution is Rare. In: Proceedings of the British Academy, vol. 88, pp. 73–93 (1996)
4. Canamero, L., Avila-Garcia, O., Hafner, E.: First Experiments Relating Behaviour Selection Architectures to Environmental Complexity. In: Proceedings of 2002 IEEE/RSJ international conference on intelligent robots and systems, pp. 3024–3029 (2002)
5. Cangelosi, A.: Evolution of communication and language using signals, symbols and words. IEEE Transactions on Evolutionary Computation 5, 93–101 (2001)
6. Cangelosi, A., Harnad, S.: The adaptive advantage of symbolic theft over sensorimotor toil: Grounding language in perceptual categories. Evolution of Communication 4, 117–142 (2002)
7. Dawkins, R.: The Selfish Gene. Oxford University Press, Oxford (1976)
8. Di Paolo, E.A.: Behavioural coordination, structural congruence and entrainment in a simulation of acoustically couped agents. Adaptive Behaviour 8, 25–46 (2000)
9. Dreyfus, H.L.: What Computers Still Can't Do: A Critique of Artificial Reason. MIT Press, Cambridge (1992)
10. Flinn, M.V., Ward, C.W., Noone, R.J.: Hormones and the Human Family. In: Buss, D.M. (ed.) The Handbook of Evolutionary Psychology, pp. 552–580. Wiley, Hoboken (2005)
11. Harnad, S.: The Symbol Grounding Problem. Phys. D 42, 335–346 (1990)
12. Humphrey, N.: The social function of intellect. In: Humphrey, N. (ed.) Consciousness Regained: Chapters in the Development of Mind, pp. 14–28. Oxford University Press, Oxford (1983)
13. Knight, C., Studdert-Kennedy, M., Hurford, J.R.: Language: A Darwinian Adaptation? In: Knight, C., Hurford, J., Studdert-Kennedy, M. (eds.) The Evolutionary Emergence of Language: Social function and the origins of linguistic form, pp. 1–15. Cambridge University Press, Cambridge (2000)
14. Marocco, D., Nolfi, S.: Emergence of communication in teams of embodied and situated agents. In: Cangelosi, A., Smith, A.D.M., Smith, K. (eds.) Proceeding of the VI International Conference on the Evolution of Language, pp. 198–205 (2006)
15. Nolfi, S.: Emergence of Communication in Embodied Agents: Co-Adapting Communicative and Non-Communicative Behaviours. Connection Science 17, 231–248 (2005)
16. Parisi, D., Cecconi, F., Nolfi, S.: Econets: Neural networks that learn in an environment. Network 1, 149–168 (1990)
17. Pfeifer, R., Iida, F.: Embodied Artificial Intelligence: Trends and Challenges. In: Iida, F., Pfeifer, R., Steels, L., Kuniyoshi, Y. (eds.) Embodied Artificial Intelligence. LNCS (LNAI), vol. 3139, pp. 1–26. Springer, Heidelberg (2004)
18. Pfeifer, R., Bongard, J.C.: How the Body Shapes the Way We Think: A New View of Intelligence. MIT Press, Cambridge (2007)
19. Plotkin, H.C.: Darwin Machines and the Nature of Knowledge. Harvard University Press, Cambridge (1997)
20. Prem, E.: Symbol Grounding Revisited. Technical Report TR-94-19. Austrian Research Institute for Artificial Intelligence. Vienna (1994)

21. Quinn, M., Smith, L., Mayley, G., Husbands, P.: Evolving Controllers for a homogeneous system of physical robots: Structure cooperation with minimal sensors. Philosophical Transactions of the Royal Society of London, Series A: Mathematical, Physical and Engineering Sciences 361, 2321–2344 (2003)
22. Roy, D.: Grounded spoken language acquisition: experiments in word learning. IEEE Transactions on Multimedia 5(2), 197–209 (2003)
23. Sims, K.: Evolving Virtual Creatures. Computer Graphics, 15–22 (1994)
24. Steels, L.: The origins of syntax in visually grounded robotic agents. Artificial Intelligence 103, 133–156 (1997)
25. Vogt, P., Divina, F.: Social symbol grounding and language evolution. Interaction Studies 8, 31–52 (2007)
26. Wagner, K., Reggia, J.A., Uriagereka, J., Wilkinson, G.S.: Progress in the Simulation of Emergent Communication and Language. Adaptive Behaviour 11, 37–69 (2003)
27. Yaeger, L.: Computational Genetics, Physiology, Metabolism, Neural Systems, Learning, Vision, and Behaviour or PolyWorld: Life in a New Context. In: Proceedings of the Artificial Life III Conference, pp. 263–298. Addison-Wesley, Reading (1994)
28. Ziemke, T.: Embodied AI as Science: Models of Embodied Cognition, Embodied Models of Cognition, or Both? In: Iida, F., Pfeifer, R., Steels, L., Kuniyoshi, Y. (eds.) Embodied Artificial Intelligence. LNCS (LNAI), vol. 3139, pp. 27–36. Springer, Heidelberg (2004)

Testing Punctuated Equilibrium Theory Using Evolutionary Activity Statistics

O.G. Woodberry, K.B. Korb, and A.E. Nicholson

School of Information Technology
Monash University
Clayton, VIC 3800

Abstract. The Punctuated Equilibrium hypothesis (Eldredge and Gould, 1972) asserts that most evolutionary change occurs during geologically rapid speciation events, with species exhibiting stasis most of the time. Punctuated Equilibrium is a natural extension of Mayr's theories on peripatric speciation via the founder effect,(Mayr, 1963; Eldredge and Gould, 1972) which associates changes in diversity to a population bottleneck. That is, while the formation of a foundation bottleneck brings an initial loss of genetic variation, it may subsequently result in the emergence of a child species distinctly different from its parent species. In this paper we adapt Bedau's evolutionary activity statistics (Bedau and Packard, 1991) to test these effects in an ALife simulation of speciation. We find a relative increase in evolutionary activity during speciations events, indicating that punctuation is occurring.

1 Introduction

Eldredge and Gould (1972) first presented the Punctuated Equilibrium (PE) hypothesis. In contrast with a possible Darwinian gradualism relying on anagenesis, PE claims that most evolutionary change occurs during relatively short bursts of activity, punctuating long eras of evolutionary stasis. Although there is suggestive paleontological evidence in support of the thesis (Cheetham et al., 1994, e.g.,), the evidence is far from definitive, and the hypothesis has stirred considerable debate (e.g., the Dennett-Gould exchange; Dennett, 1996; Gould, 1997a,b).

In work on the theory of evolutionary Artificial Life, Mark Bedau and collaborators have developed a number of "evolutionary activity statistics" which have been used to assess the extent of evolutionary change in both ALife simulations and in real biology(Bedau and Packard, 1991; Channon, 2006). Whereas these statistics have been widely used to assess the extent to which ALife simulations exhibit, or fail to exhibit, "open-ended" evolution, we judged them to be even better suited to measuring the rate of genetic change in evolutionary processes and so to assessing any association between speciation and punctuation events.

1.1 Speciation

The concept of species remains central to biology, evolutionary and otherwise, and to a host of related fields. However, the concept, just as the concept of life,

K. Korb, M. Randall, and T. Hendtlass (Eds.): ACAL 2009, LNAI 5865, pp. 86–95, 2009.
© Springer-Verlag Berlin Heidelberg 2009

resists efforts to capture it in some precise definition (Hey, 2006). There seems to be an ineradicable vagueness to it, exhibited in application to, for example, asexual species, ring species and hybrids. The most generally accepted definition of species, which we follow here, identifies them with a reproductively isolated sub-population (Mayr, 1963) — that is, a group of actually or potentially inter-breeding populations that are reproductively isolated from other such groups.

Eldredge and Gould (1972, p. 84) asserted:

> The history of life is more adequately represented by a picture of "punc-tuated equilibria" than by the notion of phyletic gradualism. The history of evolution is not one of stately unfolding, but a story of homeostatic equilibria, disturbed only "rarely" (i.e., rather often in the fullness of time) by rapid and episodic events of speciation.

They labeled the contrasting position "phyletic gradualism", the view that evo-lution can only occur gradually and, indeed, can only occur at a constant, contin-uous rate. While acknowledging that the fossil record cannot definitively settle the question of whether most evolutionary change occurs in punctuation events, they asserted that PE theory better matches the paleontological record. Whereas the fossil record is limited in its ability to support or undermine PE, our artificial life simulations can provide clear tests of whether or not PE is actually occurring within different simulations.

On a wider front, Gould used PE theory to support his anti-adaptationist views, claiming that punctuation events are associated with population bottle-necks, pre-adaptations and genetic drift rather than natural selection pressure.[1] In effect, he was saying that rapid shifts in genetics and speciation events are not specifically adaptive, but random. Our simulation, and our use of Bedau's evolutionary activity statistics, has something to say on this subject as well.

1.2   Activity Statistics

Bedau and Packard (1991) introduced statistics for quantifying evolutionary adap-tive activity (cf. also Channon, 2006). An evolving system is represented by a set of adaptive "components" which changes over time. Exactly what components consist of is left open by Bedau, but may be alleles or genetically dependent phenotypic traits. Here, we interpret them as alleles. These components may be introduced into the system via mutation and will usually persist so long as they remain adaptive.

Given a discrete design space of possible components, we can index them and associate a counter with each one. Component activity is recorded by:

$$\Delta_i(t) = \begin{cases} 1 & \text{if component } i \text{ is in use at } t \\ 0 & \text{otherwise} \end{cases} \tag{1}$$

---

[1] For an acknowledgement of the debt his PE theory owed to Mayr and the founder effect see Gould (1989, p. 56).

Note the qualification "is in use at t"; this is meant to exclude the counting of, for example, recessive alleles which are not actually *in use*. Now, $a_i(t)$ will measure the history of the ith component's activity level up to time t:

$$a_i(t) = \sum_{k \leq t} \Delta_i(k)$$

We now define various of Bedau's activity statistics based on these activity counters.

System Diversity: The system diversity, $D(t)$, is the total number of components that have been present in the system up to time t:

$$D(t) = |\{i : a_i(t) > 0\}| \tag{2}$$

Component Activity Distribution: The component activity distribution, $C(t, a)$, measures the frequency of components with exactly activity level a at t:

$$C(t, a) = \sum_i \delta(a - a_i(t)) \tag{3}$$

where $\delta(a - a_i(t))$ is the Dirac delta function, equal to one if $a = a_i(t)$ and zero otherwise.

New Evolutionary Activity: Rates of change in new evolutionary activity are what we are primarily interested in here. New evolutionary activity, $A_{new}(t)$, measures the average activity of *new* components up to time t:

$$A_{new}(t) = \frac{1}{D(t)} \sum_{i : a_i(t) \in [a_0, a_1]} a_i(t) \tag{4}$$

where $[a_0, a_1]$ is selected so as to exclude components which have too little activity, i.e., which have not established themselves in the genome and so may not be adaptively useful, and also to exclude components which have too much activity, i.e., which are fixed in the population and not new. We shall discuss exactly how we selected a_0 and a_1, as well as how we have had to adapt Bedau's statistics, below.

2   Methods

2.1   Simulation

We now briefly describe the ALife simulations we have used in this study. There were three main varieties of experiment conducted:

1. **Patch experiments** restricted agents to 10×10 patches of cells within which they could move, eat and mate freely. Attempts to interact with cells beyond the patch were redirected to the closest cell inside the patch. Interactions between patches were limited to migration actions.

2. **Niche experiments** forced populations to adapt to specific, changing niches. In particular, plants (or simply the cells containing the plants) carried 100-bit signatures which agents had to match genetically in order to be nourished.
3. **Reduced mobility experiments** kept agents very close to their parents throughout their lifetimes, thereby reducing gene flow within species and thus increasing the chance of speciation events.

World: The basic simulated world consisted of a 30×30 grid of cells, wrapped so that the edges meet, forming a torus. The agents within cells interacted with the nine cells in the Moore neighbourhood.

Cells: Each cell had an agent population, unlimited in size, and a food store. In **niche experiments** cells also had a 100-bit niche signature.

Cell updating: In each simulation cycle all cells were updated, in a random order. During cell updating, first the contents of the food store were distributed, evenly, to the nine neighbouring cells; then the food store was replenished with one unit of new food (enough to sustain a single agent). In **niche experiments**, there was a 0.01 probability that the niche signature would be updated. When updating occurred, the signature was crossed-over with a signature from a randomly selected neighbouring cell, with a 0.002 probability of mutation flipping each bit.

Agents: Each agent had an age, a health level and a 100-bit mating signature. In **niche experiments**, agents also had a 100-bit eating signature, which was matched against the cell niche signatures.

Agent updating: Each cycle all agents consumed one unit of energy (taken from their health store). They then had an opportunity, in randomly selected order, to move, eat and reproduce, after which they were tested for death conditions.

Agent moving: Movement action occurred with a 0.3 (0.0, in **reduced mobility experiments**; i.e., "movement" only occurred across generations, as with plants) probability each cycle. This caused the agent to move to a randomly selected cell from the agent's neighbourhood. In **patch experiments** agents also had an 0.002 probability of migrating between patches, which was otherwise impossible. In order to allow for founding new colonies, when migrating, a companion was randomly selected from neighbouring agents, and the two agents moved together into a neighbouring patch.

Agent eating: Eating occurred every cycle. A random neighbouring cell was selected and energy was transfered to the agent's health, which was capped at 10 units. In **niche experiments**, agents could only eat from cells with niche signatures similar to their own eating signatures, in particular, when the Hamming distance between the two signatures was less than 10. Edible cells in the neighbourhood were first identified and, if any existed, one was randomly selected for food supply.

Agent mating: Mating was possible when the agent's health was at 10 units. This meant that agents were, potentially, mature one cycle after birth, when enough food was available. A random mate was selected from the mature agents in the neighbourhood. Assuming one was found, the two agents were then checked for mating compatibility, that is, for a Hamming distance between their respective mate signatures being less than 5. Given compatibility, the agents reproduced sexually and an offspring was created, with an initial health of 5 units, via equal health donations from both parents. The offspring inherited a mate signature (and eat signature in **niche experiments**) which were formed via cross-over of parent signatures, with a 0.002 probability of mutation flipping each bit.

Agent death: The agent died whenever: its health fell below zero; its age exceeded 10 cycles; or, it died of external, accidental causes, as determined by an 0.1 accident probability each cycle.

We simulated a closed ecosystem, allowing population levels to stabilize without having to impose arbitrary limits on cell capacity. Having a closed ecosystem required us to remove dead agents from the board and recycle any remaining energy held (as health) through the growth of new plant food.

Species: Species are reproductively isolated agent groups within the simulation. Every 100 cycles the simulation agents were clustered into species based on mate compatibility. Statistics of the species' agents were collected and used to measure punctuation events and the activity levels of the agents' mate signatures. These statistics are discussed below in §2.2. In order to ensure a continued production of new species, each cycle a species had a $5 \times 10^{-6} \times Size$ probability of being killed off. To simplify the simulation, once new species were identified, they were no longer permitted to inter-mate.

2.2   Adapting the Activity Statistics

In order to measure the rate of evolutionary change in the species of our simulation we had to adapt Bedau's activity statistics described above.

Evolutionary Components: The first problem was to identify the components we should be counting. Since we were interested in rates of genetic change in diverging species and sub-species groups, it was natural to focus on the mating signatures. Since those signatures have no functional substructure (they only mattered for computing Hamming distances), any fixed substrings could be used to identify components. However, taking either too long or too short a substring of those signatures would cause problems. Too long substrings would lead to very few components showing enough activity to be counted as adaptive (i.e., greater than a_0); too short substrings would lead to components that are present too often. We found that bit strings of length four, with sixteen possible components, worked well (for a total of 400 possible components).

Activity Counters and Increment Function: Activity counters were kept for each of the species identified. When a speciation event occurred, counters were inherited from the parent species. Activity counters were incremented whenever a successful mate occurred within a species — using the mate signatures of the two reproducing agents. To compensate for the differences in species size, we report per capita activity counters in our statistics. As the simulation had bounded diversity, and in particular a fairly small space of possible components, we measured the activity counters only over a period of 200 cycles (an "epoch"), after which they were all reset.

Neutral Shadow: In order to assess activity statistics for *adaptive* components, we had to subtract background non-adaptive activity. We did this using a "neutral shadow" simulation, applying the results to determine values for a_0 and a_1 in equation 4. That is, for each epoch we kept a shadow population corresponding to the normal population. At the beginning of each epoch the two were "re-synched", with the foreground population being duplicated into a new neutral shadow world. Thereafter, births and deaths in the foreground triggered equivalent events in the shadow run, but using randomly selected agents, thereby eliminating any effect of selective pressure on genetic change. As the shadow population had no species, activity counters were kept for the entire shadow population. (For more discussion of this kind of neutral shadowing, see Channon, 2006.)

3   Results and Discussion

As outlined in Section 2.1, three different simulation environments were conducted to induce speciation. In Section 3.1 we examine the speciation effects in each of the experiments; in Section 3.2 we examine their activity statistics.

3.1   Speciation Experiments

In this set of experiments, simulations were run without any species extinction events to determine baseline average numbers of species (see Table 1). For contrast, a null run without any of the speciation triggers was included. In the null run, one species dominated, and any new species were quickly driven extinct. Note that the movement rate was selected for this to occur. If the world size were greater, or the movement rate was lower (as in the reduced mobility experiment), then we would expect a greater number of species to be sustainable. The average number of species in each of the other experiments demonstrated that the introduced mechanism was enough to maintain multiple species.

Table 1. The average number of species present in each experiment. A null run, without speciation triggers, is included for contrast.

Experiment	Avg. #Species
Null	1
Reduced Mobility	17.5
Patches	13.6
Niches	11.8

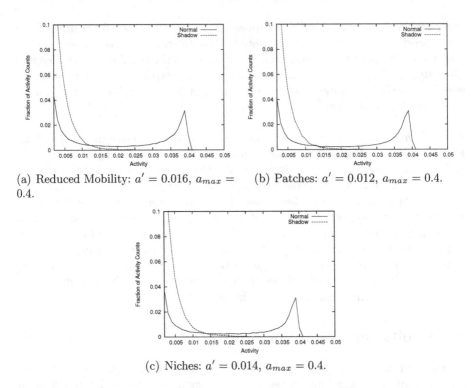

(a) Reduced Mobility: $a' = 0.016$, $a_{max} = 0.4$.

(b) Patches: $a' = 0.012$, $a_{max} = 0.4$.

(c) Niches: $a' = 0.014$, $a_{max} = 0.4$.

Fig. 1. Component activity distributions of normal and shadow runs for each experiment. Components with low activities have little adaptive value and components with high activities are fixed in the species. Components with activity levels in-between are relatively new and relatively adaptive. The shadow runs have no high end activity peak, since the components appear and disappear at random.

3.2   Activity Experiments

According to PE, we expect to see a **relative** boost in new evolutionary activity during speciation events. To test this hypothesis we used the activity statistics to measure the rate of evolutionary change lineages undergo during speciations.

First we needed to determine values for a_0 and a_1 in Equation 4 for each simulation. We found the component activity distributions for each experimental run and plotted them against the distributions from the corresponding neutral shadow. Figure 1 shows these plots, where the distributions have been summed over the epoch. The shapes of the distributions are similar, however the crossover point varies. The normal runs have two peaks: one at low activities, which are components with little adaptive value introduced by mutation and quickly removed by natural selection; and one at the high end, which are components that are well fixed in the species throughout the epoch. Components with activity levels between these peaks are relatively new and relatively adaptive. The shadow runs have no high end activity peak, since nothing is very adaptive, and an inflated low end peak, since the components appear and disappear at random.

Table 2. Selected values for a_0 and a_1, based on intersection point, a', between normal and shadow runs

Experiment	a'	a_{max}	a_0	a_1
Reduced Mobility	0.012	0.04	0.0078	0.0162
Patches	0.012	0.04	0.0078	0.0162
Niches	0.014	0.04	0.0101	0.0175

Table 3. Comparison of new activity in lineages which underwent speciation events and those that did not

Experiment	Speciation activity		No speciation activity	
	Mean	SE	Mean	SE
Reduced Mobility	0.00043	6.13308E-6	0.00035	1.1189409E-5
Patches	0.00037	7.4510413E-6	0.00026	8.948636E-6
Niches	0.00034	8.707967E-6	0.00026	8.521615E-6

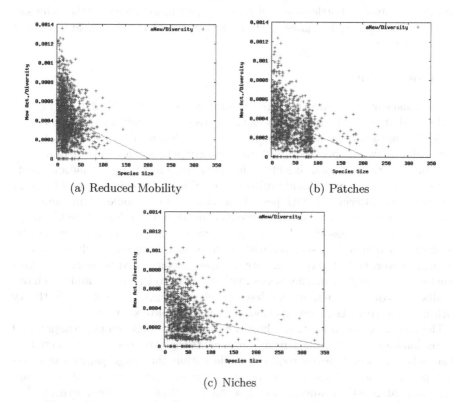

(a) Reduced Mobility   (b) Patches

(c) Niches

Fig. 2. New activity levels versus population sizes, with regression lines

The intersection point, a' between the normal and shadow distributions and the highest activity level, a_{max}, were identified and used to set values a_0 and a_1 to be $a' \pm (0.15 \times (a_{max} - a'))$, identifying a middle range of new and adaptive components per simulation (see Table 2).

With the new activity statistics defined, we compared them with speciation activity. Two tests were conducted: first, we compared new activity levels in lineages which underwent a speciation event versus those that didn't; second, we tested for a relationship between new activity levels and species' bottlenecks.

Table 3 shows a comparison of the means (and their standard errors) of new activity levels in lineages experiencing speciation versus no speciation for each experiment. It can be seen that, on average, lineages with speciation events have higher new activity levels, as predicted by PE. All results yielded a statistically significant difference, with new activity levels higher in lineages associated with speciation events (p values were well below 0.001 in all cases).

Figure 2 shows the relationship between new activity and the species bottleneck for each experiment. The founder effect predicts a greater amount of genetic change during group bottlenecks. We expected that punctuation events would correspond with these lineage bottlenecks. Fitting regression lines to the plots in Figure 2 shows that there is indeed a correlation between new evolutionary activity and lineage bottlenecks (all regression coefficents were statistically significant). Lineages with smaller bottleneck sizes tend to have a greater amount of new activity.

4 Conclusion

Our simulations illustrate the experimental potential of ALife for testing hypotheses that are difficult or impossible to test adequately in the non-virtual world. Using Bedau's evolutionary activity statistics, we have demonstrated an association between speciation and evolutionary activity in a variety of scenarios conducive to speciation. It's clear that something like the punctuation events posited by Eldredge and Gould (1972) is occurring, with periods of rapid change and speciation alternating with periods of relative stasis. Some might think it is tautologous to claim that speciation events coincide with periods of rapid change in genes specifically associated with reproductio — and the genes we specifically monitor *are* exclusively associated with reproduction. What is clearly not tautologous, however, is that periods in between speciations should be associated with genetic stasis, which is another necessary ingredient to PE theory and which our results also confirm. In short, we have developed a genuine test of PE theory with multiple speciation processes, and all of them have supported that theory.

The fact that our simulations show this association between punctuation and speciation based upon evolutionary components that are specifically selected for their adaptive value (being required to lie within the range $[a_0, a_1]$) suggests that punctuation and speciation are not occurring *independently* of the adaptive value of genetic changes. Gould seemed to think otherwise, writing that PE theory leads to a "decoupling" of macro- and microevolution, i.e., leaving macroevolution inexplicable in terms of the natural selection of adaptive traits (Gould, 1993, p. 224). Our simulations, however, while having random elements, were primarily driven by natural selection, and our activity statistics were restricted to components with demonstrated adaptive value, so they clearly suggest that a "recoupling" of macro- and microevolution is in order.

References

Bedau, M.A., Packard, N.H.: Measurement of evolutionary activity, teleology, and life. In: Langton, C.G., Taylor, C., Farmer, J.D., Rasmussen, S. (eds.) Artificial Life II, pp. 431–461. Addison-Wesley, Reading (1991)

Channon, A.: Unbounded evolutionary dybnamics in a system of agents that actively process and transform their environment. Genetic Programming and Evolvable Machines 7, 253–281 (2006)

Cheetham, A., Jackson, J., Hayek, L.-A.: Quantitative genetics of bryozoan phenotypic evolution. Evolution 48, 360–375 (1994)

Dennett, D.: Darwin's Dangerous Idea. Simon & Schuster, New York (1996)

Eldredge, N., Gould, S.J.: Punctuated Equilibria: An Alternative to Phyletic Gradualism. In: Schopf, T.J.M. (ed.) Models in Paleobiology, pp. 82–115. Freeman, Cooper and Company, San Francisco (1972)

Gould, S.: Punctuated equilibrium in fact and theory. In: Somit, A., Peterson, S. (eds.) The dynamics of evolution, pp. 54–84. Cornell University, Ithica (1989)

Gould, S.: Punctuated equilibrium comes of age. Nature 366, 223–227 (1993)

Gould, S.: Darwinian fundamentalism. New York Review of Books, 34–37 (June 12, 1997a)

Gould, S.: Evolution: The pleasures of pluralism. New York Review of Books, 47–52 (June 26, 1997b)

Hey, J.: On the failure of modern species concepts. Trends in Ecology and Evolution 21, 447–450 (2006)

Mayr, E.: Animal Species and Evolution. Harvard University Press, Cambridge (1963)

Evaluation of the Effectiveness of Machine-Based Situation Assessment

David M. Lingard and Dale A. Lambert

Defence Science and Technology Organisation,
PO Box 1500, Edinburgh, SA 5111, Australia
David.Lingard@dsto.defence.gov.au

Abstract. This paper describes a technique for measuring the effectiveness of machine-based situation assessment, one of the levels of data fusion. Using the computer to perform situation assessment assists human operators in comprehending complex situations. The evaluation technique is an iterative one that utilises a metric to measure the divergence between the situation assessment and the ground truth in a simulation environment. Different pieces of divergent information can be weighted separately using methods based on the Hamming distance, the number of antecedents, or a Bayesian approach. The evaluation technique is explored using Random Inference Networks and shows promise. The results are very sensitive to the phase of the inference network, i.e. stable, critical or chaotic phase. Key Words: situation assessment, inference network, measure of effectiveness.

1   Introduction

The most dominant model of data fusion is the Joint Directors of Laboratories (JDL) model [1]. The two levels of data fusion from that model that are most relevant to this paper are: [2], [3]

1. **Level 1**. Object assessments are stored representations of objects. The object assessment process is usually partitioned into data registration, data association, position attribute estimation, and identification.
2. **Level 2**. Situation assessments are stored representations of relations between objects. Situation assessment fuses kinematic, temporal and other characteristics of the data to create a description of the situation in terms of indications and warnings, plans of action, and inferences about the distribution of forces and flow of information.

This paper is primarily concerned with algorithms that are developed to perform machine-based situation assessments, thus assisting the human operators in comprehending complex military situations. In particular, this paper is concerned with how to assess the effectiveness of these algorithms.

In relation to machine-based situation assessment, there needs to be a way to represent the domain of interest in a meaningful way. An ontological framework

K. Korb, M. Randall, and T. Hendtlass (Eds.): ACAL 2009, LNAI 5865, pp. 96–105, 2009.

named "Mephisto" is being developed to achieve this goal [4], [5]. An ontology is the systematic specification of the concepts required to describe the domain of interest. Mephisto's layers are the metaphysical, environmental, functional, cognitive and social layers.

The ontology can be expressed in a formal logical language. A formal logic is a formal language together with an inference relation that specifies which sentences of that formal language can be inferred from sets of sentences in that formal language. A formal theory is a set of sentences expressed in a formal language. When combined with a formal logic, inferences from a formal theory can be made that describe a domain of interest in the world.

The goal of the algorithm for machine-based situation assessment is to determine the most likely state of the domain of interest. For example, [2] develops a modal model for the state of the domain of interest that comprises a set of possible worlds. The maintenance of a probability density function over the possible worlds allows the most likely possible world to be determined.

Several nations have collaborated through the Defence-orientated Technical Cooperation Program to develop and test algorithms for machine-based situation assessment. To facilitate this, they have developed a detailed "North Atlantis" simulation scenario [6]. North Atlantis is a fictional continent between Europe and Greenland and comprises six nations with various alliances and hostilities.

A complete formal theory could in principle be developed for the ground truth of the scenario. This represents an omniscient view of the scenario in the simulation environment where the state of each entity and the relationships between entities are correctly understood and articulated. In practice, this comprises a set of ground truth propositions, and becomes the standard to test the machine-based situation assessments against.

In the simulation environment, a given algorithm for machine-based situation assessment will also produce a set of propositions to describe the situation in the scenario. At this stage, to simplify the process, it will be assumed that the information that the situation assessment algorithm receives from the object assessment is perfectly accurate and fully comprehensive information. Even so, the Situation Assessment (SA) algorithm may be imperfect, and so the SA propositions may diverge from the set of Ground Truth (GT) propositions. To assess the performance of the SA algorithms, there needs to be a method for comparing the set of SA propositions with the set of GT propositions. The aim of this paper is to describe a methodology that could be employed when simulation experiments are performed, developing further the ideas in [7].

Section 2 discusses the proposed methodology for assessing the performance of machine-based situation assessment. Section 3 illustrates this methodology using a simple example. Section 4 shows how this methodology can be explored using Random Inference Networks. Section 5 contains some simulation results and discussion. Section 6 outlines conclusions.

2   Iterative Correction Process

This section will firstly describe situation assessment in a network paradigm, and will then describe the iterative correction process for measuring the effectiveness of the situation assessment algorithm.

With machine-based situation assessment, the processing takes place at the semantic level and involves logical propositions that describe the situation of interest. An Object Assessment may provide a set of atomic propositions, such as *Vessel 6 is named the Ocean Princess*. The machine-based situation assessment takes these atomic propositions from the Object Assessment (call them OA propositions for convenience) and applies logical rules to infer conclusions. A simple example of such a rule might be:

(X has recently been communicating with Y) AND (Y is a member of the secret service of Z) → (X is associated with the secret service of Z)

For example, two input propositions might be *John Brown has recently been communicating with Andrew Smith*, and *Andrew Smith is a member of the secret service of Orangeland*, and thus the rule results in the output proposition *John Brown is associated with the secret service of Orangeland*. In this way, the application of the rules to the OA propositions results in output propositions that in turn provide input to other rules in the rule set. This can be described using a network diagram that has a tree structure of logical propositions with OA propositions at the bottom as leaves, and with propositions comprising the situation assessment towards the top. Each output proposition forms a node that is linked by edges to the input propositions involved in the creation of that output proposition via a rule. Each edge is directed from the input proposition to the output proposition. A logical proposition can only have two states, TRUE or FALSE. Each rule corresponds to a Boolean function with the input propositions and output proposition as the inputs and output respectively.

An iterative correction process for comparing the set of SA propositions with the set of GT propositions was introduced in [7]. A brief description of the process will be given here, and it will be illustrated using a simple example in Section 3. The main steps in the process are:

1. Use some form of metric to measure, in an overall sense, how well the SA propositions match the GT propositions.
2. Identify all SA propositions that satisfy the following conditions:
   (a) The SA proposition is TRUE, but the corresponding GT proposition is FALSE, or vice versa. For convenience, call this a mismatch. (On the other hand, a correct match is when the SA proposition and corresponding GT proposition are both TRUE, or both FALSE.)
   (b) The input SA propositions to the rule that created the SA proposition correctly match the corresponding GT propositions.
3. Correct all the SA propositions identified in step (2) by over-riding the SA algorithm and making it produce correct matches for those propositions.
4. Re-calculate the rules to produce a new set of output propositions for the SA algorithm and go to Step (1).

Step (2) identifies cases where there is clearly a problem with a SA rule since it takes in correct inputs, but produces an incorrect output. In Step (3), the iterative correction process forces these faulty rules to produce correct results. The technical expert can carefully examine these rules offline to determine how to repair them for the next release of their SA algorithm. The corrections made in Step (3) will have a ripple effect during calculation of the remaining rules in the rule-set. Some mismatches may disappear, and some new candidates for Step (2) may surface.

One possible scenario is that the overall match between the situation assessment and ground truth will improve with each new iteration of the process described above. Eventually, after a finite number of iterations (N), Step (1) may indicate a perfect match between the SA and GT propositions. The set of values of the metric calculated in Step (1) across the N iterations provides a measure of the performance of the SA algorithm, and the algorithm developer obtains a list of faulty rules that require more development effort.

Three options have been explored for the metric in Step (1). The most basic option is to employ the Hamming distance. This compares each SA proposition with its corresponding GT proposition and counts the number of mismatches.

The other options make allowance for the fact that some mismatches may be deemed to be more important than others. Thus the second and third options allow the various mismatches to be weighted differently before they are summed to form an overall measure of mismatch between the situation assessment and ground truth.

The second option uses the number of antecedents as the weighting factor for a mismatch. For proposition A to be an antecedent of proposition B in the tree structure described above, there must be a direct path of links leading from proposition A to proposition B. The number of antecedents of proposition B is the cardinality of the set of antecedents of proposition B. The concept is that mismatches "higher" in the tree will tend to be more important, especially where they have resulted from the fusion of many pieces of information in the form of propositions. This higher level information is generally more meaningful to the user than lower-level information "lower" in the tree structure. A reasonable measure of a proposition's level of information is its number of antecedents.

The third option assigns an error value between 0 and 1 to each mismatch, and these errors propagate through the network described above in a Bayesian fashion. This is best illustrated using the simple example described in the next section.

3   A Simple Example

This section describes a simple example to illustrate the iterative correction process.

Table 1 shows the calculation of the three different types of metric (Hamming, Antecedent and Bayesian for short) for a hypothetical simple network in the third iteration of the iterative correction process. There are three OA propositions and

Table 1. A simple example illustrating calculation of the measures of effectiveness of a hypothetical SA algorithm in the third iteration of the iterative correction process. The results in boldface are the mismatches.

Node	1	2	3	4	5	6	7	8	9	10	11	12	13	
GT	1	0	0	1	0	0	1	1	**1**	**0**	**0**	**1**	0	
SA	1	0	0	1	0	0	1	1	**0**	**1**	**1**	**0**	0	Total:
Ham	0	0	0	0	0	0	0	0	**1**	**1**	**1**	**1**	0	4
Ant	0	0	0	0	0	0	0	0	**7**	**8**	**9**	**11**	0	35
Bay	0	0	0	0	0	0	0	0	**1**	**0.9571**	**0.6788**	**0.6105**	0	3.2464
Input 1									6	9	8	10		
Input 2									7	8	9	11		

10 output propositions numbered sequentially 1 to 13 in the first row. Whether each proposition is TRUE or FALSE is indicated in the second and third rows for the Ground Truth (GT) and Situation Assessment (SA) versions of the network. One indicates TRUE and zero FALSE. The measure of mismatch is shown for each proposition in rows four to six for the three types of metric: Hamming, Antecedent and Bayesian respectively. For convenience, call these measures the Hamming Divergence, Antecedent Divergence and Bayesian Divergence. The last two rows in the table show the input links for relevant nodes, e.g. nodes 9 and 8 provide input to node 10. Inputs are assigned for the other nodes (4–8, 13), and Boolean functions are assigned for nodes 4–13, but these aren't described in the table.

In Table 1, the mismatches only occur for nodes 9 to 12 in the third iteration. Node 9 is a special case because it satisfies the two conditions described above for Step 2 of the iterative correction process. This node will be corrected in this iteration and forced to provide the correct output, and this will affect the divergence measures obtained in the fourth iteration.

The Hamming Divergence is one for each mismatched node, and thus the total Hamming Divergence is 4 for the third iteration. Note that each mismatch has the same weight.

The Antecedent Divergence for nodes 9 to 12 is just the number of antecedent nodes for each node, and the total of these gives 35 as the Antecedent Divergence for the third iteration. The mismatch for node 12 is weighted more heavily than the other mismatches.

The Bayesian Divergence is obtained by propagating error values through the input links. The error values are defined on the interval $[0, 1]$. If there is a correct match for the node between the ground truth and SA algorithm, then the Bayesian Divergence is set to zero. If the two conditions described above for Step 2 of the iterative correction process are met, then the Bayesian Divergence is set to 1. This is the case for node 9 in Table 1. Otherwise a table of conditional probability values is consulted to determine the Bayesian Divergence for a node. For each node there are four conditional probabilities (assuming two inputs):

Table 2. Hypothetical conditional probability values for nodes 10, 11 and 12 in the simple example

Input 1	Input 2	Node 10	Node 11	Node 12
no error	no error	1	1	1
error	no error	0.9571	0.7208	0.0816
no error	error	0.4399	0.6788	0.2745
error	error	0.6015	0.2128	0.8675

P(error in the node | no error in Input 1 & no error in Input 2)
P(error in the node | error in Input 1 & no error in Input 2)
P(error in the node | no error in Input 1 & error in Input 2)
P(error in the node | error in Input 1 & error in Input 2)

   In this study the conditional probabilities were set randomly, and hypothetical values for nodes 10, 11 and 12 are shown in Table 2. If there is no error in the inputs to a node, but the node is mismatched, then the conditional probability is one, resulting in a Bayesian Divergence of one as described in the previous paragraph.

   The Bayesian Divergence for node 12 is calculated by:

P(error in node 12 | no error in node 10 & no error in node 11) ×
        P(no error in node 10) × P(no error in node 11) +
P(error in node 12 | error in node 10 & no error in node 11) ×
        P(error in node 10) × P(no error in node 11) +
P(error in node 12 | no error in node 10 & error in node 11) ×
        P(no error in node 10) × P(error in node 11) +
P(error in node 12 | error in node 10 & error in node 11) ×
        P(error in node 10) × P(error in node 11)

$= 1 \times (1 - 0.9571) \times (1 - 0.6788) + 0.0816 \times 0.9571 \times (1 - 0.6788)$
        $+0.2745 \times (1 - 0.9571) \times 0.6788 + 0.8675 \times 0.9571 \times 0.6788$
$= 0.6105$

This assumes that the error in Input 1 is independent of the error in Input 2. This is valid for the great majority of nodes in this study, although for a few nodes the inputs may be dependent. Independence was assumed as an approximation in this study for the sake of simplicity.

   The technical experts who develop the SA algorithm and assess its performance would need to think carefully about what values to assign to the conditional probabilities. In this study values were randomly assigned to study the generic properties of the network of propositions and Bayesian Divergence metric. However, in a realistic situation, the conditional probabilities would be set to reflect expert judgement about how the error in a node depends on the errors in its inputs. Effectively, the error in a node can be discounted by apportioning some of the cause of the mismatch for that node to the erroneous inputs.

   In Table 1, it can be seen that summing the individual Bayesian Divergences for each node gives an overall Bayesian Divergence for the third iteration of 3.2464. The individual Bayesian Divergences for each node are the weighting factors.

4   Random Inference Networks

This section describes how the iterative correction process described in the previous section can be explored in a generic manner using Random Inference Networks (RIN). The structure of the RIN model employed in this study is described below.

The RIN model consists of N_{OA} OA propositions and N_P output propositions. Each output proposition is linked to K other output propositions or OA propositions that provided the inputs to the rule that created the output proposition. K is the same for each output proposition. A Boolean function is randomly assigned to each output proposition representing the logical content of the corresponding rule.

When randomly choosing the K inputs for each output proposition, an approach is employed similar to that described in [8] for growing directed networks. The OA propositions have identification numbers $1, 2, \ldots, N_{OA}$. The output propositions are created sequentially and have identification numbers $1, 2, \ldots, N_P$. In total there are $N_{OA} + N_P$ nodes in the RIN, identified as $1, 2, \ldots, N_{OA}, N_{OA} + 1, N_{OA} + 2, \ldots, N_{OA} + N_P$, starting with the OA propositions. For the first output proposition, inputs are randomly chosen from the OA propositions according to some Probability Mass Function (PMF). For later output propositions, inputs are chosen from the previous output propositions and OA propositions according to the same PMF.

A moving window can be employed as the PMF to choose K inputs. This is a uniform distribution defined on the N_W propositions that are directly prior (according to the ordering defined above) to the node being created. That is, during creation of node J, the inputs are chosen on a uniform distribution defined on nodes $J - N_W, J - N_W + 1, \ldots, J - 1$. No duplicate inputs are allowed for a given output proposition. This is called the moving window version of the RIN. It is convenient to set $N_W = N_{OA}$.

This study used RINs to model the generic properties of situation assessment algorithms, and to study the impact of defective rules on their performance. The first step is to assign the K input links and the Boolean function for each of the N_P output propositions as described above. The N_{OA} OA propositions are randomly initialised to 0 or 1. Then the Boolean states (0 or 1) of the N_P output propositions are calculated sequentially. This forms the baseline or "ground truth" version of this specific RIN. The corrupted version of the RIN models the real-world imperfect situation assessment algorithm, and is formed by altering some of the assigned Boolean functions or input links. Then the N_P output propositions are calculated sequentially for the corrupted version of the RIN. Finally, the ground truth and situation assessment versions of the RIN can undergo the iterative correction process described in Section 2 to produce a quantitative measure of the effectiveness of the SA algorithm.

5   Results and Discussion

This section describes the simulation results when the RINs described in Section 4 were used to explore the iterative correction process described in Section 2.

A single ground truth network was created for a given value of the connectivity (K) by assigning input links and Boolean functions, and randomly initialising the OA propositions. Then N_C corruptions of this network were created to simulate SA algorithms with varying degrees of imperfection. For each corrupted network, the probability that a rule was corrupted was taken from a uniform distribution on the interval $[0, 1]$; this probability was fixed across the network. Thus there was a spectrum of corrupted networks varying from little or no corruption, to cases where nearly all rules were corrupted. When a rule was corrupted, there was 50% probability that the input links were altered; otherwise the Boolean function was changed. For this paper, the following parameter values were employed: $N_{OA} = 50$, $N_P = 5000$ and $N_C = 5000$. Then the variation of the measure of effectiveness of the SA algorithm could be examined as a function of: the degree of corruption, the type of metric (Hamming, Antecedent or Bayesian), and the connectivity (K).

Fig. 1 shows the divergence versus the number of corrupted rules in the version of the network simulating the SA algorithm. Plots A, B and C show the Hamming Divergence for connectivity values of 1, 2 and 3 respectively. Plot D shows the Bayesian Divergence for a connectivity value of 2. The left side of each plot corresponds to no corruptions, while the right side of the plot corresponds to all the rules being corrupted. The top curve in black gives the divergences for the first iteration, the next curve (in grey) gives the divergences for the second iteration, the next curve in black, third from the top, is for the third iteration, etc.

Firstly focus on plots B and D for a connectivity of 2. The greatest number of iterations required for the divergence to converge to zero, indicating a perfect match between the ground truth and corrected situation assessment, was 17. The divergence values clearly increase as the SA algorithm becomes more corrupted. This indicates that the divergence metrics provide a good measure of the degree of corruption. Plots B and D suggest, and analysis not shown here confirms, that more corrective iterations are required as the degree of corruption increases, as expected. Fig. 1 shows that for a given corrupted network, the successive divergence values across the iterations tend to diminish, at least for the early iterations.

Due to the stochastic nature in which the networks were created, the same value of divergence can be associated with a range of values of the number of the rules corrupted. This indicates that a corrupted network can have fewer corrupted rules than another network, but it may have the same divergence because its random corruptions happen to be of a more strategic nature.

The divergence metric could be used to compare two different SA algorithms in a given simulated scenario. Lower divergence values would indicate the better-performing algorithm. Also, the divergence metric could be used to confirm that the next release of a SA algorithm performs better in a given simulated scenario compared with the previous release.

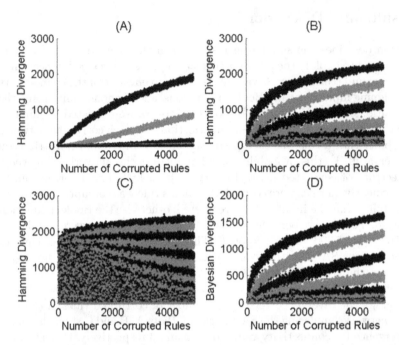

Fig. 1. The divergence versus the number of corrupted rules in the version of the network simulating the SA algorithm. Plots A, B and C show the Hamming Divergence for connectivity values of 1, 2 and 3 respectively. Plot D shows the Bayesian Divergence for a connectivity value of 2. Other parameter values employed are $N_{OA} = 50$, $N_P = 5000$, $N_C = 5000$. The divergence values for the first iteration are shown at the top in black, and then successive iterations are plotted alternately in grey and black.

It is interesting to note that plots B and D in Fig. 1, for the Hamming and Bayesian Divergences, are very similar. The corresponding curves for the Antecedent Divergence (not shown) are very similar to plots B and D. This indicates that the different weighting schemes don't have a significant effect on the divergence results in the case of Random Inference Networks. The next step will be to use a real-world SA algorithm instead of a RIN, and to compare the three types of metric. It may be feasible to use the realistic inference networks being developed for the North Atlantis simulated scenario (described in Section 1). Realistic networks will be associated with non-random conditional probabilities for the Bayesian Divergence, and this may have a significant impact.

The Hamming Divergence can be compared across different values of connectivity in plots A, B and C. Reference [9] shows that for the RIN described in Section 4, $K = 1$ is associated with the stable phase, $K = 2$ with the critical phase, and $K = 3$ with the chaotic phase. The chaotic phase is quite evident in plot C. It may be unlikely that the chaotic phase would yield a useful real-world inference network since such a network would be very sensitive to small perturbations. It is interesting to note that the stable phase in plot A converges to zero divergence much more quickly over successive iterations compared with the critical phase in plot B. These results indicate that the phase of the inference network has a very significant impact on the divergence results. Further work could examine realistic inference networks and determine which phase they are in.

6 Conclusions

This paper discusses an approach for assessing the performance of machine-based situation assessment. The iterative correction process, initially introduced in [7], has been further developed in this paper. Three metrics, named the Hamming Divergence, Antecedent Divergence and the Bayesian Divergence, have been introduced and explored using Random Inference Networks. The divergences appear to provide good measures of how defective a simulated situation assessment algorithm is. The divergence results are very sensitive to which phase the inference network is in, i.e. stable, critical or chaotic. Further work is required to confirm and build on these results through analysis of more realistic inference networks.

References

1. Llinas, J., Bowman, C., Rogova, G., Steinberg, A., Waltz, E., White, F.: Revisiting the JDL Data Fusion Model II. In: 7th International Conference on Information Fusion, pp. 1218–1230. International Society of Information Fusion (2004)
2. Lambert, D.A.: STDF Model Based Maritime Situation Assessments. In: 10th International Conference on Information Fusion, art. no. 4408055. International Society of Information Fusion (2007)
3. Smith, D., Singh, S.: Approaches to Multisensor Data Fusion in Target Tracking: A Survey. IEEE Trans. Knowl. Data Eng. 18, 1696–1710 (2006)
4. Lambert, D.A.: Grand Challenges of Information Fusion. In: 6th International Conference on Information Fusion. International Society of Information Fusion, pp. 213–219 (2003)
5. Lambert, D.A., Nowak, C.: The Mephisto Conceptual Framework. Technical Report DSTO-TR-2162, Defence Science & Technology Organisation, Australia (2008)
6. Blanchette, M.: Military Strikes in Atlantis – A Baseline Scenario for Coalition Situation Analysis. Technical Memorandum, Defence R&D Canada – Valcartier (2004)
7. Lingard, D.M., Lambert, D.A.: Evaluation of the Effectiveness of Machine-based Situation Assessment – Preliminary Work. Technical Note DSTO-TN-0836, Defence Science & Technology Organisation, Australia (2008)
8. Yuan, B., Wang, B.-H.: Growing Directed Networks: Organisation and Dynamics. New J. Phys. 9, 282 (2007)
9. Lingard, D.M.: Perturbation Avalanches in Random Inference Networks (submitted to Complexity)

Experiments with the Universal Constructor in the DigiHive Environment

Rafal Sienkiewicz

Gdansk University of Technology, Gdansk, Poland
Rafal.Sienkiewicz@eti.pg.gda.pl

Abstract. The paper discusses the performance and limitations of the universal constructor embedded in the DigiHive environment and presents the results of two simulation experiments showing the possibility of workaround the limitations.

1 Introduction

The DigiHive environment [1,2] is an artificial world aimed for modeling the various systems which a complex global behaviour emerges as a result of many simultaneous, distributed in space, local and simple interactions between its constituent entities. It is especially convenient for modeling the basic properties of self-organizing, self-modifying and self-replicating systems. A brief description of the DigiHive system is given in the first part of the paper.

The various strategies of self-replication usually involves the existence of a universal constructor, a device which can create any entity basing on its description and using surrounding building materials. The design of the universal constructor in the DigiHive environment is described in the second part of the paper.

The implemented constructor has some limitations. They are discussed and some solutions are proposed and two simulation experiments are presented.

2 The DigiHive Environment

The environment is a two dimensional space containing objects which move, collide and change their structure.[1] The constituent objects of the environment, called *particles*, are represented by hexagonal tiles. There are 256 of particles and they are marked by velocity, position, and internal energy.

Particles are able to bond together (as a result of collissions) forming a *complex* of particles. Particles are able to bond vertically up and down forming the stacks. The particles at the bottom of the stack can bond horizontally. A sample stack of particles and a complex formed by horizontal bonds are shown in Fig.1.

---

[1] The section presents an abbreviated description of the DigiHive environment. For more details see [1,2].

K. Korb, M. Randall, and T. Hendtlass (Eds.): ACAL 2009, LNAI 5865, pp. 106–115, 2009.

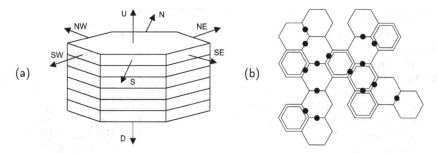

Fig. 1. Examples of complexes: (a) horizontal view of single stack of particles with directions shown, and (b) vertical view of complex formed by horizontal bonds, where hexagons drawn with single lines represent single particles, and with double lines represent stacks of particles and black dots mark horizontal bonds between particles

2.1 Functions

Besides the reactions resulting from the collision of particles, there also exists an additional class of interactions in which complexes of particles are capable to recognize and manipulate particular structures of particles in the space around them. The description of the function performed by a complex is given in the types and locations of particles in the complex. The structure of the complex is interpreted as a program written in a specially defined declarative language. Syntax of the program encoded in a complex of particles is similar to the Prolog language, using only the following predicates: program, search, action, structure, exists, bind, unbind, move, not. Predicates: program, search, action and structure help to maintain the structure of the program. The other predicates are responsible for selective recognition of the particular structure of particles (exists) and for manipulation of them (bind – create bonds, unbind – remove bonds and move – move particles).

An example of a program and a structure transformed by it are presented in Fig. 2.

The predicate program exactly consists of two predicates search and action. The first one calls the searching predicates, while the second one calls the predicates responsible for performing some actions in the environment.

The predicate search calls the predicate structure. The predicate structure consists of sequence of exists predicates and/or other structure predicates, always followed by the negation not. It provides the ability of recognizing the particular structure if some other structure does not exist. In this example, the particle of type 10101010 plays a role of the reaction inhibitor (its existence would prevents program from being executed).

With the predicate exists it is possible to check various conditions, e.g.: existence of some particular particle type (exists([0,0,0,0,1,1,1,1] ...)), check if the particle is bound to some other particle in a given direction (exists(... bound to V2 in N ...)) etc. It is also possible to mark the particle which fulfills the predicate condition with one of 15 labels (variables: V1 to V15), e.g. the

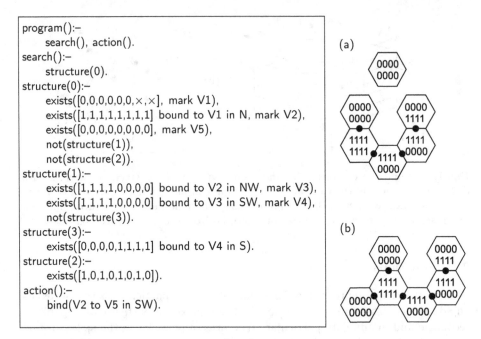

```
program():-
    search(), action().
search():-
    structure(0).
structure(0):-
    exists([0,0,0,0,0,0,×,×], mark V1),
    exists([1,1,1,1,1,1,1,1] bound to V1 in N, mark V2),
    exists([0,0,0,0,0,0,0,0], mark V5),
    not(structure(1)),
    not(structure(2)).
structure(1):-
    exists([1,1,1,1,0,0,0,0] bound to V2 in NW, mark V3),
    exists([1,1,1,1,0,0,0,0] bound to V3 in SW, mark V4),
    not(structure(3)).
structure(3):-
    exists([0,0,0,0,1,1,1,1] bound to V4 in S).
structure(2):-
    exists([1,0,1,0,1,0,1,0]).
action():-
    bind(V2 to V5 in SW).
```

Fig. 2. Example of a program (in box) recognizing the structure (a) and transforming it into the structure (b)

exists([1,1,1,1,0,0,0,0] bound to V2 in NW, mark V3) means: find the particle of type 11110000 which is bound to the particle marked as V2 in direction NW, and store the result in variable V3 (mark the found particle as V3).

The program is encoded by a complex of particles. Each predicate structure is represented by a single stack of particles. Such the stack encodes a list of predicates exists. Stack which encodes structure(0) also encodes predicates bind, unbind and/or move. Adjoining stacks encode a negative form of predicates structure.

3   The Universal Constructor

The universal constructor is a concept introduced by von Neumann in his famous work on self-replicating cellular automata [3] (see also [4] for the most recent ideas).

The universal constructor in the DigiHive environment (see [5]) is a structure A (complex of particles) being able to constructs other structure X based on its description $d(X)$. It is admissible to construct the X structure via description $d(X')$ of an intermediate structure $X' \neq X$, being able to transform itself into the X.

The universal constructor is a consistent structure (a set of programs) being able to execute the following tasks:

1. search for valid information structure (information string) – $d(X)$. The information string encodes description of some structure $d(X)$. The description may be viewed as another program written in a simple universal constructor language with the following commands: PUT (add specified particle to the stack), SPLIT (start construction of a new stack connected horizontally to the existing one), NEW (start construction of a new stack and do not connect it to the existing one), and END (the end of an information string). The information string is a stack of particles of particular types. As an example the following program is presented:

PUT(01010101) PUT(01010101) END

which describes a stack of two particles of type 01010101 that can be encoded by the stack:

11111111
01010101
00000001
01010101
00000001

2. connect itself to the found information string and start constructing the structure X. The structure X consists of horizontally joined stacks of particles. There is always exactly one stack of particles being built at the moment, called the active stack $X^\star$,
3. sequentially process the joined information string:
   (a) if a current particle in the information string encodes command PUT – find the particle of *specified* type which is on the top of stack of building material. The building material is contained in stacks of particles (named M) marked by the particle at the bottom (material header) of the type $0000\times\times10$. The newly found particle is removed from the top of the stack M, and put on the top of the stack $X^\star$ (Fig. 3a).
   (b) if current particle in the information string encodes command SPLIT – split the stack $X^\star$ into two stacks: remove particle from the top, move the trimmed stack in the *specified* direction, and create horizontal bond between $X^\star$ and removed particle. The particle becomes the bottom of active stack $X^\star$. This action is presented in Fig. 3b,
   (c) if current particle in the information string encodes command NEW – disconnect the structure X and immediately start the construction of a new structure with a *specified* particle as the beginning of $X^\star$ (a single information string can then encode various structures, e.g. both $d(B)$ and $d(A)$ – see also additional note on the use of the NEW command at the end of this paragraph),
   (d) if a current particle in the information string encodes command END – release the information string and the constructed structure.

The universal constructor was implemented as a set of cooperating 10 programs enhanced with 5 helper stacks of particles. The helper stacks are used mainly

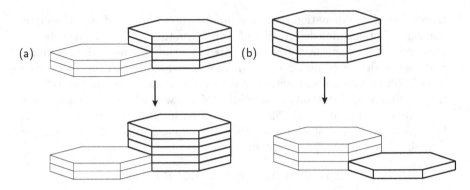

Fig. 3. Illustration of actions of the universal constructor during processing of the information string: (a) action caused by the command PUT. The particle of specified type is put on the top of active stack X^* (draw using thicker lines), (b) action caused by the command SPLIT. The particle is removed from the top of X^*, then the bond is created in a specified direction, the particle becomes a new active stack X^*.

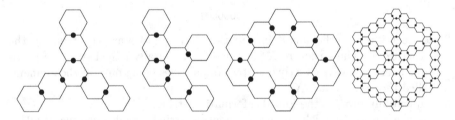

Fig. 4. Examples of structures, which cannot be directly build by the universal constructor. The construction of the most right snowflake-like structure will be discussed in the Sec. 4.1.

for performing synchronization between working programs, they also mark some characteristic part of the constructor e.g. a place where the new structure is build, a place where the universal constructor join the encoding string etc.

The universal constructor is not fully universal, i.e. it cannot build any possible spatial structure straightforwardly. Sequential adding the subsequent stacks in various directions may lead to the situation where a built structure will try to occupy a place already occupied by the constructor itself. Important limitation is also lack of possibility of create bond between built stacks in an every possible direction. The universal constructor can only then build a chain of particle stacks. Note, that the stacks of particles itself are not restricted – the constructor can build any possible stack.

Examples of problematic structures are presented in Fig. 4. There is no possibility of encode the stack (or particle) with 3 or more horizontal bonds. The universal constructor is also not able to build any closed curve of particles (or of stacks).

4   Simulations

The universal constructor as it was stated in Section 3 is not able to build any possible structure straightforwardly. In Ref. [5] two possible strategies were shortly described which can compensate its limitations:

S_1 – construction of set of programs which are able to cooperatively build the desired structure in finite time cycles – Fig. 5(a)

S_2 – construction of one intermediate structure which can transform itself into the desired structure – Fig. 5(b)

There is also possible to mix the strategies, e.g. by constructing an intermediate structure with set of programs which helps to transforms it into the desired one, etc.

In this section two simulation experiments are presented showing as examples of the above strategies. The experiments will also illuminate the large potential and universality of environment being discussed.

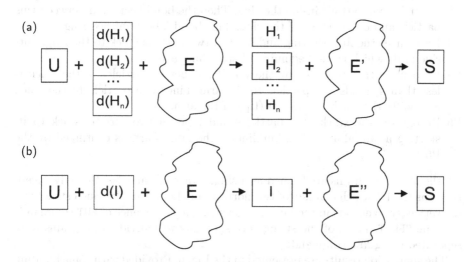

Fig. 5. Universal constructor strategies: (a) the universal constructor U joins the information string $d(H_1)$, $d(H_2)$, ..., $d(H_n)$ and, using the building material from environment E (random distributed particles and complexes), builds the set of helper $d(H_1)$, $d(H_2)$, ..., $d(H_n)$ programs. Helper programs interacts with environment E' (changed thanks to U activity, note that E' contains also both U and $d(H_2)$, ..., $d(H_n)$ which in fact becomes a part of the environment at this stage of simulation). As a result they finally build the desired structure S. Note that description of individual programs $d(H_1)$, $d(H_2)$, ..., $d(H_n)$ are separated by the NEW command. (b) the universal constructor U joins the information string $d(I)$ and builds the intermediate structure I. The structure I interacts with environment E'' and finally transforms itself into the desired structure S.

4.1   Emergent Behaviour

The S_1 strategy is particularly useful for the tasks associated with the construction of complex spatial structures. A sample construction of the snowflake-like structure illustrated in Fig. 4 is discussed.

The universal constructor is not able to build this structure straightforwardly but it is easy to build the structure by cooperative activities of a set of 6 programs (P1–P6). These programs moving and colliding with other particles, used as a building material, gradually constructs the structure.

The sequence of events triggered by programs is acting as follows:

P1 Joins the two unbound particles, then deactivates by unbinding one particle from itself

P2 Builds a ring of a six particles by adding one particle per run, (starting from two bound particles).

P3 Recognizes the finished ring of six particles by recognizing the two neighbour particles belonging to the same complex and not bound together. After recognition the program creates the bond between the particles, and puts one unbound particle inside the ring. Then the bond is created between this particle and one particle of the ring. This bond labels finished ring.

P4 Recognizes the finished ring and binds a two-particle stack to the ring, outside of it, which is the beginning of stretching arm.

P5 Recognizes the stack put by the P4. If number of particles in the stack is less than 4, binds one particle to the arm, binds one particle to the stack and shifts the stack to the end (tip) of the arm.

P6 Recognizes the stack of 4 particles and binds a two-particle stack to it, starting a lateral arm. The building of the lateral arm is continued by the P5.

All the above programs are formed as a single stack of particles, the only exception was the P5 which consists of two horizontally bound stacks. Such structures are very easy to encode in the information string as a sequence of PUT commands and one SPLIT. Parts of the string corresponding to individual programs were separated using NEW command.

The simulation results are presented in the Fig. 6. Provided with some building material, the universal constructor built the set of cooperative programs (P1–P6) which in turn built the snowflake structure.

4.2   Copying the Constructor

Case when the constructor builds its own copy (when the A works with its description $d(A)$), is a good illustration of the strategy S_2 i.e. building an intermediate structure which transforms itself into the full universal constructor. The constructor cannot directly build its copy due to the following reason:

1. it is impossible to encode the structure of bonds that is needed by a constructor (Sec. 3),

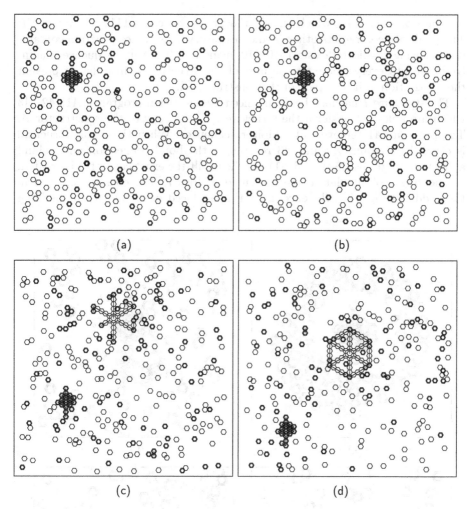

(a)

(b)

(c)

(d)

Fig. 6. Constructing the snowflake-like structure by the universal constructor: (a) step 0 – initial state; (b) step 110 – building the programs (P1–P6); (c) step 2383 – constructing the snowflake; (d) step 2910 – finished structure

2. the structure being built, should not manifest any activity before it is completely finished, in other case the simulation can become unpredictable (e.g. the unfinished constructor may start producing the copy of itself etc.)
3. the universal constructor should not recognize the structure being built as a part of itself

The intermediate structure I consists of all programs which constitute the universal constructor A but the bond structure was rearranged into the rhombus shape (see structure at the bottom of the Fig. 7(c)). To prevent the partially developed programs from being executed, they were enhanced by the inhibitor sequence of 5 particles which encode the predicate: exists 11111111, mark V1. The predicate contradicts other predicates (every constructor program contains the following predicate: exists 10101010, mark V1). The same sequence was introduced into every characteristic part of the structure I, in order to prevent the recognition of it as a part of the constructor A.

The structure I has also been enhanced by activation programs: $A1$, $A2$, $A3$ which are able to transform the structure into the copy of the constructor A. The program $A1$ removes all inhibitor sequences. When the whole structure is cleaned, programs $A2$ and $A3$ rearranges the bond connection and unbind activation programs (itself and $A1$). As the result the exact copy of A is created.

The stages of simulation are presented in Fig. 7. The constructor A is in the middle of Fig. 7(a), the encoding string is placed just above it, the building

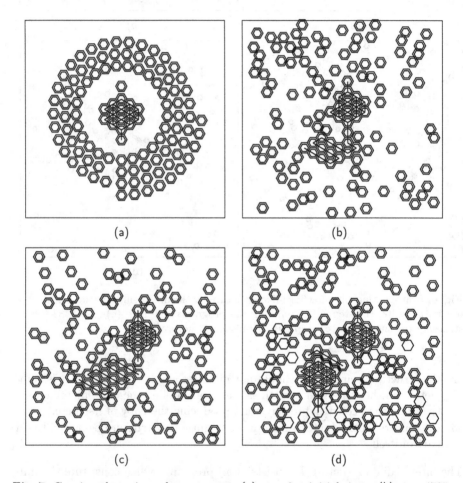

(a)

(b)

(c)

(d)

Fig. 7. Copying the universal constructor: (a) step 0 – initial state; (b) step 4735 – building the intermediate structure; (c) step 6080 – finished intermediate structure, (d) step 6100 – intermediate structure transformed into the universal constructor

materials form a circle around it. Such the "unnatural" form of an initial state was chosen due to the performance reason. The translation should start immediately and the environment size has to be as small as possible (most of its constituent object should be visible to constructor's individual programs), otherwise the time of simulation may by extended unacceptably (wasted time cycles when no program were executed takes a long time)[2]. In Fig. 7(c) the constructor has finished translation, the structure I (at the bottom of the picture) is ready for activation. In the last Fig. 7(d) the structure I has been successfully transformed into the copy of the universal constructor.

Note the small error at the end of copying process. After the structure I was transformed into the A' (copy of A), the constructor A started the new translation process but by mistake the building particle was connected to the structure A' instead of A.

5   Conclusions and Further Research

The universal constructor in the DigiHive environment opens possibilities for various simulation. The aim of further research is to implement the full self-replication structure (extend the universal constructor with the structure being able to construct copy of the encoding string). Then the various possible self-reproduction strategies will be test and compare. After allowing the small probability of random changes in the environment, due to its limited resources, we expect to observe the emergence of Darwinian evolution mechanism.

At the moment the main obstacle in performing more complex simulation in DigiHive environment is high computational complexity of searching algorithms used in embedded language, which is the price paid for a low level modeling. The solution would be the extension of the language with high level instructions e.g. direct checking equality of particles. Also the parallelization of the environment is planned.

Acknowledgments. The work was supported by The Polish Ministry of Science and Higher Education, grant no. N 516 367636.

References

1. Sienkiewicz, R.: The digihive website, http://www.swarm.eti.pg.gda.pl/
2. Sienkiewicz, R., Jedruch, W.: Artificial environment for simulation of emergent behaviour. In: Beliczynski, B., Dzielinski, A., Iwanowski, M., Ribeiro, B. (eds.) ICANNGA 2007. LNCS, vol. 4431, pp. 386–393. Springer, Heidelberg (2007)
3. von Neumann, J.: Theory of Self-Reproducing Automata. University of Illinois Press, Champaign (1966)
4. Hutton, T.J.: Evolvable self-reproducing cells in a two-dimensional artificial chemistry. Artificial Life 13(1), 11–30 (2007)
5. Sienkiewicz, R., Jedruch, W.: The universal constructor in the digihive environment. In: Proceedings of ECAL 2009 conference (accepted 2009)

---

[2] The simulation takes over 3 months of continuously calculations on Quad6600 PC.

Making a Self-feeding Structure by Assembly of Digital Organs

Sylvain Cussat-Blanc, Hervé Luga, and Yves Duthen

University of Toulouse - IRIT - UMR 5505
2 rue du Doyen-Gabriel-Marty 31042 Toulouse Cedex 9, France
{Sylvain.Cussat-Blanc,Herve.Luga,Yves.Duthen}@irit.fr

Abstract. In Nature, the intrinsic cooperation between organism's parts is capital. Most living systems are composed of organs, functional units specialized for specific actions. In our last research, we developed an evolutionary model able to generate artificial organs. This paper deals with the assembly of organs. We show, through experimentation, the development of an artificial organism composed of four digital organs able to produce a self-feeding organism. This kind of structure has applications in the mophogenetic-engineering of future nano and bio robots.

1 Introduction

Most living systems are composed of different organs. Cooperation between organs allows them to optimize the exploitation of environmental resources. Its role is crucial for survival in a complex environment. Several works on digital organs development already exist mainly based on two methods: shape generation, which is the most widely discussed, and function generation. Whereas the first is usually based on artificial Gene Regulation Networks (GRN) and, in recent years, tries to be the most biologically plausible as possible, the second method is usually based on cellular automata or block assembling and is more bio-inspired than biologically plausible.

Our previous research dealt with making isolated digital organs. We developed a bio-inspired model able to produce goal-directed organisms starting from a single cell. The aim was to make an organ library. We now present the assembly of two kinds of organs: producer-consumers and transfer systems. Assembling these organs produces a self-feeding structure and gives the organism a potentially limitless survival capacity.

The paper is organised as follows. Section 2 gives the related work about artificial development and artificial creature production. Section 3 summarizes the model, already presented in [4]. Section 4 details the experimentation of a self-feeding structure with particular emphasis on environmental parameters. Section 5 discuss the possible application of such a developmental model for morphogenetic engineering of future bio and nano systems. Finally, we conclude by outlining possible future work on this creature.

2 Related Works

Over the past few years, more and more models concerning artificial devolment have been produced. A common method for developing digital organisms is to use artificial

K. Korb, M. Randall, and T. Hendtlass (Eds.): ACAL 2009, LNAI 5865, pp. 116–125, 2009.

regulatory networks. Banzhaf [1] was one of the first to design such a model. In his work, the beginning of each gene, before the coding itself, is marked by a starting pattern, named "promoter". This promoter is composed of enhancer and inhibitor sites that allow the regulation of gene activations and inhibitions. Another different approach is based on Random Boolean Networks (RBN) first presented by Kauffman [12] and reused by Dellaert [7]. An RBN is a network where each node has a boolean state: activate or inactivate. The nodes are interconnected by boolean functions, represented by edges in the net. Cell function is determined during genome interpretation.

Several models dealing with shape generation have recently been designed such as [6,14,17,2,13,11]. Many of them use gene regulation and morphogens to drive the development. A few produce their own morphogens whereas others use environment "built-in" morphogens. Different shapes are produced, with or without cell specialisation. The well-known French flag problem was solved by Chavoya [2] and Knabe [13]. This problem shows model specialisation capacity during the multiple colour shifts.

In their models, produced organisms have only one function: filling up a shape. Other models, most often based on cellular automata or artificial morphogenesis (creatures built with blocks), are able to give functions to their organisms [16,10,8]. Here, creatures can walk, swim, reproduce, count, display... Their goals are either led by user-defined fitness objectives that evaluate the creature responses in comparison to those expected or only led by their capacity to reproduce and to survive in the environment.

The next section presents our developmental model. It is based on gene regulatory networks and an action selection system inspired by classifier rule sets. It has been presented in details in [4].

3   *Cell2Organ*: A Cellular Developmental Model

3.1   The Environment

To reduce simulation computation time, we implement the environment as a 2-D toric grid. This choice allows a significant decrease in the simulation's complexity keeping a sufficient degree of freedom.

The environment contains different substrates. They spread within the grid, minimizing the variation of substrate quantities between two neighbouring crosses on the grid. These substrates have different properties such as spreading speed or colour, and can interact with other substrates. Interactions between substrates can be viewed as a great simplification of a chemical reaction: using different substrates, the transformation will create new substrates, emitting or consuming energy. To reduce the complexity, the environment contains a list of available substrate transformations. Only cells can trigger substrate transformations.

3.2   Cells

Cells evolve in the environment, more precisely on the environment's spreading grid. Each cell contains sensors and has different abilities (or actions). An action selection system allows the cell to select the best action to perform at any moment of the simulation. Finally, a representation of a GRN is available inside the cell to allow specialization during division. Figure 1 is a global representation of our artificial cells.

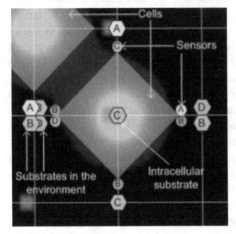

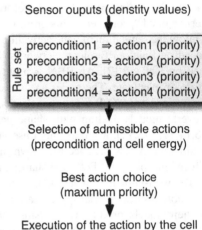

Fig. 1. The cell plan in its environment. It contains substrates (hexagons) and corresponding sensors (circles).

Fig. 2. Action selector functioning: sensors and cell energy are used to select admissible actions. The best action is chosen according to the rule priority.

Each cell contains different density sensors positioned at each cell corner. Sensors allow the cell to measure the amounts of substrates available in the cell's Von Neumann neighbourhood. The list of available sensors and their position in the cell is described in the genetic code.

To interact with the environment, cells can perform different actions: perform a substrate transformation, absorb or reject substrates in the environment, divide (see later), wait, die, etc. This list is not exhaustive. The implementation of the model enables a simple addition of actions. As with sensors, not all actions are available for the cell: the genetic code will give the available action list.

Cells contain an action selection system. This system is inspired by the rule set of classifier systems. It uses data given by sensors to select the best action to perform. Each rule is composed of three parts: (1) The *precondition* describes when the action can be triggered. A list of substrate density intervals describes the neighbourhood in which action must be triggered. (2) The *action* gives the action that must be performed if the corresponding precondition is respected. (3) The *priority* allows the selection of only one action if more than one can be performed. The higher the coefficient, the more probable is the selection of the rule. Its functioning is presented in figure 2.

Division is a particular action performable if the next three conditions are respected. First, the cell must have at least one free neighbour cross to create the new cell. Secondly, the cell must have enough vital energy to perform the division. The needed vital energy level is defined during the specification of the environment. Finally, during the environment modelling, a condition list can be added.

3.3 Action Optimisation

The new cell created after division is completely independent and interacts with the environment. During division, the cell can optimize a group of actions. In nature, this specialization seems to be mainly carried out by the GRN. In our model, we imagine a mechanism that plays the role of an artificial GRN. Each action has an efficiency coefficient that corresponds to the action optimization level: the higher the coefficient, the lower the vital energy cost. Moreover, if the coefficient is null, the action is not yet available for the cell. Finally, the sum of efficiency coefficients must remain constant during the simulation. In other words, if an action is optimized increasing its efficiency coefficient during division, another (or a group) efficiency coefficient has to be decreased.

The cell is specialized by varying the efficiency coefficients during division. A network represents the transfer rule. In this network, nodes represent cell actions with their efficiency coefficients and weighted edges representing efficiency coefficient quantities that will be transferred during the division.

3.4 Creature Genome

To find the creature best adapted to a specific problem, we use a genetic algorithm. The creature is tested in its environment. It returns the score at the end of the simulation. Each creature is coded with a genome composed of three different chromosomes: (1) the list of available actions, (2) an encoding of the action selection system and (3) an encoding of the gene regulation network.

3.5 Example of Generated Creatures

Different creatures have been generated using this model. For example:

- A *harvester*: a creature able to collect a maximum of a substrate scattered all over the environment and transform it into division material and waste. The creature has to reject the waste because of a limited substrate capacity.
- A *transfer system*: presented in [4], a creature able to move substrate from one point to another. This creature is interesting because it has to alternate its behaviour between performing its function and developing its metabolism to survive.
- Different *morphologies*: also presented in [4], such as a starfish, a jellyfish or any user-designed shape. Once again, the organism must develop its metabolism to be able to perform its function.

All creatures have a common property: they are able to repair themselves in case of injury [5]. This feature is an inherent property of the model. It shows the phenotype plasticity of produced creatures.

In the next section, we present the features obtained by producing new organisms and putting them in the same environment. We design an environment wherein the organism will be composed of four organs. Once assembled, their organs will make a self-feeding structure that will allow the organism to maintain its life endlessly. Before that, the organism must develop a sufficient metabolism to start the chain.

4   Experiments: Self-feeding Structure

In order to produce a cycle, the organism is composed of two kinds of organs: transfer systems close to the one previously presented and organs able to transform a substrate into another and to position precisely the produced substrate (to be transferred by a transfer system). The global functioning is introduced by Figure 3.

Section continuation is organised as follows. First, we will describe clearly the different organs, detailing the global environment and the different possible actions for each organ. Then we will show and discuss the organism obtained.

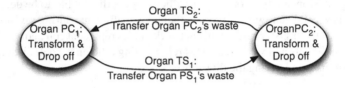

Fig. 3. Functioning diagram of the organism. It is composed of two kinds of organ: producer-consumer organs PC_1 and PC_2 able to transform substrates and to position them in a particular place; transfer organs TS_1 and TS_2 able to transfer substrates from one point to another.

4.1   Experimentation Parameters

Description of the environment

The environment is composed of 3 different substrates:

- E (represented in blue on the next figures) that will be used by the organism to develop its metabolism,
- A *and* B (respectively represented in red and yellow on the next figures) substrates that will be used by the organism to produce the self-feeding structure.

Three substrate transformations are available:

- $E \rightarrow energy$ produces energy using water,
- $A \rightarrow B + energy$ pro duces B substrate plus energy using one unit of A,
- $B \rightarrow A + energy$ produces A substrate plus energy using one unit of B.

50 units of A substrates are positioned near PC_1 and 50 units of B substrates near PC_2. Organ PC_1 has to transform the substrate A into B and must position it at the entrance of organ TS_1, which transfers the B substrate on the entrance of organ PC_2. Organ PC_2 has to permorm the opposite operation to that of organ PC_1: it transforms B substrate into A and has to put the result at the entrance of organ TS_2, which drives the A substrate back up near organ PC_1. Because all their actions provide energy to the cells, the obtained organism can work endlessly. With the purpose of producing the self-feeding structure, each organ has first been developed individually. Each kind of organ has a different list of possible actions.

Table 1. Table of possible actions of different organs of the self-feeding structure. All organs do not have all the action activate to accelerate the convergence process of the genetic algorithm.

Action	Cost	Needs	TS_1	TS_2	PC_1	PC_2
Divide to NorthEast			X	X	X	X
Divide to NorthWest	30	1 unit of E	X	X	X	X
Divide to SouthEast			X	X	X	X
Divide to SouthWest			X	X	X	X
Transform $E \rightarrow energy$	-30	1 unit of E	X	X	X	X
Transform $A \rightarrow B$	-50	1 unit of A			X	
Transform $B \rightarrow A$	-50	1 unit of B				X
Absorb E from North	2			X	X	X
Absorb E from South	2	Cell must contain less than		X	X	X
Absorb E from East	2	7 substrat units		X	X	X
Absorb E from West	2			X	X	X
Absorb A from North	-2			X	X	X
Absorb A from South	-2	Cell must contain less than		X	X	X
Absorb A from East	-2	7 substrat units		X	X	X
Absorb A from West	-2			X	X	X
Absorb B from North	-2		X		X	X
Absorb B from South	-2	Cell must contain less than	X		X	X
Absorb B from East	-2	7 substrat units	X		X	X
Absorb B from West	-2		X		X	X
Evacuate A from North	-0.5			X	X	X
Evacuate A from South	-0.5	Cell must contain at least		X	X	X
Evacuate A from East	-0.5	one unit of A		X	X	X
Evacuate A from West	-0.5			X	X	X
Evacuate B from North	-0.5		X		X	X
Evacuate B from South	-0.5	Cell must contain at least	X		X	X
Evacuate B from East	-0.5	one unit of B	X		X	X
Evacuate B from West	-0.5		X		X	X
Do Nothing	1	-	X	X	X	X

Possible actions for the organs

The table 1 gives the possible actions for the different organs TS_1, TS_2, PC_1 and PC_2. Some actions are inactivated to accelerate the convergence process of the genetic algorithm. However, organs have the possibility to divide to all directions to increase the degree of freedom. The energy cost of actions, except for division, substrate E absorption and wait actions, are negative to give the stucture to produce energy during the cycle.

4.2   Results

We compute each organ in a separate environment. Four cells containing the genetic code of their corresponding organ are then assembled to the environment. They evolve with the aim of generating a self-feeding structure. Figure 4 shows the development and the behaviour of the organism. It is worth mentionning that for each kind of organ, different strategies emerge to reach the goal. For example, organ PC_1 transfers the

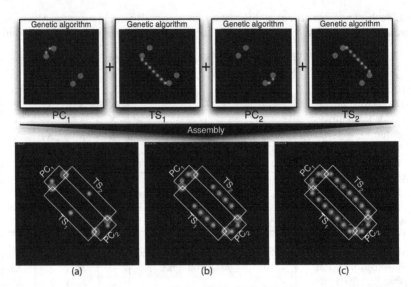

Fig. 4. Result of the assembly of the 4 different organs. (a) Beginning of the simulation: 4 cells that contain the genetic code of each organ are positioned in the environment. (b) Organ growth: while the 2 producer-consumer organs have finished their development and start their work, the transfer systems continue their growth. (c) All organs have finished their development and a self-feeding structure is made. While producer-consumer organs continue their work, transfer systems start the transfer to feed other organs with new substrates.

initial substrate near the goal before transforming it into the final substrate whereas organ PC_2 transforms the substrate before transferring the result to the right place.

The obtained organism works as expected one[1] The regulation network regulates correctly the size of the transfer systems whereas the organs that transform the substrate develop the different action selection strategies to reach their goals. Detailed functioning of organs is given by figure 5. Organ PC_1, on the top left, transfers A substrate (in red) to the second cell before transforms it into B and reject the result in the right position. Organ PC_2 adopts the opposite strategy: it absorbs substrate B, transforms it in the first cell and transfers the resulting substrate A to the final position. Organ TS_1 and Organ TS_2 use serial absorption and rejection to move the substrate from the exit of an organ to the entrance of the opposite organ.

Curves presented in figure 6 show the evolution of the number of cells and the water quantity in the environment. The quantity of E substrate strongly decreases at the beginning of the simulation, before the initialisation of the cycle (stage 1). Different organs use water to start their metabolism. When the cycle starts (stage 2), organs use the cycle as metabolism. Organs still consume water to produce energy that will be stocked for the future. The curve presented in figure 7 shows the ratio of A and B substrates. B substrate quantity slowly decreases. This proves an efficiency difference in the organs: organ PC_1 converts A into B more slowly than organ PC_2 does the opposite. Even if

---

[1] Videos of this organism development and of each organ functioning seperately are available on the website http://www.irit.fr/~Sylvain.Cussat-Blanc.

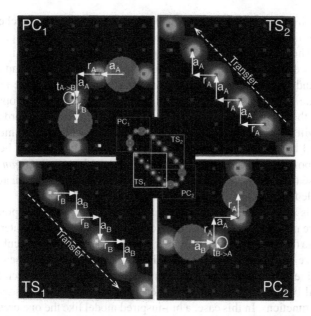

Fig. 5. Detail of the organism functioning. Acronyms on arrows correspond to actions accomplished by cells: r_A means "reject A", a_B "absorb B" and $t_{A \rightarrow B}$ "transform A into B".

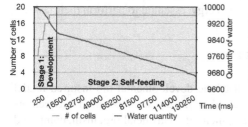

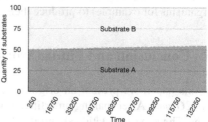

Fig. 6. Number of cells (left ordinate) and quantity of water (right ordinate) in the environment in function of time (abscissa)

Fig. 7. Evolution of A and B substrate quantity ratio. The B substrate quantity decreases: Organ PC_2 is more efficient than PC_1.

the difference is small, this curve shows that the cycle is not endless: after a long period of time, B substrate will disappear and the cycle will be broken.

5   Discussion

Developmental models that take into consideration the organism's metabolism should have many applications in the near future. Self-assembly, self-repairing and self-feeding, or more generally self-*, properties are important for the future bio and nano robots. Many works are nowadays in progress to develop artificial cells [15,9] and nano modules for the future modular robots [3,18,19].

In synthetic biology, which will produce future bio-systems, researchers are today working on modifying the genome of different bacteria to make them produce particular proteins. Those proteins are used to express a particular function in a cell. Some building blocks of this chemical self-assembly are already in place and many others have to be found [9]. In the next twenty years, it seems to be possible to build a cell able to replicate itself and to have a group of possible actions. Developmental models, especially those that include metabolism, would be necessary in order to find the genome that will allow the cell to perform its asked goal with the minimum set of actions. They will have to be more biological plausible than bio-inspired because of the constraints imposed by Nature. In other words, a model like *Cell2Organ* will have to be more precise (take into consideration physics, better simulate chemical reactions...) to be acceptable for this kind of applications.

Our model is applied more in the domain of nano robots, or more precisely meso-systems. We are now able to work on atoms to modify the molecules' material structure. We can imagine that, in some years, modular robotics like Lipson Molecube robots [19] could have a size of a couple of nanometres. Building structures composed of thousands of their units able to accomplish different actions would be possible. Such developmental models will be interesting to use in order to learn those robots to self-build their structures and functions. In this case, a bio-inspired model like the one presented in this paper could be sufficient because each robot module could include human-designed functions. Metabolism could be an interesting feature of such a robot in order to allow itself to use environmental resources (like glucose that is contained in many natural organisms for example) to produce their own energy, essential to perform their tasks.

6   Conclusion and Future Works

In this paper, we present an original result of the developmental bio-inspired model *Cell2Organ*. After making an artificial organ library, we test cooperation between organs. The experimentation shows the development of an artificial organism, composed of four digital organs. The cooperation of its organs creates a self-feeding structure. This kind of structure, with the self-repairing properties presented in [5], could be interesting for a the morphogenetic-engineering approach of future bio and nano robots.

Continuations of this work are multiple. First of all, we are currently starting the development of the organism with four cells, one for each organ. We want to develop the organism starting from only one cell. With this purpose, we are working on a "pre-organism" able to position cells on the four starting positions of the final organs. The organism will have to switch it genome to the different organs' genomes and, finally, to resorb itself so as not to interfere with the organism's evolution.

We are also working on making different layers of the simulated environment. A physical layer will allow us to develop our organism at the same time in a physical world, with all its properties and the current "chemical" world to maintain the metabolism of the creatures. A hydrodynamic layer will simulate substrate diffusions more efficiently. For example, this layer will allow a cell to expulse a substrate with a chosen strength to position it in a particular place. It will also simulate fluid flows. Cells will have to adjust their behaviour according to new data.

Acknowledgment. Experiments presented in this paper were carried out using ProActive, a middleware for parallel, distributed and multi-threaded computing (see http://proactive.inria.fr), and the Grid'5000 French experimental testbed (see https://www.grid5000.fr)

References

1. Banzhaf, W.: Artificial regulatory networks and genetic programming. Genetic Programming Theory and Practice, 43–62 (2003)
2. Chavoya, A., Duthen, Y.: A cell pattern generation model based on an extended artificial regulatory network. Biosystems (2008)
3. Christensen, D.: Experiments on Fault-Tolerant Self-Reconfiguration and Emergent Self-Repair. In: IEEE Symposium on Artificial Life, ALIFE 2007 (2007)
4. Cussat-Blanc, S., Luga, H., Duthen, Y.: From single cell to simple creature morphology and metabolism. In: Artificial Life XI, pp. 134–141. MIT Press, Cambridge (2008)
5. Cussat-Blanc, S., Luga, H., Duthen, Y.: Cell2organ: Self-repairing artificial creatures thanks to a healthy metabolism. In: Proceedings of the IEEE Congress on Evolutionary Computation, IEEE CEC 2009 (2009)
6. de Garis, H.: Artificial embryology and cellular differentiation. In: Bentley, P.J. (ed.) Evolutionary Design by Computers, pp. 281–295 (1999)
7. Dellaert, F., Beer, R.: Toward an evolvable model of development for autonomous agent synthesis. In: Artificial Life IV. MIT Press, Cambridge (1994)
8. Epiney, L., Nowostawski, M.: A Self-organising, Self-adaptable Cellular System. In: Capcarrère, M.S., et al. (eds.) ECAL 2005. LNCS (LNAI), vol. 3630, pp. 128–137. Springer, Heidelberg (2005)
9. Forster, A., Church, G.: Towards synthesis of a minimal cell. Molecular Systems Biology 2(1) (2006)
10. Garcia Carbajal, S., Moran, M.B., Martinez, F.G.: Evolgl: Life in a pond. Artificial Life XI, 75–80 (2004)
11. Joachimczak, M., Wróbel, B.: Evo-devo in silico: a model of a gene network regulating multicellular development in 3d space with artificial physics. In: Artificial Life XI, pp. 297–304. MIT Press, Cambridge (2008)
12. Kauffman, S.: Metabolic stability and epigenesis in randomly constructed genetic nets. Journal of Theorical Biology 22, 437–467 (1969)
13. Knabe, J., Schilstra, M., Nehaniv, C.: Evolution and morphogenesis of differentiated multicellular organisms: autonomously generated diffusion gradients for positional information. Artificial Life XI 11, 321 (2008)
14. Kumar, S., Bentley, P.: Biologically inspired evolutionary development. Lecture notes in computer science, pp. 57–68 (2003)
15. Service, R.: How Far Can We Push Chemical Self-Assembly (2005)
16. Sims, K.: Evolving 3d morphology and behavior by competition. ALife IV, 28 (1994)
17. Stewart, F., Taylor, T., Konidaris, G.: Metamorph: Experimenting with genetic regulatory networks for artificial development. In: Capcarrère, M.S., et al. (eds.) ECAL 2005. LNCS (LNAI), vol. 3630, pp. 108–117. Springer, Heidelberg (2005)
18. Yim, M., Shen, W., Salemi, B., Rus, D., Moll, M., Lipson, H., Klavins, E., Chirikjian, G.: Modular self-reconfigurable robot systems (grand challenges of robotics). IEEE Robotics & Automation Magazine 14(1), 43–52 (2007)
19. Zykov, V., Phelps, W., Lassabe, N., Lipson, H.: Molecubes Extended: Diversifying Capabilities of Open-Source Modular Robotics. In: International Conference on Intelligent RObots and Systems, Self-Reconfigurable Robots Workshop, IROS 2008 (2008)

Towards Tailored Communication Networks in Assemblies of Artificial Cells

Maik Hadorn[1], Bo Burla[2,3], and Peter Eggenberger Hotz[1,4]

[1] University of Zurich, Department of Informatics, Artificial Intelligence Laboratory, 8050 Zurich, Switzerland
[2] University of Zurich, Institute of Plant Biology, 8008 Zurich, Switzerland
[3] Global Research Laboratory, 790784 Pohang, South Korea
[4] University of Southern Denmark, The Mærsk Mc-Kinney Møller Institute, Campusvej 55, 5230 Odense M, Denmark
hadorn@ifi.uzh.ch, bburla@botinst.uzh.ch, eggen@mmmi.sdu.dk

Abstract. Living Technology is researching novel IT making strong use of programmable chemical systems. These chemical systems shall finally converge to artificial cells resulting in evolvable complex information systems. We focus on procedural manageability and information processing capabilities of such information systems. Here, we present a novel resource-saving formation, processing, and examination procedure to generate and handle single compartments representing preliminary stages of artificial cells. Its potential is exemplified by testing the influence of different glycerophospholipids on the stability of the compartments. We discuss how the procedure could be used both in evolutionary optimization of self-assembling amphiphilic systems and in engineering tailored communication networks enabling life-like information processing in multicompartment aggregates of programmable composition and spatial configuration.

Keywords: Living Technology, self-assembly, programmability, glycerophospholipids, vesicles, multivesicular aggregates, adhesion plaque, phase transition.

1 Introduction

Engineering Living Technology from non-living materials has attracted particular attention in minimal life and complex information systems. As part of the complex systems Future Emerging Technologies initiative, PACE (Programmable Artificial Cell Evolution) was researching novel Information Technology (IT) that makes strong use of life-like properties such as robustness, homeostasis, self-repair, self-assembly, modularity, self-organization, self-reproduction, genetic programmability, evolvability, complex systems design, and bootstrapping complexity. In this context, PACE has created the foundation for a new generation of embedded IT to build evolvable complex information systems using programmable chemical systems that converge to artificial cells [1]. Because experiments were realized both in the laboratory and in simulation, findings of *in vitro* and *in silico* experiments interacted and lead to essential additions to the evolutionary approach in design of laboratory experimentation [2].

K. Korb, M. Randall, and T. Hendtlass (Eds.): ACAL 2009, LNAI 5865, pp. 126–135, 2009.

According to the guidelines of PACE, the creation of simple forms of life from scratch in the laboratory should have pursued a bottom-up strategy choosing simple organic compounds of low molecular weight over highly evolved polypeptides. Complexity of an artificial cell featuring all aspects of a living system should have been achieved in an evolutionary process. It is hypothesized that a membrane partitioning internal constituents off the environment might have been one of the minimal requirements for living systems to arise [3, 4]. A lipid membrane represents a formidable barrier to the passage of polar molecules. It organizes biological processes by compartmentalizing them. Compartmentalization enables segregation of specific chemical reactions for the purposes of increased biochemical efficiency by restricted dissemination and efficient storage of reaction products. This partitioning is not only realized between the cell and its environment, but it is even recapitulated within the cell. Vesicles, as an instance of minimality, feature an aqueous compartment

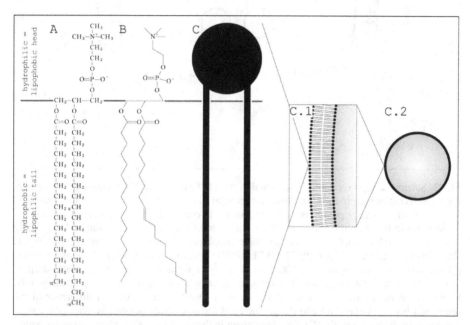

Fig. 1. Structure of glycerophospholipids and bilayer formation. The structure of 1-palmitoyl-2-oleoyl-*sn*-glycero-3-phosphatidylcholine (PC(16:0/18:1(Δ9-Cis)) is represented (*A*) as a structural formula, (*B*) as a skeletal formula, and (*C*) as a schematic representation used throughout this publication. Glycerophospholipids are amphiphilic molecules with lipophilic hydrocarbon "tails" and hydrophilic "heads". In PC(16:0/18:1(Δ9-Cis)) the headgroup is phosphatidylcholine; a saturated C_{16} palmitic acid (16:0) hydrocarbon chain and a monounsaturated C_{18} oleic acid (18:1) hydrocarbon chain occur at the C1 and C2 position of the glycerophospholipid and constitute the tail. The double bond (monounsaturation) occurs between the C9 and C10 atoms (Δ9) of the oleic acid, has cis configuration, and puts a rigid 30° bend in the hydrocarbon chain. (*C.1*) Phospholipids can form lipid bilayers (membranes) that partition an aqueous compartment off the surrounding medium. (*C.2*) In vesicles an *intra*vesicular fluid (light blue) is separated from the *inter*vesicular medium (white).

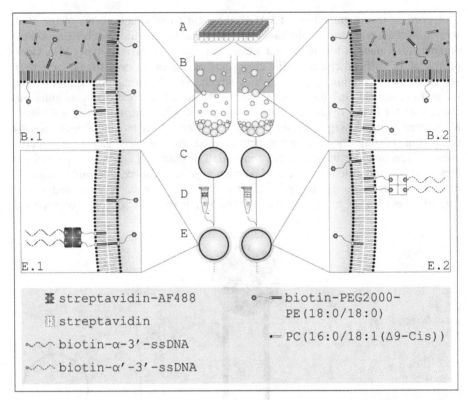

Fig. 2. Schematic representation of the parallel vesicle formation and membrane doping procedure. (*A*) Vesicles are produced in 96-well microtiter plates, providing parallel formation of up to 96 distinct vesicle populations. (*B*) The sample is composed of two parts: water droplets (light blue) in the oil phase (light gray) and the bottom aqueous phase (white), which finally hosts the vesicles. (*B.1,B.2*) Due to their amphiphilic character, glycerophospholipids (PC(16:0/18:1(Δ9-Cis)), biotin-PEG2000-PE(18:0/18:0)), solved in mineral oil, stabilize water-oil interfaces by forming monolayers. Two monolayers form a bilayer when a water droplet, induced by centrifugation, passes the interface. Glycerophospholipids are incorporated into the bilayer according to their percentage in the solution. Due to both the density difference of the *intra-* and *inter*vesicular fluid and the geometry of the microplate bottom, vesicles pelletize in the centre of the well (cp. *B*). (*C*) Vesicles remain in the same microtiter plate during formation and membrane doping. (*D*) Vesicle populations become distinct by incubating them with single stranded DNA (ssDNA) of different sequence (α: biotin-TGTACGTCACAACTA-3', α': biotin-TAGTTGTGACGTACA-3') and streptavidin differing in fluorescence labeling (Alexa Fluor 488 conjugate (AF488) or unlabeled). (*E.1,E.2*) ssDNA covalently bound to biotin is non-covalently linked to phospholipid-grafted biotinylated polyethylene glycol tethers (biotin-PEG2000-Phosphoethanolamine) using streptavidin as cross-linking agent.

partitioned off the surrounding by an impermeable lipid membrane. Like cellular membranes, vesicular membranes consist of amphiphilic phospholipids that link a hydrophilic "head" and a lipophilic "tail" (Fig. 1). Suspended phospholipids can self-assemble to form closed, self-sealing solvent-filled vesicles that are bounded by a two-layered sheet (a bilayer) of 6 nm in width, with all of their tails pointing toward the center of the bilayer. This molecular arrangement excludes water from the center of the sheet and thereby eliminates unfavorable contacts between water and the lipophilic (= hydrophobic) tails. The lipid bilayer provides inherent self-repair characteristics due to lateral mobility of its phospholipids. Wide usage of artificial vesicles is found in analytics [5-9] and synthetics, where their applications include bioreactors [10-12], and drug delivery systems [13-17].

In the laboratory work, we focused on intrinsic information processing capabilities of vesicles and multivesicular assemblies. To achieve compartmentalization we developed methods to self-assemble multivesicular aggregates of programmable composition and spatial configuration composed of distinct vesicle populations that differ in membrane and *intra*vesicular fluid composition. The assembly process of multivesicular aggregates was based on the hybridization of biotinylated single-stranded DNA (ssDNA) with which the vesicles were doped. Doping was realized by anchoring biotinylated ssDNA to biotinylated phospholipids via streptavidin as a cross-linking agent (Fig. 2). The potential of a programmable self-assembly of super-structures with high degrees of complexity [18] has attracted significant attention to nanotechnological applications in the last decade. So far, cross-linkage based on DNA hybridization was proposed to induce self-assembly of complementary monohomo-philic vesicles [5, 19] or hard sphere colloids [20-23], to induce programmable fusion of vesicles [5, 24], or to specifically link vesicles to surface supported membranes [5, 25-27]. By introducing a surface doping of distinct populations of ssDNA, as realized in the self-assembly of hard sphere colloids [28, 29], we provide n-arity to the assembly process. As a result, linkage of more than two distinct vesicle populations, as proposed by Chiruvolu *et al.* [30] and already realized for hard sphere colloids [31], becomes feasible.

Concerning procedural manageability in laboratory experimentation, we established a new protocol for *in vitro* vesicle formation and modification. It increases the versatility of the underlying vesicle formation method [11, 32, 33] by introducing microtiter plates and vesicle pelletization (Fig. 2). The potential of the vesicle formation method that provides independent composition control of the *intra-* and *inter*vesicular fluid as well as of the inner and outer bilayer leaflet was exemplified by the production of asymmetric vesicles combining biocompatibility and mechanical endurance in asymmetric vesicles [32]. Here, we exemplify the advantages of the novel protocol by carrying out a high-throughput analysis of constituents affecting vesicle formation and stability. Thereby the effect of nine different glycerophospholi-pids on vesicle formation was tested. We discuss how this procedure could be used both in evolutionary optimization of self-assembling amphiphilic systems and to realize tailored communication networks in assemblies of artificial cells by programmable localization of glycerophospholipids within the vesicular membrane.

2 Material and Methods

Major technical modifications of the vesicle formation protocol reported in Ref. [32] were: the introduction of (i) 96-well microtiter plates U96 to provide a high-throughput analysis and (ii) a density difference between *intra-* and *inter*vesicular solution to induce vesicle pelletization. For a description of the modified vesicle

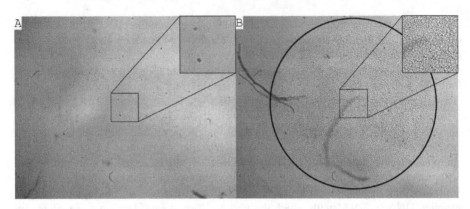

Fig. 3. Measurement of vesicle yield. (*A*) No vesicle pellet emerged for the mixture of 50% PC(24:1(Δ15-Cis)/24:1(Δ15-Cis)) and 50% PC(16:0/18:1(Δ9-Cis)). (*B*) The vesicle yield is expressed by length of circumference of the vesicle pellet (circle) emerged for the mixture 10% PC(24:1(Δ15-Cis)/24:1(Δ15-Cis)) and 90% PC(16:0/18:1(Δ9-Cis)). Fibers represent pollutants not affecting vesicular yield or handling.

protocol see Fig. 2. To analyze the effect of different glycerophospholipids on vesicle formation, data on vesicle yield of four times nine equimolar mixtures of 100%-m (m ∈ {100, 50, 10, 1}) PC(16:0/18:1(Δ9-Cis)) and m% PC(x:0/x:0) (x ∈ {12, 14, 16, 18, 24}), PC(y:1(Δ9-Cis)/y:1(Δ9-Cis)) (y ∈ {14, 16, 18}), or PC(24:1(Δ15-Cis)/24:1(Δ15-Cis)) were collected and compared to 100%-m PC(16:0/18:1(Δ9-Cis)) and m% mineral oil (solvent for all glycerophospholipids) as control. Vesicle formation was performed in duplication. Length of circumference of the vesicle pellet is used as a measure of vesicle yield (Fig. 3). Light-microscopy was performed using a Wild M40 inverted microscope equipped with a MikoOkular microscope camera. All camera settings were identical for the recordings. Confocal laser scanning microscopy was performed using an inverted Leica Confocal DMR IRE2 SP2 confocal laser scanning microscope.

3   Results and Discussion

In the literature many examples of artificially produced vesicles are reported, whose membranes are composed of PC(16:0/18:1(Δ9-Cis)) (POPC) exclusively.[1] To vary the intrinsic material properties of membranes we have to be able to alter their phospholipid content. In this study, we realized a high-throughput analysis of glycerophospholipids affecting vesicle formation. All glycerophospholipids differed in length and saturation of their hydrocarbon chains only. In the control experiment, vesicle yield was constant when POPC was present (Fig. 4). Influence of the phospholipids tested at an admixture of one percent was marginal and will not be discussed. Intergroup comparisons (cp. Fig. 4) of saturated and unsaturated glycerophospholipids of equal chain length revealed remarkable differences in vesicle yield. Whereas unsaturated glycerophospholipids of chain length up to 18 carbon atoms both could form vesicles

---

[1] In 2008 more than 150 articles were published concerning POPC vesicles or liposomes.

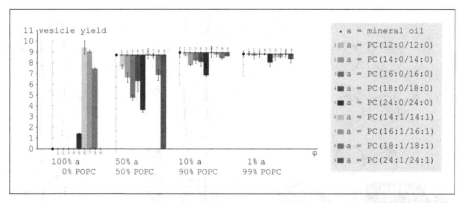

Fig. 4. Effect of different glycerophospholipids on vesicle formation. Vesicle yield of 100%-m (m ∈ {100, 50, 10, 1}) POPC = PC(16:0/18:1(Δ9-Cis)) and m% mineral oil serves as standard (XY (scatter) chart). Deviations from control for the glycerophospholipids tested are presented in absolute values as bar charts. Error bars indicate the standard deviation. The glycerophospholipids tested can be summarized according to the level of saturation of their hydrocarbon chains in two groups (saturated, PC(x:0/x:0) (x ∈ {12, 14, 16, 18, 24}); unsaturated, PC(y:1(Δ9-Cis)/y:1(Δ9-Cis)) (y ∈ {14, 16, 18}), PC(24:1(Δ15-Cis)/24:1(Δ15-Cis))).

by themselves and did not or just slightly affect the yield at 50 and 10 percent, saturated lipids seem to disturb vesicle formation down to 10 percent of admixture. For a chain length of 24 carbon atoms the situation is diametrical. A small amount of vesicles is found for solutions containing 100 percent saturated PC(24:0/24:0). Whereas vesicle yield is halved at 50 percent for this lipid, vesicle formation is inhibited totally by the unsaturated PC(24:1(Δ15-Cis)/24:1(Δ15-Cis)) up to an admixture of 10 percent (cp. Fig. 3). Intragroup comparison within saturated or unsaturated glycerophospholipids reveals a decrease in vesicle yield depending on chain length (except for PC(24:0/24:0) at 100 percent). Only a limited number of glycerophospholipid is able to form vesicles on their own or in cooperation with POPC.

By providing parallelism in vesicle formation, processing and examination (cp. Fig. 2), we not only increased the experimental throughput and the procedural manageability, but we lowered the costs and reduced the amount of contributory factors. This reduction of dimensionality and the parallelism in vesicle formation provided by the novel vesicle formation method presented herein may prove to be useful in evolutionary design of experiments by shortening the search toward the optimality region of the search space [2]. The application of microtiter plates may further enable the automatization in vesicle formation.

By self-assembling multivesicular aggregates of programmable composition and spatial configuration, composed of vesicles differing in membrane and *intra*vesicular fluid composition, an inhomogeneous distribution of information carriers is achieved (Fig. 5). To build complex information systems, the exchange of information carriers between the compartments themselves and/or between the environment and the compartments has to be realized. Biological membranes contain membrane proteins that were evolutionary optimized for catalyzing numerous chemical reactions, mediating the flow of nutrients and wastes, as well as participating in communication across the

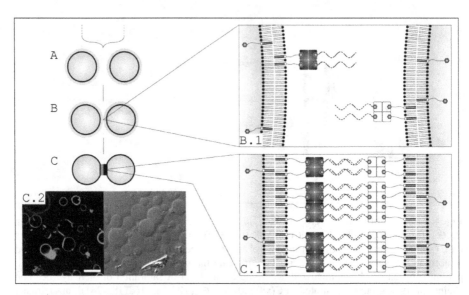

Fig. 5. Schematic representation of the self-assembly process and micrographs of adhesion plaques. (*A*) For vesicle formation, membrane doping and illustration symbols see Fig. 2.Two distinct vesicle populations are merged (brace). (*B*) The lateral distribution of linkers in the lipid membrane is homogeneous. (*B.1*) Vesicles doped with complementary single stranded DNA come into contact. (*C*) Hybridization of DNA strands results in double stranded DNA and induces the assembly process. Due to their lateral mobility, linkers accumulate in the contact zone forming an adhesion plaque – the lateral distribution of linkers in the outer leaflet becomes inhomogeneous. (*C.1*) Biotinylated phospholipids (biotin-PEG2000-PE(18:0/18:0)) colocalize with the linkers. Even though the lateral distribution of phospholipids in the inner leaflet is not affected, the membrane composition *intra*-adhesion-plaque (by accumulation) and *inter*-adhesion-plaque (by depletion) becomes different. (*C.2*) CLSM (confocal laser scanning microscope) and DIC (differential interference contrast) micrograph of a vesicular aggregate. Accumulation and depletion of linkers are clearly visible in the CLSM micrograph. Scale bar represents 10µm.

membrane. The use of such membrane proteins in information exchange was excluded by the guidelines of PACE. Thus, information processing capabilities of multivesicular aggregates had to be realized by exploiting intrinsic material properties of the phospholipid membrane. A key property of the lipid membrane is its phase [34]. Lipid bilayers can undergo phase transitions in which they become a gel-like solid and therefore lose their fluidity. The state depends on temperature; the transition temperature of a bilayer increases with the chain length and the degree of saturation of its fatty acid residues. The phase transitions are triggered externally by changing temperature; a fact exploited in exchanging reactants between the surrounding medium and a vesicular compartment setting up consecutive enzymatic reactions in a single container [35]. A selective exchange of information between compartments of a multivesicular aggregate relying on inherent material properties of phospholipid membranes has not been demonstrated so far. We recognized that the formation of adhesion plaques [36, 37] triggered by the aggregation process results in an inhomogeneous distribution of phospholipids in the lateral dimension of the membrane (Fig. 5). This inhomogeneity in phospholipid distribution causes a difference in phase transition characteristics of membrane portions *intra-* and *inter-*adhesion-plaque. Selective opening of communication channels between the compartments, whereas the membrane portions *inter-*adhesion-plaques are still impermeable, becomes conceivable. By using membrane anchors of different phospholipid composition the adhesion plaques would differ in phase transition characteristics among each other enabling programmable and triggerable communication networks in multivesicular aggregates of programmable composition and spatial configuration.

Realization of such tailored communication networks is hindered by restrictions in commercial availability of lipid anchors to dope vesicular surfaces with ssDNA. The high-throughput analysis presented herein allows for identification of lipid candidates not affecting vesicle formation and differing in phase transition characteristics. In a next step the head group of these candidates could be covalently linked to biotinylated PEG tethers therefore providing anchors for biotinylated ssDNA (via (strept-)avidin).

Acknowledgements. This work was conducted as part of the European Union integrated project PACE (EU-IST-FP6-FET-002035). Maik Hadorn was supported by the Swiss National Foundation Project 200020-118127 Embryogenic Evolution: From Simulations to Robotic Applications. Peter Eggenberger Hotz was partly supported by PACE. Wet laboratory experiments were performed at the Molecular Physiology Laboratory of Professor Enrico Martinoia (Institute of Plant Biology, University of Zurich, Switzerland).

References

1. Rasmussen, S.: Protocells: bridging nonliving and living matter. MIT Press, Cambridge (2009)
2. Forlin, M., Poli, I., De March, D., Packard, N., Gazzola, G., Serra, R.: Evolutionary experiments for self-assembling amphiphilic systems. Chemometrics Intell. Lab. Syst. 90, 153–160 (2008)
3. Griffiths, G.: Cell evolution and the problem of membrane topology. Nat. Rev. Mol. Cell Biol. 8, 1018–1024 (2007)
4. Israelachvili, J.N., Mitchell, D.J., Ninham, B.W.: Theory of self-assembly of lipid bilayers and vesicles. Biochimica Et Biophysica Acta 470, 185–201 (1977)
5. Chan, Y.H.M., van Lengerich, B., Boxer, S.G.: Effects of linker sequences on vesicle fusion mediated by lipid-anchored DNA oligonucleotides. Proc. Natl. Acad. Sci. USA 106, 979–984 (2009)
6. Hase, M., Yoshikawa, K.: Structural transition of actin filament in a cell-sized water droplet with a phospholipid membrane. J. Chem. Phys. 124, 104903 (2006)
7. Hotani, H., Nomura, F., Suzuki, Y.: Giant liposomes: from membrane dynamics to cell morphogenesis. Curr. Opin. Colloid Interface Sci. 4, 358–368 (1999)
8. Limozin, L., Roth, A., Sackmann, E.: Microviscoelastic moduli of biomimetic cell envelopes. Phys. Rev. Lett. 95, 178101 (2005)
9. Luisi, P., Walde, P.: Giant vesicles. John Wiley & Sons, Ltd., Chichester (2000)
10. Michel, M., Winterhalter, M., Darbois, L., Hemmerle, J., Voegel, J.C., Schaaf, P., Ball, V.: Giant liposome microreactors for controlled production of calcium phosphate crystals. Langmuir 20, 6127–6133 (2004)

11. Noireaux, V., Libchaber, A.: A vesicle bioreactor as a step toward an artificial cell assembly. Proc. Natl. Acad. Sci. USA 101, 17669–17674 (2004)

12. Nomura, S., Tsumoto, K., Hamada, T., Akiyoshi, K., Nakatani, Y., Yoshikawa, K.: Gene expression within cell-sized lipid vesicles. Chembiochem 4, 1172–1175 (2003)

13. Abraham, S.A., Waterhouse, D.N., Mayer, L.D., Cullis, P.R., Madden, T.D., Bally, M.B.: The liposomal formulation of doxorubicin. In: Liposomes, Pt E. Elsevier Academic Press Inc., San Diego (2005)

14. Allen, T.M., Cullis, P.R.: Drug delivery systems: Entering the mainstream. Science 303, 1818–1822 (2004)

15. Marjan, J.M.J., Allen, T.M.: Long circulating liposomes: Past, present and future. Biotechnology Advances 14, 151–175 (1996)

16. Tardi, P.G., Boman, N.L., Cullis, P.R.: Liposomal doxorubicin. J. Drug Target. 4, 129–140 (1996)

17. Sengupta, S., Eavarone, D., Capila, I., Zhao, G.L., Watson, N., Kiziltepe, T., Sasisekharan, R.: Temporal targeting of tumour cells and neovasculature with a nanoscale delivery system. Nature 436, 568–572 (2005)

18. Licata, N.A., Tkachenko, A.V.: Errorproof programmable self-assembly of DNA-nanoparticle clusters. Physical Review E (Statistical, Nonlinear, and Soft Matter Physics) 74, 041406 (2006)

19. Beales, P.A., Vanderlick, T.K.: Specific binding of different vesicle populations by the hybridization of membrane-anchored DNA. J. Phys. Chem. A 111, 12372–12380 (2007)

20. Biancaniello, P.L., Crocker, J.C., Hammer, D.A., Milam, V.T.: DNA-mediated phase behavior of microsphere suspensions. Langmuir 23, 2688–2693 (2007)

21. Mirkin, C.A., Letsinger, R.L., Mucic, R.C., Storhoff, J.J.: A DNA-based method for rationally assembling nanoparticles into macroscopic materials. Nature 382, 607–609 (1996)

22. Valignat, M.P., Theodoly, O., Crocker, J.C., Russel, W.B., Chaikin, P.M.: Reversible self-assembly and directed assembly of DNA-linked micrometer-sized colloids. Proc. Natl. Acad. Sci. USA 102, 4225–4229 (2005)

23. Biancaniello, P., Kim, A., Crocker, J.: Colloidal interactions and self-assembly using DNA hybridization. Phys. Rev. Lett. 94, 058302 (2005)

24. Stengel, G., Zahn, R., Hook, F.: DNA-induced programmable fusion of phospholipid vesicles. J. Am. Chem. Soc. 129, 9584–9585 (2007)

25. Benkoski, J.J., Hook, F.: Lateral mobility of tethered vesicle - DNA assemblies. J. Phys. Chem. B 109, 9773–9779 (2005)

26. Yoshina-Ishii, C., Boxer, S.G.: Arrays of mobile tethered vesicles on supported lipid bilayers. J. Am. Chem. Soc. 125, 3696–3697 (2003)

27. Li, F., Pincet, F., Perez, E., Eng, W.S., Melia, T.J., Rothman, J.E., Tareste, D.: Energetics and dynamics of SNAREpin folding across lipid bilayers. Nat. Struct. Mol. Biol. 14, 890–896 (2007)

28. Maye, M.M., Nykypanchuk, D., Cuisinier, M., van der Lelie, D., Gang, O.: Stepwise surface encoding for high-throughput assembly of nanoclusters. Nat. Mater. 8, 388–391 (2009)

29. Prabhu, V.M., Hudson, S.D.: Nanoparticle assembly: DNA provides control. Nat. Mater. 8, 365–366 (2009)

30. Chiruvolu, S., Walker, S., Israelachvili, J., Schmitt, F.J., Leckband, D., Zasadzinski, J.A.: Higher-order self-assembly of vesicles by site-specific binding. Science 264, 1753–1756 (1994)

31. Xu, X.Y., Rosi, N.L., Wang, Y.H., Huo, F.W., Mirkin, C.A.: Asymmetric functionalization of gold nanoparticles with oligonucleotides. J. Am. Chem. Soc. 128, 9286–9287 (2006)

32. Pautot, S., Frisken, B.J., Weitz, D.A.: Engineering asymmetric vesicles. Proc. Natl. Acad. Sci. USA 100, 10718–10721 (2003)

33. Träuble, H., Grell, E.: Carriers and specificity in membranes. IV. Model vesicles and membranes. The formation of asymmetrical spherical lecithin vesicles. Neurosciences Research Program bulletin 9, 373–380 (1971)

34. Feigenson, G.W.: Phase diagrams and lipid domains in multicomponent lipid bilayer mixtures. Biochim. Biophys. Acta-Biomembr. 1788, 47–52 (2009)

35. Bolinger, P.Y., Stamou, D., Vogel, H.: An integrated self-assembled nanofluidic system for controlled biological chemistries. Angew. Chem.-Int. Edit. 47, 5544–5549 (2008)

36. NopplSimson, D.A., Needham, D.: Avidin-biotin interactions at vesicle surfaces: Adsorption and binding, cross-bridge formation, and lateral interactions. Biophys. J. 70, 1391–1401 (1996)

37. Farbman-Yogev, I., Bohbot-Raviv, Y., Ben-Shaul, A.: A statistical thermodynamic model for cross-bridge mediated condensation of vesicles. J. Phys. Chem. A 102, 9586–9592 (1998)

A Developmental System for Organic Form Synthesis

Benjamin Porter

Centre for Electronic Media Art, Faculty of Information Technology,
Monash University, Clayton 3800, Australia

Abstract. Modelling the geometry of organic forms using traditional
CAD or animation tools is often difficult and tedious. Different models
of morphogenesis have been successfully applied to this problem; however
many kinds of organic shape still pose difficulty. This paper introduces a
novel system, the Simplicial Developmental System (SDS), which simulates morphogenetic and physical processes in order to generate specific
organic forms. SDS models a system of cells as a dynamic simplicial
complex in two or three dimensions that is governed by physical rules.
Through growth, division, and movement, the cells transform the geometric and physical representations of the form. The actions of the cells
are governed by conditional rules and communication between cells is
supported with a continuous morphogen model. Results are presented in
which simple organic forms are grown using a model inspired by limb
bud development in chick embryos. These results are discussed in the
context of using SDS as a creative system.

1 Introduction

The beautiful and complex forms found in nature are typically difficult to geometrically model using traditional CAD or animation tools. One strategy to assist the creation of these forms is to model the processes behind their generation;
in the case of organic form the processes of *biological development*. There are various *developmental systems* that successfully do this, including grammar-based
methods [1,2], rule-based methods [3,4,5], cellular automata [6], and physical
simulation systems [7,8,9] (also see surveys in e.g., [10,11].)

1.1 Organic Form

Developmental systems have traditionally focussed on modelling structurally
complex forms such as plants and architecture, and they perform this task ex-
cellently. However, the application of developmental systems to the synthesis of
organic forms with smooth, complex surfaces is still an open area of research
and the subject of this paper. This paper introduces a new system for generating organic forms with complex organic surfaces that interact in space, have
identifiable modules that repeat and vary, have high amounts of symmetry, and
that appear as though they are composed of solid organic matter. Some existing

K. Korb, M. Randall, and T. Hendtlass (Eds.): ACAL 2009, LNAI 5865, pp. 136–148, 2009.
© Springer-Verlag Berlin Heidelberg 2009

systems are able to synthesise these kinds of forms, and we briefly discuss some of them, highlighting the limitations which ultimately led to this research.

Incorporating a physical model into a simple growth model can generate surprisingly organic forms. The semi-morphogenetic system of Combaz and Neyret [7] generates rich and abstract organic forms through physical simulation and growth. The system requires a user to paint *hot-spots* onto a surface which cause an expansion of the surface geometry around that area. The system then finds a minimal-energy configuration of the geometry by incorporating physical aspects such as resistance to bending and stretching. The resultant forms have high amounts of detail in the form of folds and creases which manifest due to the physical model constraining the growth process. In this system there is no support for module and symmetry specification which makes this method tedious when constructing more complex forms. Leung and Berzins [12] and Harrison et al. [13] experimented with coupling a reaction-diffusion process to a growing surface, but both these methods have only been demonstrated to generate very simple organic shapes.

There are numerous 2D or surface-based 3D systems that can generate organic form (e.g., [14,15,9]) They are capable of synthesizing complex organic structures but their inability to model the physical effects within a growing 3D *solid* form greatly constrains the physical appearance of the generated forms (for example, modelling a tree branch that bends under its own weight is not easy to model within these systems.) The obvious generalisation of this class of systems is to 3D *volume-based* systems such as *voxel*-based [6], cell-based [16], or, as presented here, *simplex*-based. The system of Eggenberger [16] has been used to *evolve* simple solid geometries, but it is not yet clear whether it is capable of generating the type of form discussed above. The system is superficially similar to the system presented here but differs in geometric representation and application (artificial embryology versus computer graphics).

1.2   Simplicial Developmental System

The successes and limitations of the systems discussed above (and others) led to a set of requirements for a successful organic form synthesis system: the system must be able to represent smooth surfaces that interact in space; it should support the creation of modules with complex organic interfaces; and be able to generate organic symmetries with repeated, varying modules. These requirements led to the design of the *Simplicial Developmental System* (SDS): a 2D and 3D organic form synthesis system. SDS *grows* organic form through simulation of morphogenetic and physical processes. The system is capable of generating simple organic modules with repetition and variation that have organic interfaces (i.e., are geometrically coupled to their parent geometry). This paper introduces SDS, presents details for the 2D version (SDS2), and demonstrates the synthesis of the simple two-dimensional organic geometry of Figure 1. Investigation into SDS3 is ongoing.

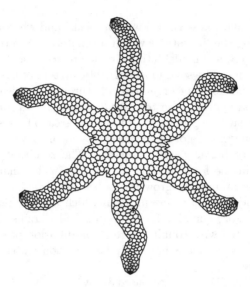

Fig. 1. A starfish–like form generated using SDS

Overview. The basic concepts in SDS are the *organism* and the *cell*. An organism is a connected collection of cells that has three different models: the *geometric, physical* and *process* models. The geometric model of the organism specifies the spatial and topological configuration of the cells (Section 2). The physical model binds a mass-spring system to the geometry, causing the organism to appear as though it consists of solid, elastic matter (Section 3). Finally, the process model (Section 5.2) specifies how cells communicate and behave. Development is driven by cellular behavior such as division, growth and movement that transform the organism over time (Section 4). This paper presents some experiments with the system, in which a model of limb bud development in chick embryos [17] was adapted and some simple organic forms were generated. Additionally, module boundary, repetition, variation, and parameterisation are discussed in this experimental context (Section 5). Finally, further research questions and directions are proposed (Section 6).

2   The Geometric Model

The geometry of an organism specifies the structural and spatial configuration of its cells. SDS2 organisms are sets of triangles joined along edges. The vertices of the geometry specify the positions of the cells, and the edges represent an adjacency relationship between cells (see Figure 2). This representation is based on Matela and Fletterick's theoretical model [18,19] and was chosen due to its conceptual simplicity, support of multi-scale detail, and ability to model arbitrary forms. SDS3 organisms are sets of non-intersecting tetrahedra joined along faces. Using the language of *simplicial complexes* we refer to triangles and tetrahedra as *2-simplices* and *3-simplices* (or just *simplex* for both). Hence an

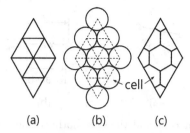

Fig. 2. The (a) geometric and (b) conceptual models of an SDS2 organism. (c) The *geometric dual* is also used in this paper as it provides a clear illustration of the cell system. Note that cells occur at the vertices in (a) but are represented by polygons in (c).

organism in SDS, whether 2D or 3D, can be said to be composed of cells and simplices.

3   The Physical Model

An SDS organism is physically modelled using a *solid mass-spring model*. This approach provides a simple approximation of the complex dynamics within a soft body and is common in physical simulation [20,21]. Using an organism's geometry the mass-spring model defines energy-minimising forces which cause the geometry to assume a more natural appearance. Mass-spring systems have been used to model cell complexes in a number of fields, e.g., computational development [16,22]. Our approach is based on that of Teschner *et al.* [23], in which the cells, edges and simplices of the geometry are modelled as point masses, edge springs, and simplex springs, respectively. Edge springs exist between every pair of adjacent cells (Figure 3) and preserve the local *structure* of the geometry. Likewise, simplex springs exist within every simplex and preserve the local *volume* of an organism. (A detailed specification of the physical model is available if required [24].) The system is solved using standard real-time numerical integration schemes with damping added to increase stability. Collision detection and response can be incorporated into the simulator to prevent the simplices from intersecting; However, the system presented in this paper did not employ collision detection, as it was not found to be necessary to achieve the results presented here.

4   Transformation Rules

The geometric and physical models are transformed by a set of operations that model processes within morphogenesis. These transformations are triggered either by the process model or as a result of cell movement and allow a simple organism to develop into a complex one. Some basic operations are defined in this section: division, growth and movement. These rules are designed to be general enough to apply to both the two and three dimensional systems. Thus growth models prototyped in SDS2 can be more easily adapted to SDS3.

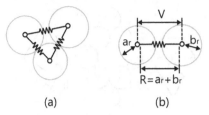

(a)   (b)

Fig. 3. (a) Three adjacent cells and the edge springs connecting them. (b) The rest length, R, of an edge spring is the sum of the radii of the two cells it joins. The actual length V is the distance between the two cells. The edge spring acts to minimize the difference between V and R.

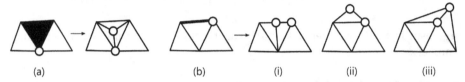

(a)   (b)   (i)   (ii)   (iii)

Fig. 4. Division in SDS2. (a) A cell (the circle) chooses a neighbouring simplex to divide into. The new daughter cell is placed in the middle of this simplex. (b) Division on the boundary proceeds by selecting an adjacent boundary edge and either (i) subdividing it as shown, or (ii,iii) attaching one or two new simplices to the boundary.

4.1   Cell Division

SDS models cell division, or *mitosis*, with an operation that replaces one cell with two. A dividing cell chooses a direction to divide into and is replaced with two adjacent daughter cells, each half the size of the parent. Cell divisions allow an organism to increase in complexity. The balanced internal cell division methodology of Ransom and Matela [25] can be applied to SDS2; however it does not easily generalise to SDS3. A simpler approach is taken in SDS (demonstrated in Figure 4(a)). The operator adds a new cell by subdividing an adjacent simplex in the direction of division. Division in the boundary is handled with a few different cases (Figure 4(b)).

4.2   Cell Growth

SDS cells control their rate of growth and as a result the sizes of simplices can change. This provides a mechanism to generate forms with detail over different scales, as demonstrated in Figure 5. Cell growth coupled with division results in *proliferation*, which allows many similar simplices to be rapidly generated. This is demonstrated in the limb bud experiment presented later.

4.3   Cell Movement

Cells move through space as a result of the physical simulation. SDS also allows cells to move with respect to the topology of their configuration. These *topological moves* occur when the topology must be changed; for instance, in SDS2 when

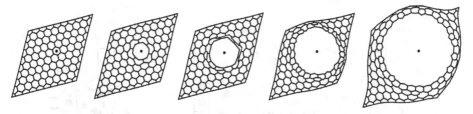

Fig. 5. A time series of a simulation of a growing (marked) cell in an SDS2 form. The physical model forces the surrounding structure to reconfigure (using topological moves) in a visually organic manner.

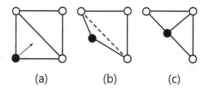

Fig. 6. (a) A cell moves through space. (b) If the cell crosses the dashed edge it will be *flipped* resulting in (c) a new configuration.

a cell crosses over an internal edge then an edge-flip is performed (Figure 6); this transformation is equivalent to Matela and Fletterick's model [18]. Boundary conditions are handled similarly. Topological transformations are applied in SDS3 if a tetrahedron is compressed to zero volume; either via a vertex moving through an opposite face or via an edge moving through an opposite edge. The transformations are understandably more complex and are the subject of ongoing research.

Topological moves allow the geometry of an organism to reconfigure in a visually organic manner. The moves, caused by internal forces, also allow stresses to distribute more evenly through an organism, resulting in a physically more stable topology. Unlike movement, the cell division, growth and death operations are controlled purely by active cellular processes. These are governed by the *process model*, presented in the next section.

5   A Study of Limb Growth

This section presents a study of shape formation in SDS and proposes a simple process model. A biological model of limb bud development in chick embryos was adapted and implemented in SDS2, resulting in the growth of simple organic geometries. The results are now presented and aspects of the model, including re-use, module boundary, repetition, and variation are discussed.

5.1   Limb Growth in Chicks

Limb growth in early chick embryogenesis is the result of a coupled reaction between the *epithelium* (a proto skin) and *mesenchyme* (free floating cells

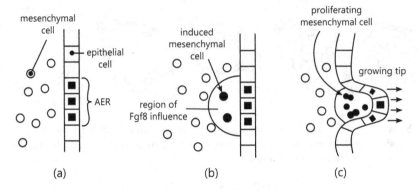

Fig. 7. Limb bud development. (a) The AER cells are initialised in the epithelium. (b) The AER releases Fgf8 which induces the nearby underlying mesenchyme to (c) grow and divide towards the AER, resulting in a feedback cycle and the proximal-distal (outwards) growth of a limb.

beneath the skin) [17]. The development involves interactions between a region on the epithelium called the *apical ectodermal ridge* (AER) and the underlying mesenchyme. Broadly, the AER diffuses a protein, *Fgf8*, which induces the underlying mesenchyme to proliferate towards it. The proliferating mesenchyme diffuses a protein, *Fgf10*, which induces the AER to produce more Fgf8. This feedback loop causes a cluster of cells to proliferate in one direction, resulting in the growth of a primitive limb (see Figure 7.)

5.2   Process Model

The limb model was adapted into SDS through the specification of a process model. The process model abstracts the aspects of development that allow cells to communicate and coordinate their actions. Previous approaches to modelling cell behaviour have included programmatic [26], GRNs [27], and the rule-based system of Fleischer [28]. In the latter approach, cells contain continuous concentrations of diffusing *morphogens* [29] and perform actions triggered by internal morphogen thresholds. This method is most suited to the limb bud model and thus was the basis for the following process model:

- There is a finite set of morphogens $\{m_1, \ldots, m_n\}$ (for this example we have two morphogens, m_{f8} and m_{f10});
- Morphogens are contained within cells as a continuous concentration value between 0 and 1;
- Morphogens diffuse isotropically between adjacent cells and decay over time;
- Morphogens are created in cells at a linear rate $\frac{dm_i}{dt}$;
- A set of threshold conditions govern the creation of morphogens and the triggering of actions;
- Cells are spherical with radius r and have a growth variable $\frac{dr}{dt}$;
- A cell can compute the local morphogen gradient;

Table 1. Conditional rules for the limb growth model

rule	condition	action
AER$_1$	$m_{f10} > $ AER$_{thres}$	$\frac{dm_{f8}}{dt} = \triangle$f8
M$_1$	$m_{f8} > $ M$_{thres}$	$\frac{dm_{f10}}{dt} = \triangle$f10, $\frac{dr}{dt} = \triangle r_M$
M$_2$	$r > $ M$_R$	divide towards f8 source
E$_1$	$m_{f8} > $ E$_{thres}$	$\frac{dr}{dt} = \triangle r_E$
E$_2$	$r > $ E$_R$	divide towards f8 source

- A cell has an internal type that distinguishes between the AER, epithelium and mesenchyme (denoted as types AER, E and M); and
- Different conditional rules apply to different cell types, for example the M_i rules apply to cells of type M.

The cell dynamics of the process are modelled as the set of rules in Table 1. For example, rule AER$_1$ dictates that a cell of type AER with an f10 concentration greater than the parameter AER$_{thres}$ produces f8 at a linear rate of $\triangle$f8. There are nine user definable parameters: AER$_{thres}$, M$_{thres}$, E$_{thres}$, $\triangle$f8, $\triangle$f10, $\triangle r_M$, $\triangle r_E$, M$_R$, and E$_R$.

5.3   Results and Discussion

The limb growth model was implemented in an SDS2 prototype. The physical model was numerically integrated using the mid-point method and morphogen diffusion was simulated using an approximation of the standard particle diffusion equation. The simulations resulted in the formation of simple organic protrusions. We explored the model in isolation, in the context of re-use, and considered the effect of changing the size of the induced region. This section presents and discusses the results of this exploration, and demonstrates the generation of organic starfish-like forms.

Limb Growth in Isolation. The model was first applied in an isolated context in order to determine its form generating potential. The experiment began with a rectangular configuration of cells, then a single cell in the complex was manually initialised with a full concentration of f10 and designated as an AER cell. The simulation sequence shown in Figure 8 demonstrates that this model results in simple limb-like forms. The limb bud model reveals a general method for directing growth: local induction with feedback and directed proliferation. The two way induction allows a cluster of cells to coordinate their actions and the form that develops is a result of this cluster forcing the epithelium outwards. The organic configuration of the cells is a direct result of the physical simulation and the dynamic reconfiguration.

During growth the limb module closely couples itself with the body it grew from resulting in an organic interface. Figure 8(g) illustrates that the coupling between the geometry of the body and limb is complex. These preliminary results indicate that SDS is capable of generating organic module boundaries. In the

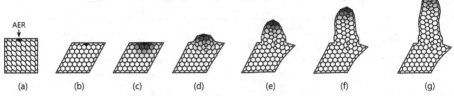

Fig. 8. Simulation of limb growth. (a) The initial form is rectangular with an AER cell at the top. The form is quickly forced into a (b) minimal energy configuration by the physical model. (c) f8 (shown as shading) begins to diffuse into the surrounding region. The nearby mesenchyme and epithelium begins to proliferate towards the AER, causing (d) a bump to appear. (e-g) During the feedback loop the growing tip is pushed away from the proliferating cluster of cells underneath it.

implemented model there is no distinction between the adhesion within the mesenchyme and epithelium; this may be the cause of the tumour-like appearance of the growths. In reality the epithelium is tightly formed whereas the mesenchyme contains free floating cells. This could be incorporated into the model by assigning spring strengths according to region, and will be a subject of future research.

Growing Multiple Limbs. In biological development the activation of a gene occasionally triggers a sequence of developmental events. The same gene can be activated in different locations resulting in the reuse and repetition of modules. The re-use of growth models in a creative system supports modularity, and is demonstrated in SDS by generating a form with multiple limbs. The simulation was initialised with a configuration of cells with multiple AER cells designated in different locations. The limb model was adapted by instructing all *non-proliferating* cells to slowly grow. This resulted in a slight tapering of the limbs because the internal cells grew at a faster rate than the growing tip.

Figure 9 presents the results of the simulation. This experiment demonstrates that SDS2 supports modularity, organic symmetry, and physicality. The arms of the starfish illustrate that modules can be designed and then applied in different locations. The different initial contexts of the arms and complex interactions during growth gives rise to subtle variation amongst the modules and results in an imperfect, organically symmetric form (see Figure 1). The starfish also has a solid appearance caused by the bending of the limbs and the low energy arrangements of the cells.

Exploring the Parameters. Adjusting the parameters of the model can lead to different characteristics in the developing form. For example, the growing tip cluster can be increased in size by decreasing the rate at which f8 decays. Experiments indicate that this leads to larger growths which are the result of a larger proliferating zone in the mesenchyme. Figure 10 illustrates a collection of different starfish forms generated using various sets of parameters. These demonstrate that qualitative aspects of the starfish model, such as limb width

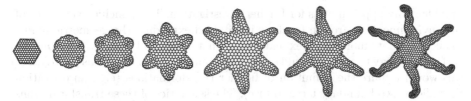

Fig. 9. Developmental sequence of a starfish-like form

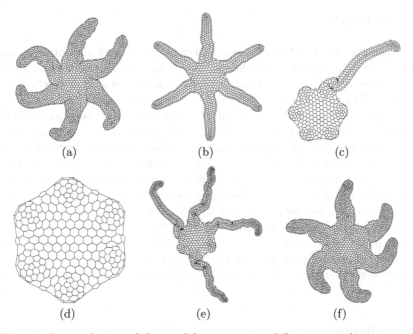

(a)   (b)   (c)

(d)   (e)   (f)

Fig. 10. Some relatives of the starfish grown using different sets of parameters

and curvature, can be changed. Additionally, images (c), (d), and (e) illustrate situations where the limb process has failed in some areas but succeeded in others, creating asymmetrical forms.

6   Future Work

There is much more research to be done in this domain. The study of SDS2 is a stepping stone to SDS3, and research is ongoing to implement these ideas into a novel 3D form generating system. Parallel to this, there is ongoing research into the limb model, for example, investigating ways to control size, taper, segmentation and length. Another current research goal is to explore the synthesis of other basic geometries such as spheres, tubes and folds. A toolbox of these *axiomatic shapes* would be highly beneficial to the artist. The synthesis of symmetries is also currently being investigated. The ideas presented in this paper provide many opportunities for further investigation. These include the effect of the environment on growth (such as in [6]) and extending the geometric model to accommodate more complex features, such as holes and creases. Allowing the surface topology to change would expand the range of potential forms greatly, but this would require the definition of further transformations (e.g., an operation that allows two boundary faces to fuse). The semantics of these transformations are unclear but this is definitely an interesting area to explore.

7   Conclusion

This paper presented initial research into synthesizing organic forms through morphogenetic and physical simulation. The Simplicial Developmental System (SDS), a system for generating organic form, was introduced and details were given for a 2D implementation. SDS can approximate smooth surfaces that interact in space due to its simplicial complex representation and physical model. The limb bud experiments resulted in the generation of basic organic geometries, and the concepts of module re-use, variation, and organic boundary were discussed in relation to SDS. The geometric transformations can be considered similar to the generative machinery of existing grammar-based or CA systems, and the addition of a physical model provides a further mechanism that constrains the possible geometries to visually organic configurations. By constructing the models and operations independent of dimension, growth models built in SDS2 should apply equally as well in SDS3. SDS shows great promise towards the goal of complex organic form generation and we hope that ongoing research will develop the techniques further into a useable, creative 3D organic modelling tool.

Acknowledgments

I am very grateful to my PhD supervisors, Jon McCormack and Alan Dorin, for seeding this project and for the numerous suggestions they have made regarding this project and paper. I would also like to thank Greg Paperin, Stephanie Porter, Ollie Bown, and the anonymous reviewers for their helpful feedback.

References

1. McCormack, J.: A developmental model for generative media. In: Capcarrère, M.S., Freitas, A.A., Bentley, P.J., Johnson, C.G., Timmis, J. (eds.) ECAL 2005. LNCS (LNAI), vol. 3630, pp. 88–97. Springer, Heidelberg (2005)
2. Prusinkiewicz, P., Lindenmayer, A.: The Algorithmic Beauty of Plants, 2nd edn. Springer, Heidelberg (1996)
3. Maierhofer, S.: Rule-Based Mesh Growing and Generalized Subdivision Meshes. PhD thesis, Vienna University of Technology (2002)
4. Todd, S., Latham, W.: Evolutionary Art and Computers. Academic Press, London (1992)

5. Hansmeyer, M.: L-systems in architecture,
   http://www.mh-portfolio.com/index1.html (retrieved January 21, 2009)
6. Greene, N.: Voxel space automata: modeling with stochastic growth processes in voxel space. In: SIGGRAPH 1989: Proceedings of the 16th Annual Conference on Computer Graphics and Interactive Techniques, pp. 175–184. ACM Press, New York (1989)
7. Combaz, J., Neyret, F.: Semi–interactive morphogenesis. In: Proceedings of the IEEE International Conference on Shape Modeling and Applications (2006)
8. Fleischer, K.W., Laidlaw, D.H., Currin, B.L., Barr, A.H.: Cellular texture generation. In: Computer Graphics. Annual Conference Series, pp. 239–248 (1995)
9. Kaandorp, J.A., Kübler, J.E.: The Algorithmic Beauty of Seaweeds, Sponges and Corals. Springer, Heidelberg (2001)
10. Prusinkiewicz, P.: Modeling and visualization of biological structures. In: Proceeding of Graphics Interface 1993, Toronto, Ontario, May 1993, pp. 128–137 (1993)
11. Stanley, K.O., Miikkulainen, R.: A taxonomy for artificial embryogeny. Artificial Life 9(2), 93–130 (2003)
12. Leung, C.H., Berzins, M.: A computational model for organism growth based on surface mesh generation. J. Comput. Phys. 188(1), 75–99 (2003)
13. Harrison, L.G., Wehner, S., Holloway, D.M.: Complex morphogenesis of surfaces: theory and experiment on coupling of reaction diffusion patterning to growth. Nonlinear Chemical Kinetics: Complex Dynamics and Spatiotemporal Patterns, Faraday Discuss. 120, 277–294 (2001)
14. Smith, C.: On Vertex-Vertex Systems and their use in geometric and biological modelling. PhD thesis, The University of Calgary (April 2006)
15. Dobrowolski, T.: Modeling geometry and growth of artificial multicellular organisms. Master's thesis, Gdańsk University of Technology (2005)
16. Eggenberger, P.: Genome-physics interaction as a new concept to reduce the number of genetic parameters in artificial evolution. In: Sarker, R., Reynolds, R., Abbass, H., Tan, K.C., McKay, R., Essam, D., Gedeon, T. (eds.) Proceedings of the IEEE 2003 Congress on Evolutionary Computation, Piscataway, NJ, pp. 191–198. IEEE Press, Los Alamitos (2003)
17. Gilbert, S.F.: 16. In: Developmental Biology, 8th edn. Sinauer Associates, Inc., Sunderland (2006)
18. Matela, R.J., Fletterick, R.J.: A topological exchange model for cell self-sorting. Journal of Theoretical Biology 76, 403–414 (1979)
19. Matela, R.J., Fletterick, R.J.: Computer simulation of cellular self-sorting: A topological exchange model. Journal of Theoretical Biology 84, 673–690 (1980)
20. Müller, M., Stam, J., James, D., Thürey, N.: Real time physics: class notes. In: SIGGRAPH 2008: ACM SIGGRAPH 2008 classes, pp. 1–90. ACM, New York (2008)
21. Turini, G., Pietroni, N., Ganovelli, F., Scopigno, R.: Techniques for computer assisted surgery. In: Eurographics Italian Chapter Conference 2007 (2007)
22. Streichert, F., Spieth, C., Ulmer, H., Zell, A.: Evolving the ability of limited growth and self-repair for artificial embryos. In: Proceedings of the 7th European Conference on Artificial Life, pp. 289–298 (2003)
23. Teschner, M., Heidelberger, B., Müller, M., Gross, M.: A versatile and robust model for geometrically complex deformable solids. In: Proceedings of Computer Graphics International, Heraklion, June 2004, pp. 312–319 (2004)
24. Porter, B.: A developmental system for organic form synthesis. Technical Report (June 2009)

25. Ransom, R., Matela, R.J.: Computer modelling of cell division during development using a topological approach. Journal of Embryology and Experimental Morphology 83, 233–259 (1984) (suppl.)
26. Agarwal, P.: The cell programming language. Artificial Life 2(1), 37–77 (1994)
27. Dellaert, F., Beer, R.D.: Toward an evolvable model of development for autonomous agent synthesis. In: Maes, P., Brooks, R. (eds.) Artificial Life IV, Proceedings of the Fourth International Workshop on the Synthesis and Simulation of Living Systems. MIT Press, Cambridge (1994)
28. Fleischer, K.: A Multiple-Mechanism Developmental Model for Defining Self-Organizing Geometric Structures. PhD thesis, California Institute of Technology, Pasadena, California (1995)
29. Turing, A.: The chemical basis of morphogenesis. Phil. Trans. R. Soc. London B 237, 37–72 (1952)

Layered Random Inference Networks

David M. Lingard

Defence Science and Technology Organisation,
PO Box 1500, Edinburgh, SA 5111, Australia
David.Lingard@dsto.defence.gov.au

Abstract. Random Boolean Networks (RBN) have been used for decades to study the generic properties of genetic regulatory networks. This paper describes Random Inference Networks (RIN) where the aim is to study the generic properties of inference networks used in high-level information fusion. Previous work has discussed RIN with a linear topology, and this paper introduces RIN with a layered topology. RIN are related to RBN, and exhibit stable, critical and chaotic dynamical regimes. As with RBN, RIN have greatest information propagation in the critical regime. This raises the question as to whether there is a driver for real inference networks to be in the critical regime as has been postulated for genetic regulatory networks. Key Words: situation assessment, inference network, information propagation, criticality

1   Introduction

1.1   Random Boolean Networks

Kauffman [1] [2] originally proposed Random Boolean Networks (RBN) as a simple model of genetic regulatory networks. RBN consist of N nodes in a cyclic directed network. Each node is controlled by K randomly chosen nodes where K is the connectivity parameter. In more complex RBN K can vary from node-to-node, and is characterised by a discrete probability distribution. A randomly chosen Boolean function is assigned to each node. For simple RBN, the nodes are updated synchronously at each discrete time step. Since the total number of possible network states (2^N) is finite, the dynamics of the network must eventually fall into a limit cycle.

The dynamical regimes of RBN can be classified according to how the network responds to a perturbation. The order parameter λ is the expectation value of the number of perturbations in the time step directly after the perturbation is introduced, given that the network has reached the bias-map fixed point before the perturbation [3]. If $\lambda < 1$, the network is in the stable regime where the effect of the perturbation tends to die out quickly. On the other hand, if $\lambda > 1$ the perturbation tends to produce very different dynamics, and this is called the chaotic regime. There is a critical regime between the stable and chaotic regimes where $\lambda = 1$ and the network exhibits a broad range of possible responses to a perturbation that is described using a scale invariant power law distribution. Kauffman postulated that genetic regulatory networks will tend

K. Korb, M. Randall, and T. Hendtlass (Eds.): ACAL 2009, LNAI 5865, pp. 149–158, 2009.

towards the critical regime, since natural selection requires significant responses to some perturbations, but low likelihood of a chaotic large response [2]. For the simple RBN described above, the critical regime ($\lambda = 1$) corresponds to $\bar{K} = 2$ where $\bar{K}$ is the mean connectivity in the network, and $\bar{K} = 1, 3, 4, 5$ correspond to $\lambda = 0.5, 1.5, 2$ and 2.5 respectively.

Reference [3] addresses the question of how to measure information propagation in RBN. They employ the annealed approximation that makes analytical treatment more tractable, assuming $N \to \infty$. They address the spread of perturbations in RBN using the theory of branching processes.

For the case where the number of outgoing links from a node obeys a Poisson distribution, [3] derives the Probability Mass Function (PMF) for the number of nodes that will be affected, at some stage during the dynamical play-out of the system, by an initial perturbation and its subsequent spread. The set of nodes affected by a perturbation is called a perturbation avalanche, and thus the PMF describes the size of the avalanche. Reference [3] postulate the entropy of the avalanche PMF as a measure of information propagation in Boolean networks. They interpret the initial perturbation as the signal, and the perturbation avalanche as the system's response to the signal. The entropy illustrates the breadth of the scale of responses of the system to the initial signal. Reference [3] shows that the entropy is maximised in the critical regime ($\lambda = 1$). They note that in the critical regime there is the broadest range of responses of the network to a perturbation, and hence the information propagation is maximised there.

In the chaotic regime the probability that the perturbation avalanche will be infinite is greater than zero [3]. Infinite avalanches are more likely as λ increases in the chaotic regime.

1.2   Random Inference Networks

A Random Inference Network (RIN) is a simple model of the inference networks employed for information fusion and is related to RBN, exhibiting similar behaviour [4] [5].

A popular model of data fusion is the Joint Directors of Laboratory (JDL) model [6]. Two of the key layers in this model are: (a) Object Assessment that is usually partitioned into data registration, data association, position attribute estimation, and identification, and (b) Situation Assessment that fuses the kinematic and temporal characteristics of the data to create a description of the situation in terms of indications and warnings, plans of action, and inferences about the distribution of military forces and flow of information [7].

Work is currently underway to produce algorithms that perform machine-based Situation Assessment to assist military operators in complex environments [8] [9]. Part of this work is the production of rules framed in a formal logical language that can be used to perform inference. The processing takes place at the semantic level and involves logical propositions that describe the situation of interest. An Object Assessment may provide a set of atomic propositions, such as *Vessel 6 is named the Ocean Princess*. The machine-based Situation Assessment takes these atomic propositions from the Object Assessment

 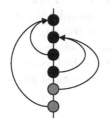

In1	In2	Out
0	0	1
0	1	1
1	0	0
1	1	1

Fig. 1. Comparison of a linear Random Inference Network with a Random Boolean Network. On the left is a small RBN with $N = 6$ nodes, and mean connectivity $\bar{K} = 2$. In the middle is part of a linear RIN, with two OA propositions in grey and four output propositions in black. Input connections are shown for two of the output propositions. On the right is a hypothetical Boolean function for a node with two inputs.

(call them OA propositions for convenience) and applies logical rules to infer conclusions. A simple example of such a rule might be:
(X has been recently communicating with Y) AND (Y is a member of the secret service of Z) → *(X is associated with the secret service of Z)*

For example, two input propositions might be *John Brown has been recently communicating with Andrew Smith*, and *Andrew Smith is a member of the secret service of Orangeland*, and thus the rule results in the output proposition *John Brown is associated with the secret service of Orangeland*.

Just as RBNs are used to gain an understanding of the generic behaviour of genetic regulatory networks, [5] conjectured that RINs may be used to gain an understanding of the generic behaviour of the inference networks described above. A RIN can be arranged in a linear topology with N_{OA} nodes representing the OA propositions, followed by N_P nodes representing the output propositions. Each output proposition is randomly assigned a Boolean function that takes its inputs from previous nodes in the linear topology — refer to Fig. 1. Reference [5] assumed that the number of inputs was fixed across the output propositions, and for each output proposition, the inputs were chosen randomly from the N_W previous nodes in the linear topology. (N_W was set to N_{OA}.) The states of the OA propositions were randomly initialised to 1 or 0 indicating TRUE or FALSE respectively. Then the output propositions were evaluated sequentially to obtain the state of the network.

Even though RBNs are cyclic networks, while RINs are acyclic networks, the RINs exhibit the same dynamical regimes as the RBNs, namely stable ($\lambda < 1$), critical ($\lambda = 1$) and chaotic ($\lambda > 1$) [5]. In addition, the entropy of the PMF of the perturbation avalanches in RINs is maximised in the critical regime, and infinite avalanches occur in the chaotic regime becoming more likely as λ increases. Using the postulate of [3], information propagation in RINs is maximised in the critical regime. In light of these findings, [5] raised the question as to how the paradigm of the stable, critical and chaotic regimes may be relevant to real inference networks, and in particular whether there may be an inherent advantage in inference networks being in the critical regime to maximise information propagation as has been postulated for genetic regulatory networks.

Section 2 introduces a layered topology for the RIN instead of the linear topology described above. Some simulation and analytical results are described in Section 3 where it is shown that the layered topology behaves similarly to the linear topology. Conclusions are presented in Section 4.

2   Layered Topology

Whereas [5] presents a RIN with a linear topology, this section describes a variation comprising a layered topology.

The layered RIN has N_L layers, and each layer has N_{OA} nodes. The first layer contains N_{OA} OA propositions, and the other layers contain the output propositions that are governed by rules. Each of the nodes corresponding to an output proposition is randomly connected to nodes on the previous layer from which it obtains its inputs, and is randomly assigned a Boolean function representing the logical content of the corresponding rule. For example if $N_{OA} = 50$, node 20 on layer 10 might be connected to nodes 39, 3 and 15 on layer 9, and node 20 would be randomly assigned a Boolean function with three inputs. A specific node on the previous layer is chosen as an input for a given node on the next layer with probability P_{input}. For example if $N_{OA} = 50$ and $P_{input} = 0.04$, then the mean the number of inputs per node over many layers will be 2. In general the number of inputs to a node will obey a binomial distribution that approaches a Poisson distribution for large values of N_{OA}. Some nodes may have no inputs assigned, in which case they remain in the state they receive when the network is initialised. Whereas the linear RIN introduced in [5] kept the number of inputs fixed across the network, the number of inputs is allowed to vary across the layered RIN introduced in this paper. As with the linear version of the RIN, the layered version is a directed acyclic network. The linear and layered versions of the RIN are both simple models, and provide two complementary avenues for exploring the generic properties of real inference networks employed for Situation Assessment.

As $N_{OA} \rightarrow \infty$, the layered RIN actually corresponds to the annealed approximation [10] used with RBN. In this approximation, the size of the RBN approaches infinity, and at each time step the links and Boolean functions are randomly re-assigned. This corresponds perfectly to the layered RIN with the state of the RBN at a given time step mapping onto a layer in the RIN, and the state of the RBN at the next time step mapping onto the next layer etc. The first layer of the RIN corresponds to the initial state of the RBN.

The approach employed in this study was to create a baseline layered RIN, and to compare it with a perturbed version of the layered RIN. Firstly the baseline RIN was created including random assignment of input links, Boolean functions, and initial states. Then each layer of the RIN was calculated sequentially to obtain the final state of the RIN. Next the initial layer of OA propositions was perturbed in some way, and the final state of the perturbed version of the RIN was calculated. The Hamming distance was calculated for each layer; this is just the number of mismatches when comparing the final states of corresponding nodes in the baseline and perturbed networks. It was noted whether or not the perturbation became extinct, and if so, at which layer this occurred. A perturbation becomes extinct at a given layer if the Hamming distance reduces to zero at that layer. Once a perturbation becomes extinct, the Hamming distance is zero for all successive layers as well. By undertaking this process for many baseline and perturbed RINs, a Cumulative Distribution Function (CDF) was obtained for the layer at which a perturbation becomes extinct, for a given set of network parameters $\{N_{OA}, \bar{K}\}$ where $\bar{K} = P_{input}N_{OA}$ is the nominal value of the mean connectivity of the network. This CDF was named the cumulative probability of extinction. Finally the entropy corresponding to the cumulative probability of extinction was calculated to test the hypothesis that this entropy, and hence information propagation, are maximised in the critical regime of the layered RIN.

Two different types of perturbation were used. The first type was a single perturbation where one of the OA propositions in the first layer was randomly chosen and its state was changed. This resulted in a Hamming distance of one for the first layer, the smallest possible perturbation. The other type of perturbation involved the first layers of the baseline and perturbed networks being complementary, that is, the state of each OA proposition was changed so that the Hamming distance for the first layer was N_{OA}. This permitted exploration of the effects of the largest possible perturbation in the first layer of OA propositions. The first type of perturbation was identified by the term 'single perturbation', and the latter approach by 'complementary inputs'.

Using a Markov random walk model, [3] derives an analytical expression that is relevant to perturbation spread for the case of a single perturbation in the limit of a large layered RIN:

$$P_{i,j} = P(Z_{l+1} = j|Z_l = i) = \frac{(\lambda i)^j}{j!}e^{-\lambda i} \qquad (1)$$

where $P_{i,j}$ is a conditional probability, Z_l and Z_{l+1} are the Hamming distances of the lth and $(l+1)^{\text{th}}$ layers respectively, λ is the order parameter, and $0 \geq i, j \geq N_{OA}$. This equation can be expressed in a matrix format:

$$Q_{l+1} = Q_l P \qquad (2)$$

where Q_l and Q_{l+1} are the PMFs of Hamming distance for the lth and $(l+1)^{\text{th}}$ layers respectively. Q_l and Q_{l+1} can be considered as row vectors of probabilities, each defined on the Hamming distances $0, 1, 2, 3, \ldots, N_{OA}$. The $(N_{OA} + 1) \times (N_{OA} + 1)$ matrix P has elements $P_{i,j}$ defined in eqn 1, with $i = i' - 1$ defining the $(i')^{\text{th}}$ row of the matrix, and $j = j' - 1$ defining the $(j')^{\text{th}}$ column.

For the case of a single perturbation, Q_1 is all zeros except for the second element being one; it is certain that the Hamming distance will be one. For the case of complementary inputs, Q_1 is all zeros except for the last element being one; it is certain that the Hamming distance will be N_{OA}. Once the PMFs of

Hamming distance have been calculated for each layer in the layered RIN using eqn 2, it is straightforward to obtain the analytical cumulative probability of extinction and the corresponding entropy.

3   Results and Discussion

This section describes simulation results for the two cases, a single perturbation and complementary inputs, and the former are compared with analytical results.

Fig. 2 shows the cumulative probability of extinction for the case of a single perturbation, as a function of the layer number. Both simulation and analytical results are shown for a range of values of the order parameter from 0.5 up to 2.5. For these simulation results, the dimension of each layer (N_{OA}) was 10000. Except for $\lambda = 1$, the results were obtained by analysing 70 independent networks, each with 70 independent sets of initial conditions, giving a total of 4900 samples for estimation of the CDF. For $\lambda = 1$, 200 networks and 200 sets of initial conditions were employed.

For the stable regime ($\lambda = 0.5, 0.75$ in Fig. 2), the perturbations die out quickly within about 10–20 layers.

For the chaotic regime ($\lambda = 1.25, 1.5, 2, 2.5$), some of the perturbations die out quickly within about 5–30 layers. The remaining perturbation avalanches are infinite ones; these are represented in Fig. 2 by the gap between the asymptotic cumulative probability and a cumulative probability of one. Note that the infinite avalanches become more likely as λ increases in the chaotic regime.

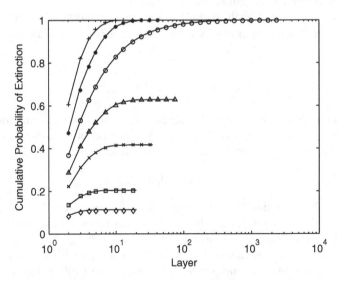

Fig. 2. The cumulative probability of extinction versus the layer number for the case of a single perturbation. The symbols with no connecting lines show the analytical results for λ equals 0.5 (+), 0.75 (*), 1 (○), 1.25 (△), 1.5 (×), 2 (□) and 2.5 (◇). Simulation results for a layer dimension (N_{OA}) of 10000 are shown using solid lines.

For the critical regime ($\lambda = 1$), a small proportion of the perturbation avalanches are extinguished much deeper into the layered RIN compared with extinguished perturbation avalanches in the stable and chaotic regimes.

In general in Fig. 2, there is good agreement between the simulation results for $N_{OA} = 10000$, and the analytical results that assume $N_{OA} \rightarrow \infty$.

Fig. 3 shows the entropies corresponding to the CDFs in Fig. 2, plotted verses the order parameter. There is good agreement between the simulation and analytical results in Fig. 3. The shape of the curve has similar form to the result in [5] (Fig. 3) that shows entropy versus order parameter where the entropy corresponds to the CDF of the total size of the perturbation avalanche from a single perturbation in the case of a linear RIN topology.

Adapting the postulate of [3], information propagation is maximised in the critical regime of the layered RIN since the entropy corresponding to the cumulative probability of extinction is maximised there. The critical regime can be viewed as providing the broadest range of responses to a small external signal. This response can be measured by how far into structure of the layered RIN that the external signal has an impact. In the critical regime there is some chance that the impact will occur deep into the layered structure where there are propositions that have resulted from the fusion of many pieces of information. The deeper layers that result from more fusion are likely to be more meaningful for the user of the inference network. However, there is also robustness against stochasticity and noise since most small external signals only impact on the layered RIN in a relatively shallow manner. Thus the critical regime provides a compromise between the need for response to selected small external signals, and the need for robustness against

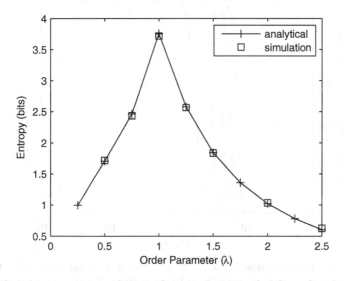

Fig. 3. The entropy corresponding to the cumulative probability of extinction versus the order parameter for the case of a single perturbation. Simulation and analytical results are compared. The simulation results are for a layer dimension (N_{OA}) of 10000.

stochasticity and noise. The stable regime offers minimal response to small external signals, and the chaotic regime is quite sensitive to noise and stochasticity.

In [5] and this paper, linear and layered RINs have been proposed as complementary simple models of inference networks used for Situation Assessment, similar to how RBNs are used to model the generic properties of genetic regulatory networks. Future work will assess whether the dynamical regimes observed in RINs are relevant to real inference networks, and indeed whether there is a driver for real inference networks to be in the critical regime in order to maximise information propagation.

Figs. 4 and 5 show the simulation results for complementary inputs. Fig. 4 shows the cumulative probability of extinction for several values of order parameter in the stable regime, and for the critical regime, and Fig. 5 shows the corresponding entropies. For these results, the dimension of each layer (N_{OA}) was 10000 nodes. Except for $\lambda = 1$, the results were obtained by analysing 70 independent networks, each with 70 independent sets of initial conditions, giving a total of 4900 samples for estimation of the CDF. For $\lambda = 1$, 100 networks and 100 sets of initial conditions were employed.

With complementary inputs, for a given value of the order parameter in the stable and critical regimes, the perturbation avalanches tend to become extinct within a band of layers. As the order parameter increases, this band has greater breadth, and the associated entropy increases.

Networks with complementary inputs and layer dimension $N_{OA} = 10000$ were examined in the chaotic regime for $\lambda = 1.05$ and 1.5. No extinctions were

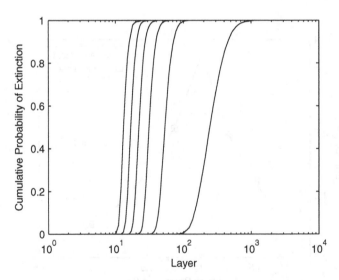

Fig. 4. The cumulative probability of extinction versus the layer number for the case of complementary inputs. The curves from left to right are for the order parameter (λ) equal to 0.5, 0.6, 0.7, 0.8, 0.9 and 1 respectively. These results were obtained through simulation, employing a layer dimension (N_{OA}) of 10000.

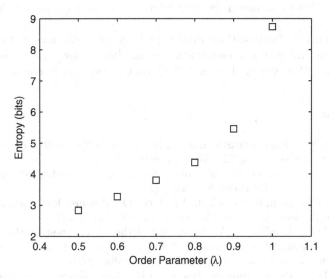

Fig. 5. The entropy corresponding to the cumulative probability of extinction shown in Fig. 4, versus the order parameter

detected out to 5000 layers over 20 independent networks, each with 20 independent sets of initial conditions, giving a total of 400 trials for each value of λ. The layers beyond 5000 haven't been explored in simulation due to the prohibitive computational requirements. It may be the case that all perturbation avalanches are infinite in the chaotic regime with complementary inputs in the limit of large networks, and this would have an associated entropy of zero. Hence the entropy corresponding to the cumulative probability of extinction would be maximised in the critical regime even for complementary inputs. Maximum information propagation would again be associated with the critical regime that would provide the broadest range of responses to the external signal. Further work is required to confirm whether or not all avalanches are infinite in the chaotic regime, preferably by proving it analytically.

4   Conclusions

This paper conjectures that layered and linear RIN are two complementary ways of studying the generic properties of real inference networks used for high-level information fusion, just as RBN have been used to study the generic properties of genetic regulatory networks.

In terms of single perturbations in the limit of large networks, layered RIN exhibit the same dynamical regimes as linear RIN and RBN, i.e. stable, critical and chaotic. As with linear RIN and RBN, information propagation is maximised in the critical regime for layered RIN. Further work could aim to determine the relevance of the stable, critical and chaotic regimes for real inference networks, and whether there is a driver for real inference networks to be in the critical regime.

There is some indication that, even with large perturbations in the limit of large networks, information propagation is maximised for layered RIN in the critical regime. However further analytical work is required to confirm this.

References

1. Kauffman, S.A.: Metabolic Stability and Epigenesis in Randomly Constructed Genetic Nets. J. Theor. Biol. 22, 437–467 (1969)
2. Kauffman, S.A.: The Origins of Order: Self-Organization and Selection in Evolution. Oxford University Press, New York (1993)
3. Rämö, P., Kauffman, S., Kesseli, J., Yli-Harja, O.: Measures for Information Propagation in Boolean Networks. Physica D 227, 100–104 (2007)
4. Lingard, D.M., Lambert, D.A.: Evaluation of the Effectiveness of Machine-based Situation Assessment – Preliminary Work. Technical Note DSTO-TN-0836, Defence Science & Technology Organisation, Australia (2008)
5. Lingard, D.M.: Perturbation Avalanches in Random Inference Networks (submitted to Complexity)
6. Llinas, J., Bowman, C., Rogova, G., Steinberg, A., Waltz, E., White, F.: Revisiting the JDL Data Fusion Model II. In: 7th International Conference on Information Fusion, pp. 1218–1230. International Society of Information Fusion (2004)
7. Smith, D., Singh, S.: Approaches to Multisensor Data Fusion in Target Tracking: A Survey. IEEE Trans. Knowl. Data Eng. 18, 1696–1710 (2006)
8. Lambert, D.A., Nowak, C.: Mephisto I: Towards a Formal Theory. In: Orgun, M.A., Meyer, T. (eds.) AOW 2006. ACM International Conference Proceeding Series, vol. 72, 238, pp. 25–30. Australian Computer Society, Inc. (2006)
9. Lambert, D.A., Nowak, C.: The Mephisto Conceptual Framework. Technical Report DSTO-TR-2162, Defence Science & Technology Organisation, Australia (2008)
10. Derrida, B., Pomeau, Y.: Random Networks of Automata: A Simple Annealed Approximation. Europhys. Lett. 1, 45–49 (1986)

Modelling Hepatitis B Virus Antiviral Therapy and Drug Resistant Mutant Strains

Julie Bernal[1], Trevor Dix[1,*], Lloyd Allison[2], Angeline Bartholomeusz[3], and Lilly Yuen[3]

[1] Faculty of Information Technology, Monash University, Clayton, 3800, Australia
[2] National ICT Australia, Victorian Research Laboratory, University of Melbourne
[3] Victorian Infectious Diseases Reference Laboratory, North Melbourne, Australia

Abstract. Despite the existence of vaccines, the Hepatitis B virus (HBV) is still a serious global health concern. HBV targets liver cells. It has an unusual replication process involving an RNA pre-genome that the reverse transcriptase domain of the viral polymerase protein translates into viral DNA. The reverse transcription process is error prone and together with the high replication rates of the virus, allows the virus to exist as a heterogeneous population of mutants, known as a quasispecies, that can adapt and become resistant to antiviral therapy. This study presents an individual-based model of HBV inside an artificial liver, and associated blood serum, undergoing antiviral therapy. This model aims to provide insights into the evolution of the HBV quasispecies and the individual contribution of HBV mutations in the outcome of therapy.

1   Introduction

The Hepatitis B virus (HBV) is not cytotoxic to hepatocytes, the liver cells. Even so, the host immune response damages the liver because of the virus's extensive and ongoing replication [1]. Most infections are cleared naturally. However, about 5% of infections in healthy adults, 40% to 50% in children and 90% in newborns do not resolve but develop into chronic infections [2,3].

HBV has an unusual replication process involving an RNA pre-genome, that is reverse transcribed to yield viral DNA [4]. Antiviral drugs are potent inhibitors of HBV replication that target this process via the reverse transcriptase (rt). Short term antiviral therapy is, however, insufficient to eliminate the virus [5], and long term therapy leads to the development of drug resistance [6].

Given that the HBV polymerase has functional regions homologous to the HIV polymerase, some antiviral drugs that were initially developed for the treatment of HIV have been used against HBV. Unlike HIV, study of the replication and drug sensitivity of HBV have been limited by the lack of *in vitro* cell cultures that support HBV replication. In addition, animal models with HBV-like viruses, such as the woodchuck and duck models, are often inconvenient [7,8].

For this reason, it has been necessary to develop statistical models that provide insights into HBV replication and its response to therapy [9,10,11]. However,

---

* Corresponding author: Trevor.Dix@infotech.monash.edu.au

K. Korb, M. Randall, and T. Hendtlass (Eds.): ACAL 2009, LNAI 5865, pp. 159–168, 2009.

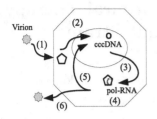

Virion

(1) HBV enters hepatocyte
(2) HBV DNA enters the nucleus and is transformed into cccDNA
(3) Cell translates cccDNA into viral proteins and RNA pre-genome
(4) Polymerase attaches to RNA and reverse transctiption begins
(5) Intracellular amplification of cccDNA
(6) Assembly and export of virions

Fig. 1. HBV life cycle

existent statistical models of HBV are simple and have been developed from small data sets. In addition, these models do not take into account the interaction of multiple mutations that ultimately lead to antiviral therapy failure.

This paper describes an individual-based model developed from biological knowledge of HBV. The model uses parameters estimated in viral dynamic studies of HBV [9,10,12] and others derived from clinical data provided by the Victorian Infectious Diseases Reference Laboratory. The data set contains the antiviral therapy outcomes of chronic HBV patients. This includes the viral load in IU/ml and HBV genomic information recorded in serological tests at different dates throughout antiviral therapy. The dominant HBV strain present at a given date in a patient is recorded as a sequence of HBV rt amino acid substitutions. Lamivudine was the first antiviral drug approved for the treatment of chronic HBV [13]. This paper restricts the data set to Lamivudine for which sufficient data is available.

2   Background

2.1   The Hepatitis B Virus

HBV is an enveloped, mostly double-stranded DNA virus. The envelope surrounds the nucleocapsid that contains the viral genome and the polymerase protein [14]. The HBV genome is circular and is $\sim$ 3.2 kilo bases long. There are many HBV strains, called wild types, however infection occurs with one strain.

The first stage in the HBV life cycle is the attachment to a cell that supports its replication. The virus enters the cell and the core is transported to the nucleus where the viral DNA is transformed into a covalently closed circular form known as cccDNA [15]. (See figure 1 for numbered stages.)

During viral replication, the cccDNA is transcribed into an RNA pre-genome, as well as shorter transcripts serving as messenger RNAs (mRNAs). The mRNAs are translated into viral proteins and the RNA pre-genome is translated into the polymerase protein, which binds to its own RNA creating a polymerase-RNA structure. A viral capsid forms around this structure and reverse transcription begins [4].

The resulting viral cores are either enveloped and exported from the cell, or transported back to the nucleus, where their genomes are converted back to cccDNA to maintain an intracellular pool for viral replication [15]. This is crucial for HBV survival [16].

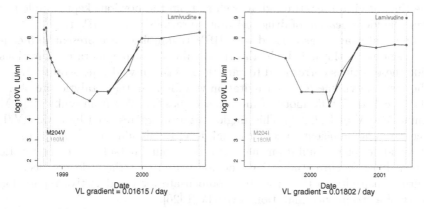

Fig. 2. Viral load vs Date plots of two lamivudine patients. Vertical gray lines indicate the dates at which the HBV genome was sequenced. Straight lines where fitted through data points corresponding to viral breakthroughs.

2.2 Viral Quasispecies

The high turn over rate of virions coupled with the error prone replication of HBV allow the virus to exist as a heterogeneous population, known as a quasispecies [11]. A cloud of potentially favorable mutants allow the quasispecies to evolve and adapt to the selective pressures of antiviral drugs [17,18].

Experiments on viruses isolated from the brain have demonstrated that a quasispecies is a group of interacting variants and that selection occurs at the population level. The viral population decreases when the mean fitness of the virus is low [19].

2.3 Expected Response to HBV Antiviral Therapy

Patients reach very high viral loads when viremia is established, commonly there are 10^7 to 10^9 virions per millimiter in the blood. At the start of therapy, the viral load drops to low, undetectable levels. However, the virus is not eliminated and continues to replicate and mutate. The added selective pressures of antiviral drugs targeting the wild strain favour mutations in the HBV rt (see figure 2 viral load reduction on different drug therapies).

The resulting mutants are less fit than the wild type strain in the absence of antiviral drugs but are strongly selected in their presence [18]. Viral breakthroughs start at low undetectable HBV DNA levels ranging from 10^1 to 10^5 IU/ml (depending on the HBV DNA detection kit) and end at high, similar to pre-treatment, viral loads, 10^7 to 10^9 IU/ml [20].

2.4 Antiviral Drug Resistance

Drug resistance is defined by the increase of the viral load of HBV DNA in serum while the patient is still undergoing therapy. Sequencing the HBV polymerase at the time of viral breakthrough as well as from the previous kept sample has allowed the identification of drug resistant mutations in the HBV rt.

Amino acid changes associated with HBV rt inhibitors are present in 5 conserved regions, A through E, in the rt domain of the polymerase. Mutations outside these regions correspond to variations between wild type strains [21].

The first lamivudine resistant mutation was found in the catalytic site of the rt, the so called YMDD motif: a mutation at position 204 from amino acid M to amino acid V or I [22,23]. This single mutation, represented by rtM204V/I, is necessary and sufficient to confer a high level Lamivudine resistance [24].

Lamivudine related mutations have also been found outside the catalytic site of the rt, [25]. *In vitro* studies have demonstrated that these do not decrease the virus sensitivity to the drug, but compensate for the defective replication capacity of a strain with mutation rtM204V/I [26].

2.5   Previous Viral Dynamics Model of HBV

Our individual-based simulation follows a deterministic computational model of HBV spread in an infected liver [27]. The purpose of that viral dynamics model (VDM) was to explore the roles of viral replication and hepatocyte death rates in the spread of two viruses: wild type and mutant.

The purpose of the individual-based simulation is, on the other hand, to investigate the interactions of HBV rt mutations in the development of drug resistance. Each viral particle in the simulation has an associated genotype, which can have multiple variations, giving rise to quasispecies. This permits the investigation of the set of mutations observed in real life patients undergoing a given antiviral drug therapy, such as lamivudine. This is not possible under the VDM approach which is not an individual-based simulation.

In VMD, a full liver is modelled by dividing hepatocytes it into partitions $c(i, j, n)$, where i is the number of wild type particles, j the number of mutants and n the maximum number of cccDNA particles of cells in the given partition.

VDM calculates the kinetics of spread of two viral strains, wild type and drug resistant using the following equations:

$$\lambda_w = i \times r_1(1 - q) + j \times r_2(q) \text{ and } \lambda_m = i \times r_1(q) + j \times r_2(1 - q)$$

where λ_w and λ_m are the rates of production and export of wild type and drug resistant mutant strains, respectively; r_1 and r_2 are their respective replication rates; and q determines the fraction of the wild type that mutates to drug resistant mutant and vice versa.

At each time step, a fraction of hepatocytes die but the liver size is kept constant through cell division. Next, the proportion of the liver that is full, $i+j = n$ exports viral particles and a fraction of these infect uninfected hepatocytes, $c(0, 0, n)$. During viral replication each liver partition with i wild type and j mutant is distributed among partitions with $i+u$ and $j+v$ following the Poisson distribution $P(u|\lambda_w) \, P(v|\lambda_m)$ where $i + u + j + v \leq n$.

Table 1. Coefficients of lamivudine resistant HBV rt mutations

Intercept	M204I/V	A200V	L80V/I	V207I	L180M	V173L	L82M
-0.0484	0.05982	0.01467	0.01304	-0.0019	-0.0033	-0.0044	-0.0072

3   Modelling HBV Antiviral Therapy

3.1   Estimating the Effects of HBV rt Mutations on Lamivudine Therapy

The automated search for resistance patterns of individual HBV rt mutations in the data is hampered by the noisy nature of clinical data and missing values. Viral loads are recorded at irregular time intervals and the different detection kits used throughout the years have generated censored data outside varying detection limits. In addition, sequences of HBV rt mutations are sparse in the data. For most patients, the viral genome is only sequenced when they stop responding to therapy.

Multivariate linear regression was used to estimate the effects of HBV rt mutations on lamivudine therapy outcome. This linear model predicts the rate of viral load increase following the emergence of HBV rt mutants. The linear coefficients are required as parameters in the individual-based simulation of HBV.

The viral load throughout therapy was plotted for all the patients in the data. These graphs were inspected to select patients that developed resistance to lamivudine therapy. Patients resistant to lamivudine have viral breakthroughs that start at low levels, $\sim 10^5$ IU/ml, and end at high levels, $\sim 10^7$ IU/ml (see figure 2).

Detecting Lamivudine Resistant Mutations. Lamivudine related HBV rt mutations were extracted from the data by listing the HBV rt mutations that emerged before viral breakthroughs. Only mutations in the conserved domains (A–E) of the HBV rt were included. The HBV rt mutations identified are: {M204V/I, L180M, L80I/V, A200V, V173L, L82M}.

Viral Load Rate Linear Coefficients for HBV rt Mutations. In this analysis, 78 lamivudine patients were included. The rate of viral load increase per day was estimated for each patient from the log_{10} VL IU/ml vs date graphs by fitting straight lines through the data points representing viral breakthroughs. The slope of diagonal line fitted in figure 2 estimates the rate for this patient.

The statistical package R was used to perform multivariate linear regression on these rates [28]. The matrix for regression analysis contains the rate of viral load increase per day observed in each patient with corresponding HBV rt mutations as regressor variables. The linear coefficients found for Lamivudine resistant mutations are listed in table 1.

The linear coefficients for mutations M204V/I, A200V and L80I/V have significance levels < 0.0001. The coefficient of M204V/I correlates with the observation that a single mutation at M204V/I confers resistance to the inhibitory

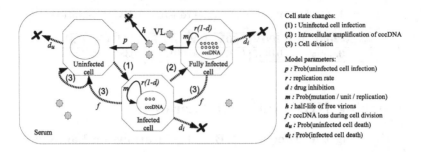

Fig. 3. Diagram of the individual-based model of HBV

effect of lamivudine. In addition, the positive coefficients of mutations A200V and L80I/V reproduce the effects of compensatory mutations that increase the replication capacity of a viral strain with mutation M204V/I [29,26].

The linear coefficients for mutations L82M, V173L, L180M and V207I are negative but are not significant. In reality these mutations are also known to increase the replication capacity of a lamivudine resistant strain. These coefficients indicate that in the current data, patients with these mutations had a slower rate of viral load increase than patients without these mutations.

3.2   Individual-Based Simulation of HBV

The individual-based simulation of HBV models the evolution of the HBV quasispecies throughout therapy. The complexity of the quasispecies is determined by the length of viral genotype in the simulation. The model is composed of a liver with a fixed number of hepatocytes, associated blood serum and viral particles either as virions in the serum or cccDNA particles inside hepatocytes; see figure 3.

The purpose of the individual-based simulation is to investigate the interactions of drug related HBV rt mutations extracted from clinical data. This is performed by representing the viral genome as sequences of HBV rt mutations positions. For example, the viral genome would be given a length of 7 to represent only the mutations in table 1. This level of coarseness in the viral genotype representation allows the simulation of a larger portion of the liver and decreases the complexity of the model by focusing on the mutations that influence therapy outcome.

An adult liver has $\sim 2 \times 10^{11}$ hepatocytes [30]. Due to the granularity of the simulation only a small fraction of the liver is modelled.

The maximum number of cccDNA particles in cells is determined from a lower and upper limit and can be either Poisson of Uniformly distributed in a similar manner to VMD. Estimates of cccDNA particle numbers in the literature range from 5 to 50 per hepatocyte [12].

The simulation models the HBV replication during antiviral therapy, focusing on the reverse transcription process. The simulation is governed by stochastic events, and it runs in discrete time steps of one day. Infection begins by infecting a small number of hepatocytes with the wild type strain. At each time step, keeping the liver size constant: cccDNA particles yield an updated generation of viral particles; a fraction of virions in the serum infect healthy hepatocytes and others are cleared by the immune system; and some hepatocytes die and others replicate. Figure 3 illustrates the model.

HBV replication. The number of new viral particles produced for each viral genotype in a cell is taken from the binomial distribution using its replication capacity and sensitivity to the drug: $B(n_v, 1 - d_v)$. Where d_v is the drug inhibition, n_v is the replication of viral phenotype v that is Poisson distributed $P(n_v | r_v * cccDNA_v)$, given the replication rate r_v and number of cccDNA particles of genotype v in the cell.

The length of viral genotypes is an input parameter of the simulation. These genotypes are represented as binary sequences of mutation positions, which are either present or absent. The mutation rate, m, causes point mutations during the reverse transcription of the new RNA pre-genomes, n_v, yielding new viral genotypes.

At the level of hepatocytes, HBV replicates in two stages. First, the virus increases the number of pooled cccDNA particles. Second, virions are assembled and exported from cells.

HBV spread and clearance. Only uninfected cells can be infected with virions and the probability of infection depends on the concentration of the virus per milliliter of the simulated serum. The half life of free virions is estimated at about 1 day [9].

Hepatocyte turn over. Hepatocyte turn over is influenced by the probabilities of uninfected and infected cell death, d_u and d_i. Values from the literature are entered in the simulation. The half life of healthy hepatocytes is ~ 100 days and the half life of infected hepatocytes is estimated to be 40 days [12].

After the death of hepatocytes, mother cells are selected at random from a uniform distribution. Mother cells divide into two daughter cells, wherein a fraction of cccDNA particles is lost. The remaining cccDNA particles, r_v of a viral genotype v, are divided among daughter cells using the binomial distribution, $B(r_v, 0.5)$.

The replication of mutants is less than that of the wild type [31,26]. Antiviral drugs inhibit the replication of the wild type and the virus is cleared from the liver due to infected hepatocyte turn over. Later, drug resistant mutants can infect these.

VMD outputs the fraction of the liver that is infected with the wild type and mutant virus. The individual-based simulation was validated against VMD by inspecting the state of the liver throughout therapy with appropriate input parameters and by limiting the length of the viral genotype to one because VMD only models two viruses: wild type and mutant. These experiments also demonstrated that even a small fraction of the liver ($\sim 10,000$ cells) can yield results that represent the state of infection in a full liver. The individual-based simulation also outputs the viral load in the serum. The expected therapy response is observed in the serum when running the simulation with viral genotypes of different lengths.

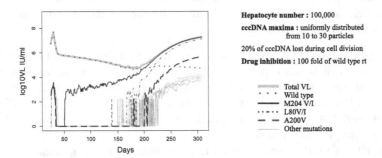

Fig. 4. Output of the simulation: graph of log_{10} VL IU/ml vs days

4   Results

The aim of the individual-based HBV model is to study the combinatory effects of the HBV rt mutations detected in serological tests in real patients. This section presents a typical experiment with the HBV simulation.

The replication of the lamivudine resistant mutant with a single mutation at rtM204V/I is set 0.2 of that for the wild-type, as suggested by the literature, [31,26]. The replication of other lamivudine resistant mutants are computed by adding the linear coefficients of their HBV rt mutations. The output of the simulation is very similar to real life responses to lamivudine therapy (see figure 2). First, the viral load drops and reaches levels below detection limits ($< 10^5$). Therapy is successful for ~ 200 days before lamivudine resistant mutants fill the liver and lead to a viral breakthrough.

Sequencing the HBV genome at the time of viral breakthough will detect only the dominant mutations. For example, sequencing the genome at day 250 will detect mutations rtM204V/I and L80V/I, whereas sequencing the viral genome at day 300 will also detect mutation A200V. The HBV data and other experiments demonstrate that this mutation is detectable with prolonged treatment.

The simulation also illustrates the HBV quasispecies during therapy. The results indicate that the virus population consists of a cloud of mutants, not a just few dominant strains, and that the wild type strain persists even when drug resistant mutants become dominant. Results with VMD also indicate the existence of wild type particles when the mutant has spread in the liver during therapy.

5   Conclusion

An individual-based simulation of the HBV quasispecies throughout antiviral therapy has been presented. The aim is to study the evolution of HBV quasispecies inside the host and the individual contributions of HBV rt antiviral mutations. Our model encapsulates known biological knowledge about HBV replication and uses parameters extracted from HBV viral dynamic studies, as well as values calculated from clinical data.

Experimental runs allowing the major mutations produce results consistent with observed patient response to antiviral therapy. Future work will attempt to model longer therapy responses and replicate real life patient data. The simulation also offers a mechanism to compare different hypotheses on how the HBV rt mutations interact in the development of drug resistance.

Experience in the study of HIV infections provides clues on the development of more efficient HBV antiviral therapies. For instance, lamivudine was developed for HIV and is also a potent inhibitor of HBV replication. Management of HIV infections has improved with the use of drug combination therapies. As more data becomes available for other anti-HBV drugs, the individual-based simulation presents a new avenue for the investigation of responses to monotherapy and drug combination therapies. This can be achieved by infecting artificial patients with HBV and subjecting them to different treatment protocols.

Acknowledgments. Thanks to S. Litwin, E. Toll, A.R. Jilbert and W.S. Mason for providing their HBV viral dynamics model implementation.

References

1. Bertoletti, A., Gehring, A.J.: The immune response during hepatitis B virus infection. J. Gen. Virol. 87, 1439–1449 (2006)
2. Lok, A.S., McMahon, B.J.: Chronic hepatitis B. Hepatology 34(6), 1225–1241 (2001)
3. Previsani, N., Lavanchy, D.: Hepatitis B: World Health Organization, Department of Communicable Diseases Surveillance and Response,
   http://www.who.int/csr/en/
4. Summers, J., Mason, W.S.: Replication of the genome of a hepatitis B-like virus by reverse transcription of an RNA intermediate. Cell 29(2), 403–415 (1982)
5. Nevens, F., Main, J., Honkoop, P., Tyrrell, D.L., Barber, J., Sullivan, M.T., Fevery, J., Man, R.A.D., Thomas, H.C.: Lamivudine therapy for chronic hepatitis B: a six-month randomized dose-ranging study. Gastroenterology 113(4), 1258–1263 (1997)
6. Honkoop, P., de Man, R.A., Zondervan, P.E., Schalm, S.W.: Histological improvement in patients with chronic hepatitis B virus infection treated with lamivudine. Liver 17(2), 103–106 (1997)
7. Summers, J.: Three recently described animal virus models for human hepatitis B virus. Hepatology 1(2), 179–183 (1981)
8. Dandri, M., Lutgehetmann, M., Volz, T., Petersen, J.: Small animal model systems for studying hepatitis B virus replication and pathogenesis. Semin. Liver Dis. 26(2), 181–191 (2006)
9. Nowak, M.A., Bonhoeffer, S., Hill, A.M., Boehme, R., Thomas, H.C., McDade, H.: Viral Dynamics in Hepatitis B Virus Infection. Proceedings of the National Academy of Science 93, 4398–4402 (1996)
10. Lewin, S.R., Ribeiro, R.M., Walters, T., Lau, G.K., Bowden, S., Locarnini, S., Perelson, A.S.: Analysis of hepatitis B viral load decline under potent therapy: complex decay profiles observed. Hematology 34(5), 1012–1020 (2001)
11. Locarnini, S.: Molecular virology of hepatitis B virus. Semin. Liver Dis. 24, (S1)3–10 (2004)
12. Zhu, Y., Yamamoto, T., Cullen, J., Saputelli, J., Aldrich, C.E., Miller, D.S., Litwin, S., Furman, P.A., Jilbert, A.R., Mason, W.S.: Kinetics of hepadnavirus loss from the liver during inhibition of viral DNA synthesis. J. Virol. 75(1), 311–322 (2001)

13. Dienstag, J.L., Perrillo, R.P., Schiff, E.R., Bartholomew, M., Vicary, C., Rubin, M.: A preliminary trial of lamivudine for chronic hepatitis B infection. N. Engl. J. Med. 333(25), 1657–1661 (1995)

14. Summers, J., O'Connell, A., Millman, I.: Genome of hepatitis B virus: restriction enzyme cleavage and structure of DNA extracted from Dane particles. Proc. Natl. Acad. Sci. USA 72(11), 4597–4601 (1975)

15. Tuttleman, J.S., Pourcel, C., Summers, J.: Formation of the pool of covalently closed circular viral DNA in hepadnavirus-infected cells. Cell 47(3), 451–460 (1986)

16. Summers, J., Smith, P.M., Horwich, A.L.: Hepadnavirus envelope proteins regulate covalently closed circular DNA amplification. J. Virol. 64(6), 2819–2824 (1990)

17. Eigen, M., Mccaskill, J., Schuster, P.: Molecular quasispecies. J. Phys. Chem. 92, 6881–6891 (1988)

18. Bartholomeusz, A., Tehan, B.G., Chalmers, D.K.: Comparisons of the HBV and HIV polymerase, and antiviral resistance mutations. Antiviral therapy 9(2), 149–160 (2004)

19. Wilke, C.O.: Quasispecies theory in the context of population genetics. BMC Evol. Biol. 5, 44 (2005)

20. Ganem, D., Prince, A.M.: Hepatitis B virus infection - natural history and clinical consequences. New England Journal of Medicine 350(11), 1118–1129 (2004)

21. Stuyver, L.J., Locarnini, S.A., Lok, A., Richman, D.D., Carman, W.F., Dienstag, J.L., Schinazi, R.F.: Nomenclature for antiviral-resistant human hepatitis B virus mutations in the polymerase region. Hematology 33(3), 751–757 (2001)

22. Tipples, G.A., Ma, M.M., Fischer, K.P., Bain, V.G., Kneteman, N.M., Tyrrell, D.L.: Mutation in HBV RNA-dependent DNA polymerase confers resistance to lamivudine in vivo. Hematology 24(3), 714–717 (1996)

23. Ling, R., Mutimer, D., Ahmed, M., Boxall, E.H., Elias, E., Dusheiko, G.M., Harrison, T.J.: Selection of mutations in the hepatitis B virus polymerase during therapy of transplant recipients with lamivudine. Hepatology 24(3), 711–713 (1996)

24. Ono-Nita, S.K., Kato, N., Shiratori, Y., Masaki, T., Lan, K.H., Carrilho, F.J., Omata, M.: YMDD motif in hepatitis B virus DNA polymerase influences on replication and lamivudine resistance: A study by in vitro full-length viral DNA transfection. Hepatology 29(3), 939–945 (1999)

25. Bartholomew, M.M., Jansen, R.W., Jeffers, L.J., Reddy, K.R., Johnson, L.C., Bunzendahl, H., Condreay, L.D., Tzakis, A.G., Schiff, E.R., Brown, N.A.: Hepatitis-B-virus resistance to lamivudine given for recurrent infection after orthotopic liver transplantation. Lancet 349(9044), 20–22 (1997)

26. Warner, N., Locarnini, S., Kuiper, M., Bartholomeusz, A., Ayres, A., Yuen, L., Shaw, T.: The L80I substitution in the reverse transcriptase domain of the hepatitis B virus polymerase is associated with lamivudine resistance and enhanced viral replication in vitro. Antimicrob Agents Chemother 51(7), 2285–2292 (2007)

27. Litwin, S., Toll, E., Jilbert, A.R., Mason, W.S.: The competing roles of virus replication and hepatocyte death rates in the emergence of drug-resistant mutants: Theoretical considerations. J. Clinical. Virology 34, 96–107 (2005)

28. Venables, W.N., Ripley, B.D.: Modern Applied Statistics with S, 4th edn. Springer, New York (2002)

29. Sheldon, J., Rodes, B., Zoulim, F., Bartholomeusz, A., Soriano, V.: Mutations affecting the replication capacity of the hepatitis B virus. J. Viral. Hepat. 13(7), 427–434 (2006)

30. Sohlenius-Sternbeck, A.K.: Determination of the hepatocellularity number for human, dog, rabbit, rat and mouse livers from protein concentration measurements. Toxicol. In. Vitro. 20(8), 1582–1586 (2006)

31. Fu, L., Cheng, Y.C.: Role of additional mutations outside the YMDD motif of hepatitis B virus polymerase in L-SddC (3TC) resistance. Biochem. Pharmacol. 55(10), 1567–1572 (1998)

Multivesicular Assemblies as Real-World Testbeds for Embryogenic Evolutionary Systems

Maik Hadorn[1] and Peter Eggenberger Hotz[2]

[1] University of Zurich, Department of Informatics, Artificial Intelligence Laboratory, 8050 Zurich, Switzerland
[2] University of Southern Denmark, The Mærsk Mc-Kinney Møller Institute, 5230 Odense M, Denmark
hadorn@ifi.uzh.ch, eggen@mmmi.sdu.dk

Abstract. Embryogenic evolution emulates *in silico* cell-like entities to get more powerful methods for complex evolutionary tasks. As simulations have to abstract from the biological model, implicit information hidden in its physics is lost. Here, we propose to use cell-like entities as a real-world *in vitro* testbed. In analogy to evolutionary robotics, where solutions evolved in simulations may be tested in real-world on macroscale, the proposed vesicular testbed would do the same for the embryogenic evolutionary tasks on mesoscale. As a first step towards a vesicular testbed emulating growth, cell division, and cell differentiation, we present a modified vesicle production method, providing custom-tailored chemical cargo, and present a novel self-assembly procedure to provide vesicle aggregates of programmable composition.

Keywords: Embryogenic evolution, real-world testbed, vesicles, hard sphere colloids, DNA, programmability, self-assembly.

1 Introduction

As evolvability is one of the central questions in artificial life (ALife), artificial evolution (AE) is one of its main aspects. AE is widely used in optimization problems [1]. Evolutionary algorithms (EAs) participating in an AE base on ideas from our current understanding of biological evolution and use mechanisms like reproduction, mutation, recombination, and selection. Since direct encoding schemes where the phenotype of an individual is directly encoded in its genome no longer work for complex evolutionary tasks [2-4], an increasing number of researchers started to mimic biological developmental processes in artificial systems. The genotype-phenotype-mapping becomes indirect by interposing developmental processes. To mimic the natural model in more detail, embryogenic evolution (EE) [5], also called computational evolution [2], simulates cell-like entities that develop into an organism, a neural network etc. Thereby the developmental processes crucially depend on the interactions between the entities typically relying on chemical signaling molecules, membrane receptors, or physical interactions between neighboring cells. EE systems are therefore testbeds to study information guided processes that create higher-order assemblies of cell-like entities, whose overall function emerges from the interaction of their entities.

K. Korb, M. Randall, and T. Hendtlass (Eds.): ACAL 2009, LNAI 5865, pp. 169–178, 2009.

Banzhaf *et al.* [2] stressed that AE should incorporate physics either in simulation or in real-world experimentation instead of trying to evolve problem solutions in a symbolic way. Although Miller and Downing [6] argued that non-conventional hardware, rather than computer chips, may be more suitable for computer controlled evolution, no real-world experimentation are realized in AE so far. If real biological cells are used to realize a real-world testbed, two arguments have to be considered: Their structure is complex and they consist of a myriad of interacting molecules. In contrast, vesicles are chemically well-defined and therefore easier to analyze and understand. Like real biological cells, vesicles feature an aqueous compartment partitioned off the surrounding by an impermeable lipid membrane. Like cellular membranes, vesicular membranes consist of amphiphilic phospholipids that link a hydrophilic head and a lipophilic tail (Fig. 1.A). Suspended phospholipids can self-assemble to form closed, self-sealing solvent-filled vesicles that are bounded by a two-layered sheet (a bilayer) of 6 nm in width, with all of their tails pointing toward the center of the bilayer (Fig. 1.C.1). This molecular arrangement excludes water from the center of the sheet

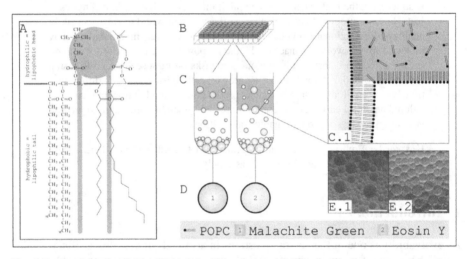

Fig. 1. Production of vesicles differing in their content. (*A*) The chemical structure of the phospholipid POPC (= PC(16:0/18:1(Δ9-Cis)). PC(16:0/18:1(Δ9-Cis) is represented as a structural formula (front, left), as a skeletal formula (front, right), and as a schematic representation (back). The schematic representation is used throughout this publication. For a discussion of the relevance of the amphiphilic character of phospholipids in the formation of biological and artificial membranes see text. (*B*) Microwell plate in which the vesicles are produced. (*C*) The sample is composed of two parts: distinctly stained water droplets (Malachite Green and Eosin Y) in the oil phase and the bottom aqueous phase, which finally hosts the vesicles. (*C.1*) Due to their amphiphilic character, phospholipids, dissolved in mineral oil, stabilize water-oil interfaces by forming monolayers. Two monolayers form a bilayer when a water droplet, induced by centrifugation, passes the interface. Due to both the density difference of the *inter*- and *intrave-sicular* fluid and the geometry of the microplate bottom, vesicles pelletize in the centre of the well. (*D*) The two vesicle populations differ only in intravesicular staining. (*E.1, E.2*) Light microscopic visualization of the two distinct vesicle populations stained by Malachite Green (*E.1*) and Eosin Y (*E.2*) that emerged in real-world experimentation. Staining persisted for days. Scale bars represent 100μm.

and thereby eliminates unfavorable contacts between water and the lipophilic (= hydrophobic) tails. The lipid bilayer provides inherent self-repair characteristics due to lateral mobility of its phospholipids [7]. As vesicles vary in size from about 50 nm to 100 μm, mechanisms of interest such as self-assembly or information processing can be tested over a wide range of scale. This scale range, called mesoscale, is not yet very well understood scientifically. Many of the biological concepts fall in this mesoscale range; the proposed testbed may therefore be of special interest in ALife. Their minimality and self-assembling and self-sealing properties make vesicles an excellent candidate for a testbed in which concepts of EEs can be set up and tested in the real world. Elaborated *in vitro* vesicle formation protocols [8-10] provide independent composition control of the *inter-* and *intra*vesicular fluid as well as of the inner and outer bilayer leaflet. Asymmetry in the inner and outer bilayer leaflet was realized by the production of phospholipid and polymer vesicles combining biocompatibility and mechanical endurance [9]. As a result of the analogy to natural systems and the compositional simplicity, artificial vesicles are the most studied systems among biomimetic structures [11] providing bottom-up procedures in the analysis of biological processes. Starting with distinct vesicle populations differing in content, size, and/or membrane composition, self-assembly of vesicle aggregates providing programmability of composition and structure may emulate growth, cell division, and cell differentiation. The potential of a programmable self-assembly of superstructures with high degrees of complexity [12] has attracted significant attention to nanotechnological applications in the last decade. The high specificity of binding between complementary sequences and the digital nature of DNA base coding enable the programmable assembly of colloidal aggregates. So far, cross-linkage based on DNA hybridization was proposed to induce self-assembly of complementary monohomophilic vesicles [13, 14] or hard sphere colloids [15-18], to induce programmable fusion of vesicles [14, 19], or to specifically link vesicles to surface supported membranes [14, 20-22].

Here, we present a new protocol for *in vitro* vesicle formation and membrane modification that increases the versatility of the underlying vesicle formation method by introducing microtiter plates and vesicle pelletization. Asymmetry in the *inter-* and *intra*vesicular fluid was realized by vesicle staining. We contrast assemblies of hard sphere colloids and multivesicular aggregates to research the influence of material properties on the creation of higher-order assemblies of cell-like entities. We discuss how an asymmetry in intravesicular fluids in combination with vesicular assemblies of preprogrammed structure and composition may emulate cell differentiation and may be used as a real-world testbed for EE systems on mesoscale.

2 Material and Methods

We performed self-assembly experiments of colloidal particles and vesicles in real-world. The self-assembly process was based on the hybridization of single-stranded DNA (ssDNA) with which the surfaces were doped (Figs. 2, 3). By introducing a surface doping of distinct populations of ssDNA, as realized in the self-assembly of hard sphere colloids [23, 24], n-arity may be provided to the assembly process.

Technical modifications of the vesicle formation protocol reported by Pautot *et al.* [9] were: (i) the introduction of 96-well microtiter plates U96 to increase procedural manageability in laboratory experimentation and (ii) a density difference between *inter-* and *intra*vesicular solution to induce vesicle pelletization. Solutions of the vesicle lumen and the surrounding medium were equal in osmolarity but differed in the size of dissolved saccharides (lumen: disaccharides, environment: monosaccharides) providing density differences between the lumen and the environment. For a description of the modified vesicle protocol see Figure 1. Vesicles were either made of 100 percent PC(16:0/18:1(Δ9-Cis)) (staining experiments, Fig. 1) or 99 percent PC(16:0/18:1(Δ9-Cis)) and one percent biotin-PEG2000-PE(18:0/18:0) (1,2-Distearoyl-sn-Glycero-3-Phosphoethanolamine-N-[Biotinyl(Polyethylene Glycol) 2000]) (self-assembly experiments, Fig. 3). Doping of the vesicular surface was realized by anchoring biotinylated ssDNA to biotinylated phospholipids via streptavidin as a cross-linking agent. The mean particle size of the latex beads constituting the hard sphere colloid was 0.25 or 1.0 μm (Fig. 2). Their surface was streptavidin labeled off the shelf. The sequences of the DNA single strands were the same for the colloidal and vesicular self-assembly experiments. For a detailed protocol of the surface doping procedure see Figures 2 and 3. The sequence of complementary biotinylated ssDNA strands (α: biotin-TGTACGTCACAACTA-3', α': biotin-TAGTTGTGACGTACA-3') were produced by a genetic algorithm. Light and confocal laser scanning microscopy was performed using a Wild M40 inverted microscope and an inverted Leica DMR IRE2 SP2 confocal laser scanning microscope.

3 Results and Discussion

3.1 Asymmetry in the Inter- and Intravesicular Fluid

Vesicles were found to sediment and hence to be easily available for inverse microscopy. Asymmetry of *inter-* and *intra*vesicular fluid (vesicle staining, Fig. 1.E) persisted for days (data of long-term observation not shown). Although vesicles are not able to differentiate, they can be prepared with specific content and then positioned specifically in space (see section 3.2). This allows to mimic the final result of a cell differentiation process.

3.2 Assemblies of Hard Sphere Colloids and Multivesicular Aggregates

In assemblies of two bead populations of equal concentration and size, large clusters of beads exhibiting regular patterns of red and green emerged (Fig. 2.F.2). If the bead populations differed in size and if the concentration of the larger beads exceeded the concentration of the smaller ones by a factor 10, small clusters were observed (Fig. 2.f.2).

When vesicles doped with complementary ssDNA came into contact, linkers accumulated in the contact phase forming an adhesion plaque (Fig. 3.G). Thus, the lateral distribution of linkers in the outer leaflet becomes inhomogeneous as a result of the self-assembly process (cp. [25]). It is conceivable that an accumulation of linkers

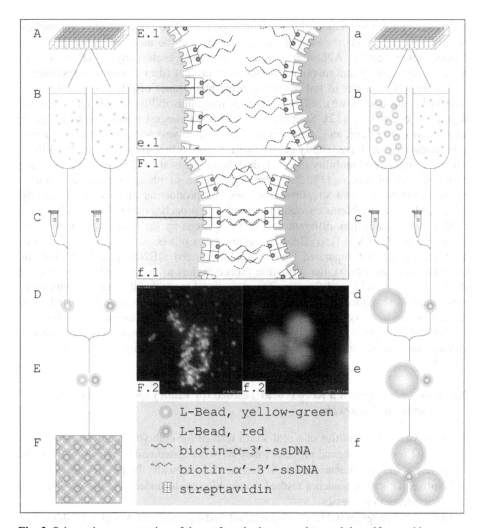

Fig. 2. Schematic representation of the surface doping procedure and the self-assembly process of mono- and bidispersed spherical colloids. (*A,a*) Surface doping and self-assembly are realized in 96-well microtiter plates, providing parallel modification of up to 96 distinct bead populations. (*B,b*) The two populations of latex beads differ in fluorescent labeling (yellow-green, red). (*C,c*) The surfaces of the latex beads are homogenously labeled with streptavidin. The two populations of beads are doped with single stranded DNA (ssDNA-α and ssDNA-α') of complementary sequence (α, α'). (*D,d*) ssDNA covalently bound to biotin is non-covalently linked to the streptavidin labeled surface. (*E,e*) The two bead populations become merged (brace). (*E.1,e.1*) Beads come into contact. (*F,f*) Hybridization of DNA strands results in double stranded DNA and induces the assembly process. The self-assembly of monodispersed bead populations results in non-terminating aggregates (*F*), whereas aggregates are terminating in assemblies of bidispersed bead populations (*f*). (*F.1,f.1*) The lateral distribution of linkers is not affected by the assembly process (cp. Fig. 3). (*F.2,f.2*) Confocal laser scanning microscope visualization of the aggregates that emerged in real-world experimentation.

may result in self-termination of the assembly process. Thus, in contrast to assemblies of hard sphere colloids the size of the aggregates may be adjusted by variations in surface linker density. Adhesion plaques were found exclusively, if DNA strands were complementary and monovalent ions were present (data of control experiments not shown). No transfer of linkers between the membranes of different vesicles was observed (data not shown). ssDNA provides programmability, specificity, and high degrees of complexity [12]. Streptavidin offers the strongest noncovalent biological interaction known [26], an extensive range of possible vesicle modifications, component modularity, and availability off the shelf. Phospholipid-grafted biotinylated PEG tethers feature lateral mobility [7], high detachment resistance [27], and no intermembrane transfer of linkers [13, 20]. The combination of phospholipid-grafted biotinylated PEG tethers and streptavidin allows fast production of vesicles differently doped and avoids problems encountered in other approaches using cholesterol-tagged DNA to specifically link different vesicle populations by the hybridization of membrane-anchored DNA [14]: (i) Because the processes of vesicle formation and vesicle modification are not separated (the cholesterol-tagged ssDNA have to be present during vesicle formation), the formation procedure has to be adjusted anew for each change in the vesicle modification. The procedural manageability in laboratory experimentation is reduced therefore. (ii) As discussed by Beales and Vanderlick [13] the cholesterol anchors of the cholesterol-tagged ssDNA spontaneously leave the lipid bilayer and incorporate randomly into (other) lipid bilayers. Thus, in contrast to our linking mechanism specificity of linking is lost when using cholesterol-tagged ssDNA.

3.3 How to Implement a Real-World Testbed for Embryogenic Evolutionary Systems on Mesoscale

By using vesicles as entities of a real-world testbed for embryogenic evolution, some aspects of the developmental processes may be investigated and compared at the right scale (nano- and micrometer scale). At this mesoscopic scale, the physics are no longer intuitive. This makes such a testbed a valuable tool to understand the workings of (implicit) physical processes.

Due to the programmable DNA-mediated self-assembly of vesicles of designer-specified content, the emulation of cell differentiation becomes feasible. Although vesicles are not able to divide or grow, they can be prepared with specific content (see section 3.1), positioned in space (see section 3.2), and their content can be released on an external trigger [28] and may serve as signaling molecule triggering other processes. Already Noireaux and Libchaber [8] incorporated a transcription-translation cell-free system into vesicles and were able to induce protein synthesis. The exchange of material between the vesicles is provided by inducing pores in the membrane. To induce pores several methods are available such as electroporation [29], phase transitions [30, 31], or protein channels/transporters [32].

We think that such a vesicular system will provide main aspects of intercellular communication and cell differentiation, but in contrast to biological systems, it would be simpler and better defined and therefore easier to be understood.

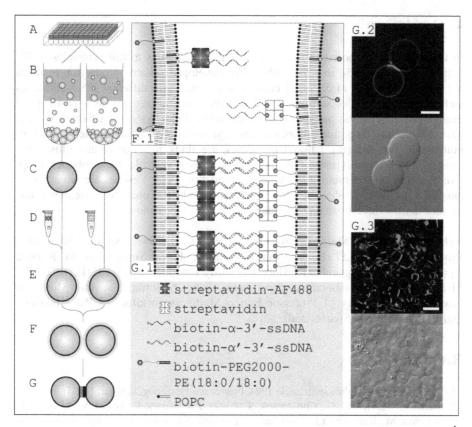

Fig. 3. Schematic representation of the membrane doping procedure, the vesicular self-assembly process and micrographs of adhesion plaques. (*A-C*) Vesicle formation: for details see Fig. 1. Vesicles remain on the same microtiter plate during formation and membrane doping. (*D*) Vesicle populations become distinct by membrane doping. The membrane of the vesicles is doped with two complementary populations of single stranded DNA (ssDNA). (*E*) ssDNA covalently bound to biotin is non-covalently linked to phospholipid-grafted biotinylated poly-ethylene glycol tethers using streptavidin as cross-linking agent (cp. *F.1*). The vesicle populations differ in fluorescence labeling of the streptavidin (Alexa Fluor 488 conjugate (AF488)). (*F*) The vesicle populations become merged (brace). (*F.1*) The lateral distribution of linkers in the lipid membrane is homogeneous. Vesicles come into contact. Hybridization of DNA strands results in double stranded DNA and induces the assembly process. (*G*) Due to their lateral mobility, linkers accumulate in the contact zone forming an adhesion plaque – linker density *inter* adhesion plaques is reduced due to depletion. (*G.1*) Biotinylated phospholipids of the outer leaflet colocalize with the linkers. (*G.2,G.3*) Confocal laser scanning microscope (CLSM) and differential interference contrast micrographs of vesicular aggregates that emerged in real-world experimentation. Accumulation and depletion of linkers are clearly visible in the CLSM micrograph. Scale bars represent 10μm.

4 Conclusion

In this work we proposed to use multivesicular assemblies to test principles of EE systems in a real-world *in vitro* testbed. We developed suitable production methods for vesicles differing in chemical content and doped with different DNA-addresses and presented results of self-assembled multivesicular aggregates, as a prototypical example of information processing in distributed cellular systems. Although we still have a long way to go for self-assembling and working vesicular clusters with programmable and designer tailored properties, we think that our results illustrate a promising step towards interesting applications such as vesicular nano-robotics, adaptive materials or programmable chemical fabrication tools. Vesicles are scale invariant and it is easy to produce vesicles from 50 nm up to 100 μm. Thus, one may transfer mechanisms investigated on the microscale also to the nanoscale.

Acknowledgements. Maik Hadorn was supported by the Swiss National Foundation Project 200020-118127 Embryogenic Evolution: From Simulations to Robotic Applications. Peter Eggenberger Hotz was partly supported by PACE (EU-IST-FP6-FET-002035). Wet laboratory experiments were performed at the Molecular Physiology Laboratory of Professor Enrico Martinoia (Institute of Plant Biology, University of Zurich, Switzerland).

References

1. Rechenberg, I.: Evolutionsstrategie 1994. Frommann-Holzboog, Stuttgart (1994)
2. Banzhaf, W., Beslon, G., Christensen, S., Foster, J.A., Kepes, F., Lefort, V., Miller, J.F., Radman, M., Ramsden, J.J.: Guidelines - From artificial evolution to computational evolution: a research agenda. Nat. Rev. Genet. 7, 729–735 (2006)
3. Ruppin, E.: Evolutionary autonomous agents: A neuroscience perspective. Nat. Rev. Neurosci. 3, 132–141 (2002)
4. Stanley, K.O., Miikkulainen, R.: A taxonomy for artificial embryogeny. Artif. Life 9, 93–130 (2003)
5. Bentley, P., Kumar, S.: Three ways to grow designs: A comparison of embryogenies for an evolutionary design problem. In: Genetic and Evolutionary Computation Conference (GECCO 1999) at the 8th International Conference on Genetic Algorithms/4th Annual Genetic Programming Conference, pp. 35–43. Morgan Kaufmann Pub. Inc., Orlando (1999)
6. Miller, J.F., Downing, K.: Evolution in materio: Looking beyond the silicon box. In: NASA/DOD Conference on Evolvable Hardware, pp. 167–176. IEEE Computer Soc., Alexandria (2002)
7. Singer, S.J., Nicolson, G.L.: Fluid mosaic model of structure of cell-membranes. Science 175, 720–731 (1972)
8. Noireaux, V., Libchaber, A.: A vesicle bioreactor as a step toward an artificial cell assembly. Proc. Natl. Acad. Sci. USA 101, 17669–17674 (2004)
9. Pautot, S., Frisken, B.J., Weitz, D.A.: Engineering asymmetric vesicles. Proc. Natl. Acad. Sci. USA 100, 10718–10721 (2003)

10. Träuble, H., Grell, E.: Carriers and specificity in membranes. IV. Model vesicles and membranes. The formation of asymmetrical spherical lecithin vesicles. Neurosciences Research Program bulletin 9, 373–380 (1971)

11. Wang, C.Z., Wang, S.Z., Huang, J.B., Li, Z.C., Gao, Q., Zhu, B.Y.: Transition between higher-level self-assemblies of ligand-lipid vesicles induced by Cu2+ ion. Langmuir 19, 7676–7678 (2003)

12. Licata, N.A., Tkachenko, A.V.: Errorproof programmable self-assembly of DNA-nanoparticle clusters. Physical Review E (Statistical, Nonlinear, and Soft Matter Physics) 74, 041406 (2006)

13. Beales, P.A., Vanderlick, T.K.: Specific binding of different vesicle populations by the hybridization of membrane-anchored DNA. J. Phys. Chem. A 111, 12372–12380 (2007)

14. Chan, Y.H.M., van Lengerich, B., Boxer, S.G.: Effects of linker sequences on vesicle fusion mediated by lipid-anchored DNA oligonucleotides. Proc. Natl. Acad. Sci. USA 106, 979–984 (2009)

15. Biancaniello, P.L., Crocker, J.C., Hammer, D.A., Milam, V.T.: DNA-mediated phase behavior of microsphere suspensions. Langmuir 23, 2688–2693 (2007)

16. Mirkin, C.A., Letsinger, R.L., Mucic, R.C., Storhoff, J.J.: A DNA-based method for rationally assembling nanoparticles into macroscopic materials. Nature 382, 607–609 (1996)

17. Valignat, M.P., Theodoly, O., Crocker, J.C., Russel, W.B., Chaikin, P.M.: Reversible self-assembly and directed assembly of DNA-linked micrometer-sized colloids. Proc. Natl. Acad. Sci. USA 102, 4225–4229 (2005)

18. Biancaniello, P., Kim, A., Crocker, J.: Colloidal interactions and self-assembly using DNA hybridization. Phys. Rev. Lett. 94, 058302 (2005)

19. Stengel, G., Zahn, R., Hook, F.: DNA-induced programmable fusion of phospholipid vesicles. J. Am. Chem. Soc. 129, 9584–9585 (2007)

20. Benkoski, J.J., Hook, F.: Lateral mobility of tethered vesicle - DNA assemblies. J. Phys. Chem. B 109, 9773–9779 (2005)

21. Yoshina-Ishii, C., Boxer, S.G.: Arrays of mobile tethered vesicles on supported lipid bilayers. J. Am. Chem. Soc. 125, 3696–3697 (2003)

22. Li, F., Pincet, F., Perez, E., Eng, W.S., Melia, T.J., Rothman, J.E., Tareste, D.: Energetics and dynamics of SNAREpin folding across lipid bilayers. Nat. Struct. Mol. Biol. 14, 890–896 (2007)

23. Maye, M.M., Nykypanchuk, D., Cuisinier, M., van der Lelie, D., Gang, O.: Stepwise surface encoding for high-throughput assembly of nanoclusters. Nat. Mater. 8, 388–391 (2009)

24. Prabhu, V.M., Hudson, S.D.: Nanoparticle assembly: DNA provides control. Nat. Mater. 8, 365–366 (2009)

25. NopplSimson, D.A., Needham, D.: Avidin-biotin interactions at vesicle surfaces: Adsorption and binding, cross-bridge formation, and lateral interactions. Biophys. J. 70, 1391–1401 (1996)

26. Green, N.M.: Avidin and streptavidin. Method Enzymol. 184, 51–67 (1990)

27. Burridge, K.A., Figa, M.A., Wong, J.Y.: Patterning adjacent supported lipid bilayers of desired composition to investigate receptor-ligand binding under shear flow. Langmuir 20, 10252–10259 (2004)

28. Torchilin, V.: Multifunctional and stimuli-sensitive pharmaceutical nanocarriers. Eur. J. Pharm. Biopharm. 71, 431–444 (2009)

29. Weaver, J.C., Chizmadzhev, Y.A.: Theory of electroporation: A review. Bioelectrochem. Bioenerg. 41, 135–160 (1996)

30. Monnard, P.A., Deamer, D.W.: Preparation of vesicles from nonphospholipid amphiphiles. In: Liposomes, Pt B. Academic Press Inc., San Diego (2003)
31. Bolinger, P.Y., Stamou, D., Vogel, H.: An integrated self-assembled nanofluidic system for controlled biological chemistries. Angew. Chem.-Int. Edit. 47, 5544–5549 (2008)
32. Abraham, T., Prenner, E.J., Lewis, R.N., Marwahal, S., Kobewka, D.M., Hodges, R.S., McElhaney, R.N.: The binding to and permeabilization of phospholipid vesicles by a designed analogue of the antimicrobial peptide gramicidin S. In: 51st Annual Meeting of the Biophysical-Society Biophysical Society, Baltimore, MD (2007)

Designing Adaptive Artificial Agents
for an Economic Production and Conflict Model

Behrooz Hassani-M and Brett W. Parris

Department of Econometrics and Business Statistics,
Monash University, VIC 3800, Australia
{Behrooz.Hassani.Mahmooei,Brett.Parris}@buseco.monash.edu.au

Abstract. Production and conflict models have been used over the past 30 years to represent the effects of unproductive resource allocation in economics. Their major applications are in modelling the assignment of property rights, rent-seeking and defense economics. This paper describes the process of designing an agent used in a production and conflict model. Using the capabilities of an agent-based approach to economic modelling, we have enriched a simple decision-maker of the kind used in classic general equilibrium economic models, to build an adaptive and interactive agent which uses its own attributes, its neighbors' parameters and information from its environment to make resource allocation decisions. Our model presents emergent and adaptive behaviors than cannot be captured using classic production and conflict agents. Some possible extensions for future applications are also recommended.

Keywords: Production and Conflict Models, Agent-based Modelling, Adaptive Behavior.

1 Introduction

According to classic economic theories, self-interested economic agents allocate their resources for production and acquiring wealth. But empirical evidence shows that economic agents also have options to allocate their resources unproductively. Unproductive activities like theft, piracy, predation and unmerited lobbying for special treatment, do not increase the total wealth of society but instead transfer economic products and rights artificially from one agent to the other. In the last three decades, different models have been suggested to study unproductive allocations, which can be categorized in two main groups: Production and Conflict (P&C) models and Rent-Seeking models [1]. In this paper we chiefly concentrate on the P&C models.

The main component of any P&C model is an economic agent who can allocate its resources to productive or unproductive activities. This kind of agent was proposed in two papers by Hirshleifer [2, 3] and then developed further during 1990's and early 2000's in different studies [4-11]. More recent papers have mainly focused on using the P&C model to analyze the assignment of property rights [12-15].

In almost all of these models an agent can allocate its resources to three main activities: production, predation and protection (sometimes called 3P). *Production* means that the agent uses its resources to produce goods or services. Output is

K. Korb, M. Randall, and T. Hendtlass (Eds.): ACAL 2009, LNAI 5865, pp. 179–190, 2009.

governed by a production function which is usually similar for all agents. *Predation* is the process of appropriating other agents' belongings (resources, output, wealth or any combination of them). Agents have a predation technology which governs the effectiveness of resources allocated to predation. Finally, *protection* is the process of defending against possible predation by providing defenses.

In this paper, we first review the structure of classic P&C decisions. Then in section three we introduce our agent describing its attributes and procedures. In the last section we present some results from our agent implementation in a real model and offer some possible extensions and applications for the designed agent.

2   Classic P&C Decision

In order to explain the motivation for designing an artificial agent for P&C models, we first describe the attributes and procedures of agents found in classical equation-based models. It is worth mentioning that before being studied in economics, these models were considered in biology. For example Smith and Price [16] used a computer simulation to model conflict among animals using five strategies titled "Mouse", "Hawk", "Bully", "Retaliator" and "Prober-retaliator" after animal behaviors.

In economics, Hirshleifer's seminal work [2] presented a two-party interaction for the first time where agents had "two alternatives ways for generating income: (1) "peaceful" production and exchange, versus (2) "appropriative" efforts designed to seize resources previously controlled by others (or to defend against such invasions)." Later, a new aspect was added to the structure which was the possibility of protection against attacks, which led to a "3P" framework including Production, Predation and Protection options. Anderton [17] added "trade" as an option, but for simplicity, we neglect that alternative in this paper.

In addition to the possible choices, the sequence of allocations is important as well. Models can be categorized into two groups according to the arrangement of their allocation decisions: 1) parties make their decisions simultaneously, which usually happens when they make their decisions independently [5, 12, 18] or 2) parties make their decisions sequentially, and defensive decisions are made first [15, 19].

Reviewing different 3P models, it can be seen that although the analysis has been extended and improved, two main problems remained unsolved. Firstly, agents need to choose between predator or producer/protector strategies. In other words, due to the limitations in general-equilibrium modelling, authors have not been able to build a player which is able to allocate resources to all three options at the same time. Modelers have always studied the P&C behavior at a meso-level, avoiding the consideration of micro-behavior of individual agents and instead studying their aggregate decision outputs. Secondly, agents' strategies are fixed or limited. For example in two-player predator/prey conflicts, which usually happen in one round, there is no chance for a player to switch its strategy during the running time. Only in a two player/role game like [20] can agents switch between cooperation and anticipation which leads to four different states based on a Hobesian structure. In two or more player general-equilibrium models the ratio of predators to producers is usually governed by a single exogenous variable. This is in contrast to theory and empirical evidence [21] that the desirability of being a predator or a producer is an endogenous function of the population of each group.

The second issue in classic P&C models is that agents are homogeneous in their characteristics. For example they have the same productivity and similar predation ability. This follows from the widespread use of representative agents in economic modelling which can give quite misleading results. For instance, we know that in a real-world perspective, one of the reasons that people become thieves is the lack of adequate production facilities. This fact has been neglected in most of the earlier studies, apart from papers like Grossman's [6] which considered this topic by adding the ratio of poorly-endowed people to well-endowed people to his model and using two new variables which indicate the productivity of each group. It should be mentioned that even in Grossman's paper, the productivity parameter is a constant, as agents' accumulated experience in their allocation does not affect their skills and return to that allocation. Muthoo [22] nominated this limitation as the first constraint in his model: "First, a simplifying but somewhat restrictive assumption that underlies my model is that the players cannot make any (further) investment in their fighting and productive skills."

The third common feature in P&C models is that the agents have little interactive effects on each other's decisions. The only possible route for an interaction is when an agent makes a protection decision based on its opponent's predation allocation, although in models like Grossman's [6] the variable G, which stands for guarding resources, is determined exogenously. The other limitation in the interaction among agents is that classic models only concentrate on predator/producer dealings or what Anderton [17] calls "dove/hawk" interaction. Clearly though, predator/predator and producer/producer interaction is also possible and the first one in particular can affect the outcomes of the model, since there is no inherent reason why stronger predators would not target weaker predators.

Returning to the agent itself, earlier studies did not usually consider the personal characteristics of the agents. Only Grossman and Kim [23] appear to have studied the morality of agents, including moral decay and revival in P&C models. In other models, our player is a simple undefined person without any personal distinctive properties and preferences.

Classic P&C models also treat information as complete and symmetric. In other words, any agent knows about all the other agents, their decisions, allocations and properties. The agent is also aware of global indicators and variables like the total number of agents, the ratio of predators to producers and so on. As we know, many of us do not know the real wealth of our close friends and relatives, their expenditures and even their monthly salaries, so this assumption is far from real world contexts.

The next common feature of classic P&C models is the lack of evolutionary methods in their optimization and learning processes. Agents usually use simple marginal comparisons to allocate their resources in any round. But no-one wakes up every morning and calculates the marginal return for every unit of resources in order to decide daily resource allocations. Past models also do not generally make any connection between the current period and previous decisions, so agents cannot optimize their outcome using past experiences.

The final issue is the *subject* of the conflicts – that which is going to be predated. There are two main views: The first is that conflicts happen over an *object* and the second is that conflicts happen over gaining a *right* [24]. In this paper we mainly concentrate on the object view, though the model can be adapted to consider conflicts over *rights* too. In the *object* view, different models have selected different objects which become subject to predation. In some models, agents contest an insecure re-source [7, 25, 26] and in others the conflict happens over access to produced goods [10, 11]. Kolmar [14] has reviewed both kinds of models and suggested a compound model. In his model, agents have two different resources: time, which cannot be pre-dated, and capital which can. The last option is for predators to steal another agent's *endowment*. Muthoo [22] alluded to this possibility as a limitation in his P&C model, because in current models agents cannot predate each other's production or predation technologies.

3   Our P&C Decision Maker

In the last section we outlined some of the main features and limitations in designing the agents which operate in classic general-equilibrium P&C models. In this section we try to develop a more capable and flexible agent using an agent-based modelling approach. Agent-based modelling uses an object-oriented simulation environment in which individual 'agents' posses both attributes and behaviours. Agents interact with each other dynamically according to certain rules, which may themselves evolve.

In our model agents have a life-cycle including birth, life, and death. Each agent, when born, receives an integer value as its maximum age. These integer values are distributed normally for the society and can easily be changed. For example, if we are speaking about a developed society, we can use a higher mean for our distribution compared with a model for agents in a less developed country. In this model we have assumed that our society's population growth rate is zero for simplicity, but it can also be easily changed. An agent will die if its total wealth is less than zero, which may occur after it has been predated, if it runs out of resources, or if it loses its production technology.

The second key attribute of our agent is its strategy. We present each agent's strat-egy using one of four possible two-bit strings: [0 0], [0 1], [1 0] and [1 1]. As men-tioned in the previous section, the classic models have two main types of agents, the *producer* which is assigned the [0 1] strategy, and the *predator*, which is assigned the [1 0] strategy. Our producer is a moral player who will never predate the others and our predator is a pure predator who only allocates resources to predating others.

Beyond these two classic combinations however, we also have the [1 1] strategy used by *3P* agents (who can produce, predate and protect at the same time) and the [0 0] strategy, which is adopted by *inactive* individuals. An inactive agent is one who does nothing for a specific period of time. We can think an inactive agent as someone unemployed, out of resources for either production or predation or maybe someone who is migrating to find a new place. Showing the strategy of agents by a bit pattern gives us the opportunity to use a genetic algorithm for optimizing our agents' behav-iors. All agents are born with the [0 1] producer strategy.

Each agent also has a *wealth* variable which represents its income accumulated over time and which will be inherited by its child when the agent dies. In our model each agent has a specific spatial location, which influences the agent's productivity. It can be thought as its work place, like a farm. Each agent also has a cost of living which must be paid by the agent every time-step, which here is equal to one week, with a model 'year' consisting of 52 time-steps. Standard short-run processes take one week to happen. For example, agents receive their income and spend it every week. Long-run processes like changing strategy happen on a yearly basis. Our agent also has a set of control variables, like its production history, which are used for managing its processes.

Agents can be linked to one to five other agents who are close by, which give us structures like those shown in Figure 1, in a 20 by 20 spatial grid (400 places) containing 100 agents.

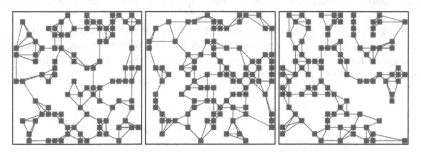

Fig. 1. Three sample arrangements of agents in our environment

In our model, information is neither complete, nor symmetric. When an agent is not linked with other agents, it means it does not know those agents at all and has no interaction with them, but when it is linked to another agent it may have some information about that other agent. Our agent cannot know all the attributes of the other agent, like its allocations, but it can partly be informed about its wealth or strategy.

Up to now, we have an agent who has its own attributes and location and it dies when it gets old or very poor. When an agent dies, a new agent is born on the same location. As discussed earlier, the new-born agent will be a producer ([0 1]) by default, but before starting its activities, its strategy at birth will also interact with the strategy of its parent agent through a cross-over (as outlined in Figure 4). The probability of this cross-over for each bit is 50%. Using this method we have been able to transfer the strategy from one generation to the other.

Production and predation technologies can be in any form, but we have used a simple function in the form of $f(x) = a\sqrt[b]{x}$ where a and b will be positive and non-zero to ensure that $f'(x) \geq 0$, $f''(x) \leq 0$, thereby maintaining a direct relation between input and output and ensuring decreasing returns. The subject of the conflict can be any object, including the resources, wealth or even technology, but in order to present some new features in our model we have used the combination of wealth and production and predation technologies.

Two other issues remain: how should our agent make decisions about the allocation of its resources and how it learns from its experience in order to optimize decision process? Answering these questions form the core activity of agents, which is the allocation of resources.

At this stage, we need a new concept which can connect an agent's characteristics to its decisions, so we introduce *danger* to our model. *Danger* is a variable between zero and one, which is higher when the environment becomes more dangerous. The concept of danger is calculated in the same way for all agents and each agent perceives danger based on the equation below:

$$Danger = \frac{1}{2}\sum_{t=0}^{n}\left(\frac{\alpha}{2}\right)^{n-t}$$ (1)

where α is 1 if the agent has been predated in the round t and 0 if it has not been. So for example, if after five rounds, an agent has a thread such as 0,0,1,0,1 which means it has been predated during the last round and the third round, its danger will be 0.3125. Danger is calculated in the same way for all agents and each agent may be predated only once during each round. Since $\sum_{1}^{n}\left(\frac{1}{2}\right)^{n}$ is always less than 2, danger is always less than 1. This equation reduces the effects of older experiences and limits *danger* to a range of (0, 1). To personalize this variable for our agents, we designed a simple fuzzy function. This function is determined by three values a, b and c as in Figure 2 below, where $a > 0$, $a < b$, and $0 < c < 1$.

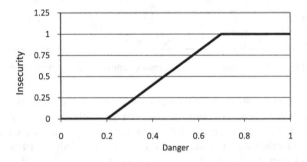

Fig. 2. A sample Danger and Insecurity Fuzzy Function

Figure 2 helps us relate agent's observations of predation to a personal interpretation of the danger which we call it *insecurity*. Every agent has its own a, b, and c which are assigned when an agent is initialized by a random normal distribution. For example Figure 2 shows that when danger is less than 0.2, this agent does not have any sense of insecurity. As soon as danger becomes more than 0.2, the agent's sense of insecurity increases linearly with increasing until it reaches maximum of 1 at a danger level of 0.7.

In Figure 3, we have shown the danger-insecurity diagram for two different agents. As can be seen, the solid line leaves the zero level at 0.2 and reaches its maximum value, 1, when danger is 0.3. But the dotted line increases with a shallower slope and finally it remains on 0.5 at its maximum insecurity. According to Equation 1, and using Figure 3 we can say that, if the associated agent of the first line is predated during the previous round – which will cause its danger to be at least 0.5 – it will feel completely insecure, while in the same situation the second (braver) agent's insecurity will be only 0.4, so in this situation the first agent feel around twice as insecure as the second agent. This fuzzy function enables us to add the characteristic of "appetite for risk" to our agents, which is defined by the parameters a, b, and c.

Now, using the available features in our agent we define its allocation process. The first decision which is made by any agent is its protection. For our current agent, we

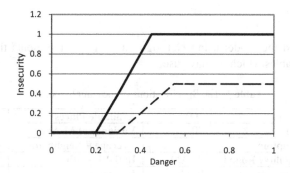

Fig. 3. Two sample Danger-Insecurity Functions

assume that it will protect itself in proportion to its sense of insecurity. So, if its insecurity is 0.6, it will allocate 60% of its resources for protection. Then if it is a [0 1] type agent, it will allocate the remaining resources to production and if it is a [1 0] type agent, it will predate using the rest of its resources. Now the question is about the [1 1] agent who both produces and predates. For making another connection between the personal attributes of an agent and its decision we have decided to allow [1 1] agents to be able to predate and produce as below, which means our agent predates more when it has a greater appetite for risk:

$$Predation = (1 - insecurity) \cdot (1 - c) \qquad (2)$$
$$Production = (1 - predation) \qquad (3)$$

The final features are the learning and optimization processes which are presented in Figure 4. Each agent's strategy can be mutated every five years on average (the probability is one in 260 weeks).

If its strategy is set to [0 0] it will do nothing for one round and then it will continue its previous strategy. The agents keep a history of their best outcomes and associated strategies over time so they always know which strategy has been the most advantageous. During each run there may be a cross-over between its best and its current strategy. The probability of crossover can be set by the modeler and we allow agents to revise their strategies every year.

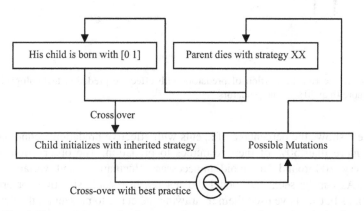

Fig. 4. The complete cycle of an agent's strategy selection and optimization

4 Results

We implemented the model using NetLogo 4.0.3. Table 1 presents the main initial
and control variables which we have used.

Table 1. The Specifications of the sample model

Number of Agents	100	Available Places	150
Mean of M-Age	20	Variance of M-Age	5
Weekly Consumption	1	Average # Neighbours	4
Mutation Probability / Round	0.0020	Initial Wealth	1
Initial Type of Agents	[0 1]	Resources	10 (Constant)
a (random distribution)	(0.5, 0.15)	b (random distribution)	(1.5, 0.15)
c (random distribution)	(0.5, 0.15)	Pop Growth Rate	0
Death Wealth Condition	<= 0	Probability of Crossover	0.5 / bit

Implementing the model and executing a sample run 50 times, we have found some
emergent and adaptive behaviour that cannot be captured in classic P&C models. The
first result that can be considered is the tendency of different types of models to select
different strategies. The results show that usually agents with lower productivity and
higher appetites for risk are more interested in allocating their resources to predation,
which seems to be close to what happens in real-world societies. For example, according to our outputs, on average around 65% of agents who allocate more than 50% of
their resources to predation are risk-takers and the effectiveness of their predation is
more than 0.5.

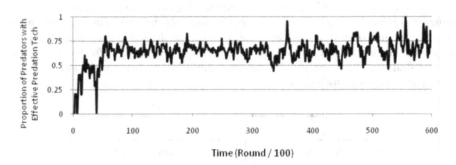

Fig. 5. A sample run. Proportion of predators with effective predation technology is usually
consists more than 50% of the predators.

Figure 5 shows the percentage of agents with effective predation technologies who
allocate more than 50% of their resources to predation. The observation has been
made every 100 rounds to avoid unnecessary fluctuations and variations in the
diagram. As can be seen, the model did not reach a stable situation for around 50
observations but we have used them in drawing the curve to present all the data.

The other observation which shows adaptive decisions is the reaction of agents to increases in returns to production. As we expected, when returns to production become more than returns to predation, there is a sudden rise in the ratio of resources allocated to predation. Consequently, as some productive agents allocate more to predation, the production growth rate decreases. This result may not be captured using classic models easily as it happens based on the interaction of agents. Two samples are shown in figure 6.

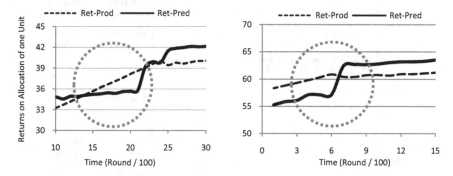

Fig. 6. Two sample runs. *Ret-Prod*: Return to Production. *Ret-Pred*: Return to Predation. As can be seen in both diagrams when returns to production are more than returns to predation, the number of predators increases very fast which causes sudden returns to predation and decrease in returns to predation.

In a long-run prespective these adaptive decisions lead to a dynamic cyclic variation which continuously happens in allocation processes. Figure 7 shows that constant consideration of returns and optimization of strategies by agents avoids any domination by a specific strategy even in the long run. As can be seen, although the diagram shows the average results for 10 runs, the cyclical allocation of production and predation resources is centred around the 40% line. The remaining resources are allocated to protection, which is not shown here. Figure 8 shows more details about the production and predation average allocation data.

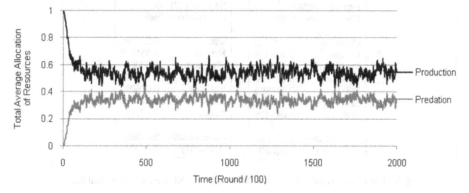

Fig. 7. A sample cyclic variation of average resource allocation in 10 runs simulation

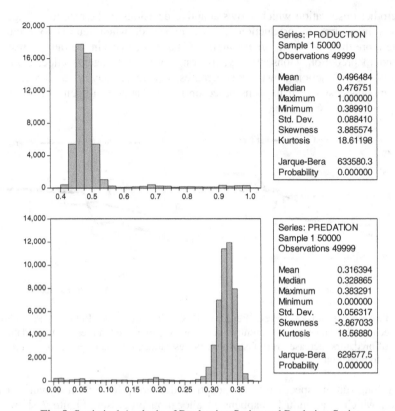

Fig. 8. Statistical Analysis of Production Series and Predation Series

Based on reviewers' suggestion we also run another type of the model where new-born agents directly inherit their parent strategy with a following mutation with the probability of 0.5 for each bit. The observations can be shown as below:

Series: PREDATION		Series: PRODUCTION	
Sample 1 50000		Sample 1 50000	
Observations 49999		Observations 49999	
Mean	0.396120	Mean	0.372336
Median	0.402937	Median	0.361279
Maximum	0.461822	Maximum	1.000000
Minimum	0.000000	Minimum	0.275583
Std. Dev.	0.048559	Std. Dev.	0.076040
Skewness	-6.035172	Skewness	6.142672
Kurtosis	44.41935	Kurtosis	45.00874
Jarque-Bera	3877539.	Jarque-Bera	3990886.
Probability	0.000000	Probability	0.000000

Fig. 9. The results from 10 run simulation with direct inter-generational transfer

The results illustrate that inter-generational transfer of strategy among the agents based on our definition in this paper can decrease the intensity of predation in our environment. As an application, if we think about the transfer as an external means like effective education, the results of a 10 run simulation shows that on average, production allocated resources may increase from 37% to 49% – a 32% increase – and at the same time the same index for predation decreased from 39% to 31%. So, finding an useful method for preventing conflict-based behavior from being inherited by children can lead to more production and economic growth.

5 Conclusions and Suggestions

While the classic Production and Conflict models usually concentrate on physical aspects of the theory like technologies of production and predation and recognize them as the most effective factors when an agent makes its allocation decisions, we have attempted to also consider some new features in these models.

Our results show that our simulated agent is responsive to its environment signals and optimizes its decision according to the other agents which it knows and also its own characteristics and initial conditions. The decisions are made at the local level in our simulation based on our definitions of the neighborhood, which makes the analysis closer to real-world events.

Using an agent-based approach, we have been able to make the agents more interactive, which led to emergent behavior and more adaptive decision making compared with classic P&C agents, which have been studied using their aggregate performance.

In general terms, in this paper we showed that in a simulated environment where there are no property rights, a significant proportion of resources will be allocated to unproductive activities which will not be useful for economic growth.

The agent designed here and its associated environment can be extended for use in different theoretical and applied areas such as the estimation of resources wasted in economic conflict and rent-seeking models, assessing the efficiency of different methods for simulating economic expectations, and modelling collective behavior, especially regarding protection and conflicts over natural resources.

Acknowledgement

The authors thank anonymous referees of this conference for their useful comments.

References

1. Hausken, K.: Production and Conflict Models versus Rent-Seeking Models. Public Choice 123(1/2), 59–93 (2005)
2. Hirshleifer, J.: The analytics of continuing conflict. Synthese 76(2), 201–233 (1988)
3. Hirshleifer, J.: Conflict and rent-seeking success functions: Ratio vs. difference models of relative success. Public Choice 63(2), 101–112 (1989)

190   B. Hassani-M and B.W. Parris

4. Garfinkel, M.R., Skaperdas, S.: The political economy of conflict and appropriation. Cambridge University Press, New York (1996)
5. Garfinkel, M.R., Skaperdas, S.: Economics of Conflict: An Overview. In: Sandler, T., Hartley, K. (eds.) Handbook of Defense Economics: Defense in a globalized world, Elsevier, Amsterdam (2007)
6. Grossman, H.I.: Producers and Predators. Pacific Economic Review 3(3), 169–187 (1998)
7. Grossman, H.I., Kim, M.: Swords or Plowshares? A Theory of the Security of Claims to Property. The Journal of Political Economy 103(6), 1275–1288 (1995)
8. Grossman, H.I., Kim, M.: Predation and accumulation. Journal of Economic Growth 1(3), 333–350 (1996)
9. Hirshleifer, J.: The dark side of the force: economic foundations of conflict theory. Cambridge University Press, Cambridge (2001)
10. Skaperdas, S.: Cooperation, Conflict, and Power in the Absence of Property Rights. The American Economic Review 82(4), 720–739 (1992)
11. Skaperdas, S., Syropoulos, C.: The Distribution of Income in the Presence of Appropriative Activities. Economica 64(253), 101–117 (1997)
12. Annen, K.: Property Rights Assignment: Conflict and the Implementability of Rules. Economics of Governance 7(2), 155–166 (2006)
13. Hafer, C.: On the Origins of Property Rights: Conflict and Production in the State of Nature. Review of Economic Studies 73(1), 119–143 (2006)
14. Kolmar, M.: Goods or resource contests? Public Choice 131(3), 491–499 (2007)
15. Kolmar, M.: Perfectly Secure Property Rights and Production Inefficiencies in Tullock Contests (cover story). Southern Economic Journal 75(2), 441–456 (2008)
16. Smith, J.M., Price, G.R.: The Logic of Animal Conflict. Nature 246(5427), 15–18 (1973)
17. Anderton, C.H.: Conflict and Trade in a Predator/Prey Economy. Review of Development Economics 7(1), 15–29 (2003)
18. Skaperdas, S.: Conflict and Attitudes Toward Risk. The American Economic Review 81(2), 116–120 (1991)
19. Anderton, C.H., Anderton, R.A., Carter, J.R.: Economic activity in the shadow of conflict. Economic Inquiry 37(1), 161–166 (1999)
20. Vanderschraaf, P.: War or Peace?: A Dynamical Analysis of Anarchy. Economics and Philosophy 22(02), 243–279 (2006)
21. Romer, D.: Advanced macroeconomics. Mass. McGraw-Hill, Boston (2006)
22. Muthoo, A.: A model of the origins of basic property rights. Games and Economic Behavior 49(2), 288–312 (2004)
23. Grossman, H.I., Kim, M.: Predators, moral decay, and moral revivals. European Journal of Political Economy 16(2), 173–187 (2000)
24. Sánchez-Pagés, S.: On the social efficiency of conflict. Economics Letters 90(1), 96–101 (2006)
25. Grossman, H.I.: The Creation of Effective Property Rights. The American Economic Review 91(2), 347–352 (2001)
26. Meza, D.d., Gould, J.R.: The Social Efficiency of Private Decisions to Enforce Property Rights. The Journal of Political Economy 100(3), 561–580 (1992)

Emergent Societal Effects of Crimino-Social Forces in an Animat Agent Model

Chris J. Scogings and Ken A. Hawick

Massey University, Private Bag 102 904, NSMSC, Auckland, New Zealand
{c.scogings,k.a.hawick}@massey.ac.nz

Abstract. Societal behaviour can be studied at a causal level by perturbing a stable multi-agent model with new microscopic behaviours and observing the statistical response over an ensemble of simulated model systems. We report on the effects of introducing criminal and law-enforcing behaviours into a large scale animat agent model and describe the complex spatial agent patterns and population changes that result. Our well-established predator-prey substrate model provides a background framework against which these new microscopic behaviours can be trialled and investigated. We describe some quantitative results and some surprising conclusions concerning the overall societal health when individually anti-social behaviour is introduced.

Keywords: animat agent; societal impact; social behaviour; emergence; complexity; adaptive system.

1 Introduction

Modelling sociological phenomena at a macroscopic level is a non-trivial problem generally, and understanding anti-social behaviours such as criminal tendencies is particularly difficult [1]. Some attempts have been made to develop computer simulations of criminal phenomena [2] and these have had some success in studying spatial patterns of criminal activity [3,4]. Intelligent agent systems work in the area of criminal social simulation [5] has focused on dynamically generated beliefs and desires and on the use of general social reasoning mechanisms and models [6,7] which have been difficult to scale up to large systems.

We have attempted to incorporate simplified anti-social or criminal behaviours at a microscopic level in a large scale Artificial Life (ALife) model system with the expectation that such behaviours would modify emergent macroscopic behaviours of the system as a whole. Several sophisticated artificial ALife simulation models exist, including [8,9,10]. These concentrate almost exclusively on the evolution of "digital organisms" and the corresponding emergent macro-behaviours. They are less concerned with the microscopic details of the lives of individual "animats" [11].

Our predator-prey model [12] has been refined over a period of several years and has proved a useful substrate model against which to study new behaviours such as trading [13] and species segregation [14]. Instead of noting evolutionary behaviour

K. Korb, M. Randall, and T. Hendtlass (Eds.): ACAL 2009, LNAI 5865, pp. 191–200, 2009.

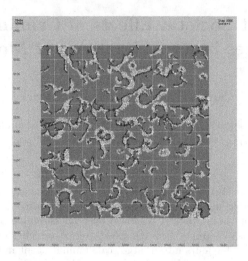

Fig. 1. The situation at step 3000 of a typical run (without criminal or law-enforcement agents) showing animats on a square grassed area. Predators are black and prey are white. Various macro-clusters, including spiral formations, have emerged.

(which is often difficult to measure) we have concentrated on making small, well-defined adjustments to the model and then analysing new animat behaviours.

The predator-prey mix of animats has allowed us to make relatively simple modifications to predator behaviours which give rise to amplified responses in the animat population as a whole and which we can study numerically through the use of simulations of multiple independent runs of model systems comprising $10^5 - 10^6$ individual animats. Typical emergent formations can be seen in the screenshot of our model shown in Figure 1.

Few Artificial Life models have been used to study the effects of "higher-order" social interactions. In this article, we study the effects on animat population and behaviour of introducing criminals into their community. We also introduce "police" that can control the criminals. In this paper we focus on how the collective animat behaviours change when first a criminal sub-population is introduced and secondly when some sort of law-enforcement animats are introduced. We describe the workings of our numerical simulation model in section 2. We present details on how to introduce criminals and police into the animat community in sections 3 and 4 and we give some selected results – in terms of model system snapshots as well as population and behaviour measurements, in section 5. Finally we offer some tentative conclusions and areas for further work in section 6.

2   The Animat Model

Our model consists of two species of interacting animats – the predators and the prey. At each time step of the model, every animat updates and records its current state which consists of: current health; current age; and an x-y location in their 2-dimensional world. When an animat is first created the current age is set to zero. The age is incremented at every time step and when it reaches a pre-set maximum for the species the animat "dies of old age" and is removed. All animats in the model have a "current health" value. This value is reduced each time-step and if it reaches zero the animat "starves to death". When a new animat is created it is allocated the average of the health of its parents. If an animat eats something then the current health value will be increased by a certain amount, although it may never be increased past the maximum health value which is predetermined for each animat species. The concepts of health values and animats eating behaviours are discussed in [15].

Prey eat only "grass" which is placed at specific locations on the map – usually in a contiguous area. Grass has a fixed "nutritional value" and this is the number of health points that prey receive when executing the graze rule. In these experiments all grass had a value of 60% of maximum prey health. This means that prey with zero health and executing the graze rule would then have 60% health. If prey with 70% health executed the graze rule they would then have 100% health as the maximum may not be exceeded. Grass is assumed to be continually replenished although "overgrazing" is simulated in that prey can only execute the graze rule if not crowded (i.e. if less than 10 immediate neighbours). Prey are also provided with a rule to move away from other prey in order to relieve overcrowding. For practical purposes prey (and with them, the predators) are contained within the "grassy area". Containing the animats is useful as it prevents the populations becoming large and unmanageable and also limits the area of the (otherwise unbounded) grid in which the animats exist. In previous work [16] we have demonstrated that these limitations do not affect the emergent macro-behaviours of the model. The experiments discussed in this article take place on a large square "grassed area" which explains why the animat locations have a fairly distinct edge in the diagrams.

Predators eat only prey and other things being equal we can reproduce the well known boom-bust limit cycles predicted by predator-prey models such as the Lotka-Volterra coupled differential equations [17,18] and their spatial variants [19]. Although the model is synchronous, animats are updated in a random order which we found adequate to remove any spatial artifacts from sweep order. The process is thus a two-phase system in which the variables for all animats are updated after all checks have been made and all rules have been executed. This system was developed in order to ensure fairness across all the animats in the model and a full discussion of alternative updating systems is available in [20].

Every animat carries a small set of rules that govern its behaviour and this rule set is passed on unchanged to any offspring. It is possible to allow mutations and to introduce genetic algorithms into the model but an important feature of our work is to make small, well-defined changes to the microscopic model and measure the effects of those changes. We have experimented with changing the order of the rules and have investigated which rule sets generate the most successful animat groups [12]. Table 1 summarises the animat rules.

Table 1. Rules currently used in the model. A rule is only executed if the conditions are met; otherwise that rule is ignored. The birth conditions are discussed below.

Rule	Predator	Prey	Conditions
Move Away	No	Yes	overcrowded prey
Breed	Yes	Yes	female with adjacent male; birth conditions
Eat Prey	Yes	No	health < 50%; adjacent prey
Flee Predator	No	Yes	adjacent predator
Graze	No	Yes	health < 50%; not crowded; requires grass
Seek Mate	Yes	Yes	health > 50%; no adjacent mate
Seek Prey	Yes	No	health < 50%; no adjacent prey
Random Move	Yes	Yes	

Rules are considered in a strict priority order. Each time-step, every animat attempts to execute the first rule in its rule set. However, most rules have conditions so can often not be executed. For example, prey will only move away from a predator if a predator is actually adjacent. If the conditions for the first rule can not be satisfied, the animat attempts to execute the next rule in the set and so on. When a new animat is produced the gender is selected at random. Males never execute the Breed rule. They carry the rule only in order to pass it on to their daughters. Females can execute the Breed rule but may only do so if a male is adjacent. Breeding only has a certain chance of success. This is a simple alternative to factoring in a host of complicated parameters including birth defects, nutrition, adequate shelter and so on. Changing the chances of a successful birth can dramatically alter the number of animats and can sometimes cause the extinction of all animats. For these experiments the chance of a successful birth was set to 15% for predators and 80% for prey, as experience gained from many simulations has shown that these values produce stable populations.

The interaction of the animats as they execute their individual rules has produced interesting emergent features in the form of macro-clusters often containing many hundreds of animats. We have analysed and documented these emergent clusters in [21]. The most fascinating cluster that consistently appears is a spiral and several spirals are visible in Figure 1. One benefit of our model is that it can handle very large numbers of animats. Several of the experiments discussed in this article contain of the order of 200,000 animats and other simulations have produced over a million animats. These substantial populations enable the study of unique emergent macro-behaviours that would not develop with smaller numbers of animats.

3   Simulation Experiment 1 - Introducing Criminals

We conducted some experiments to see if any macro-behaviours emerged when criminals appeared in predator society. It is interesting to study the predators, as the higher-level artificial life-form, and to use prey simply as the food resource or "wealth" for the predator society. During these experiments all references to society, criminals, etc refer only to the predators.

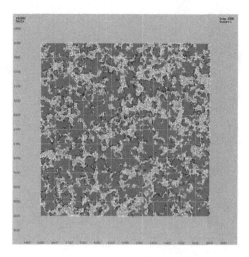

Fig. 2. The situation at step 3000 of a typical run when criminals have been introduced. Predators are black and prey are white. The predator population is 26,024 which is less than the 30,980 shown in the control situation in Figure 1 and approximately 25% of the predators are criminals. The prey population of 151,262 is almost exactly double the 75,494 prey in Figure 1. Spirals and other formations continue to emerge.

The first experiment was to introduce criminals into predator society. A criminal is defined as a predator that attacks an adjacent predator and removes some of the neighbour's health points. Health points are essential to life so in many cases the removal of health points can lead to an early death for the victim. In order to keep things simple, the following rules were used. Criminals are "born". It is not possible for a predator to change from non-criminal to criminal or vice versa. A non-criminal parent has a 15% chance of producing criminal offspring whereas a criminal parent has an 80% chance of producing criminal offspring. A neighbour is attacked at random - the neighbour could be "wealthy" (have maximum health points) or "poor" or, in fact, a criminal themselves. Attacks always succeed. The victim always loses 40% of its current health points.

In future work it may prove interesting to vary some of these parameters, e.g. criminals may target wealthy victims or not all attacks may succeed. Because some of the victims died (due to lack of health points) the overall population of predators dropped and this, in turn, caused an increase in the prey population. A typical situation at step 3000 in one of these runs can be seen in Figure 2. During these runs the number of criminals remained stable at approximately 25% of the total predator population.

4   Simulation Experiment 2 - Introducing Police

The second experiment was to introduce police into the predator society which already contained criminals (introduced in first experiment). Police are not "born".

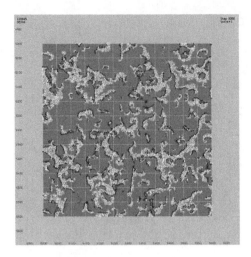

Fig. 3. The situation at step 3000 of a typical run when both police and criminals have been introduced. Predators are black and prey are white. The predator population is 32,316 which is higher than the 26,024 shown in Figure 2. Approximately 6% of the predators are police and 2.5% are criminals. The prey population is 103,945 which is lower than the 151,262 in Figure 2.

Instead police are ordinary predators that are appointed as police under particular circumstances. The rules governing the appointment of a police (predator) are as follows. The predator must be adjacent to a predator that is the victim of a criminal attack. The predator must be neither criminal nor police. There is then only a 30% chance that the predator is appointed as police.

Once a predator is appointed as police it remains police for the rest of its existence. If a police predator is adjacent to a criminal, the criminal is destroyed. Future work may investigate alternatives to this. Because criminals are destroyed the runs in this experiment tended to have higher populations of predators than those when criminals were introduced without a police presence. A typical situation is shown in Figure 3. During these runs the number of police and criminals remained stable at approximately 5.9% and 2.5% of the total predator population respectively. Thus the presence of the police reduced the criminal population from 25% to 2.5%.

5   Simulation Results

In our previous work we have found that measurements on the bulk animat population gave a good metric for the success of a particular agent society in terms of the mean sustainable population level. Predator-prey systems typically exhibit boom-bust periodic fluctuations, and our model populations usually settle down to oscillate about a mean representative value. Predator populations lag prey and the grosser uncertainties in these measurements due to microscopic spatial

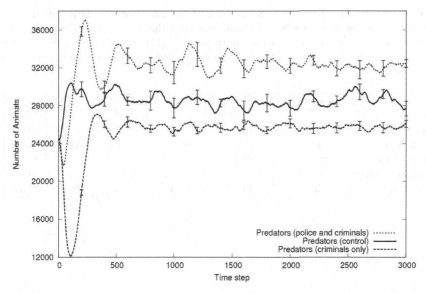

Fig. 4. Predator populations averaged over 10 independent runs of the model in three different types of simulation – the control, the introduction of criminals and the introduction of both criminals and police. Note the initial dramatic drop in the population when criminals are introduced.

fluctuations can be dampened out upon averaging over independent model runs that were started with microscopically different but statistically similar initial conditions.

The introduction of criminals and police had a marked effect on the predator populations. When criminals were introduced the predator population suffered an initial dramatic drop but then recovered to almost the same levels as in the control simulations – see Figure 4. An unexpected result was that the predator population with both criminals and police present is higher than the control population. This occurs because the initial attacks by criminals cause a drop in predator population which enables an increase in the prey population which, in turn, can sustain a higher number of predators.

Table 2 shows some of the measured predator activities during the simulations. When criminals are present and stealing health points, predators are forced to spend more time eating prey in order to regain the lost health points. This means that the Eat Prey rule is executed more often and usually this, in turn, causes the Random Move rule to be executed less.

Figure 5 shows the typical age distribution of animats. When criminals are introduced into the population, the number of older animats drops. This is due to the fact that in a "lawless" society where health points are constantly stolen, the chances of surviving successfully into old age diminish. Introducing police agents mostly restores the distribution to its control level before anti-social agents were introduced at all.

Table 2. This table shows what percentage of the predator population used a particular rule during time step 3000. "Crim" indicates that criminals are present. "Pol" indicates that police and criminals are both present. (M) and (F) denote male and female respectively. "new born" refers to animats created in this time step that therefore do not execute a rule.

Rule	Control(M)	Control(F)	Crim(M)	Crim(F)	Pol(M)	Pol(F)
Breed		4.5%		5.8%		5.1%
Eat Prey	9.3%	9.2%	17.0%	19.3%	10.3%	10.2%
Seek Mate	4.9%	4.2%	8.5%	6.4%	5.9%	4.6%
Seek Prey	20.8%	20.9%	20.0%	20.6%	19.8%	20.7%
Random Move	32.8%	30.9%	27.3%	23.4%	32.0%	29.8%
new born	32.2%	30.3%	27.3%	24.5%	32.1%	29.5%

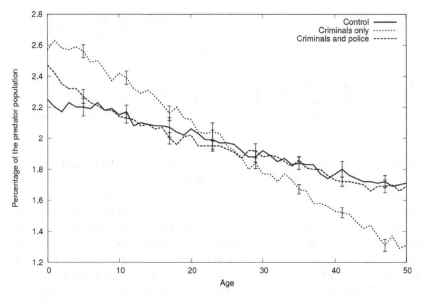

Fig. 5. Graph showing the spread of ages across the predator population at time step 3000. These figures are based on an average of 10 runs for each situation.

6   Discussion and Conclusions

We have reported upon simulation experiments to introduce criminal and law-enforcement behaviour into animat agent models. We have introduced anti-social and police enforcement microscopic behaviours amongst the predator agents which are the higher-level artificial life-forms in our system.

Our data shows some surprising results concerning the overall animat population when first criminals and subsequently police agents are introduced. Introducing criminals appears to *improve* the general level of predator animat population and subsequently introducing police agents improves it even further.

An examination of the agent age distribution and rule execution frequencies suggests that criminal agents are serving to remove old and unhealthy predator animats from the population. While this might be politically unappealing it is an emergent numerical consequence of the model rules. Happily the introduction of police does yield a sociologically more intuitive and politically appealing result and further increases the sustainable mean predator populations whilst restoring the age distribution to its control experiment levels.

We believe numerical experiments such as these, involving dispassionate numerical microscopic parameters, are a useful means to exploring sociological phenomena at a bulk level in large scale systems that approach the size of realistic social systems.

References

1. Bartol, C.R.: Criminal Behaviour: A Psychosocial Approach, 6th edn. Prentice-Hall, Englewood Cliffs (2002)
2. Brantingham, P.L., Brantingham, P.J.: Computer simulation as a tool for environmental criminologists. Security Journal 17, 21–30 (2004)
3. Furtado, V., Melo, A., Menezes, R., Belchior, M.: Using self-organization in an agent framework to model criminal activity in response to police patrol routes. In: Proc. FLAIRS Conference, pp. 68–73 (2006)
4. Furtado, V., Melo, A., Coelho, A., Menezes, R.: A crime simulation model based on social networks and swarm intelligence. In: Proc. 2007 ACM symposium on Applied computing, pp. 56–57 (2007)
5. Bosse, T., Gerritsen, C., Treur, J.: Cognitive and social simulation of criminal behaviour: the intermittent explosive disorder case. In: Proc. 6th International Joint Conference on Autonomous Agents and Multiagent Systems, Honolulu, Hawaii, pp. 1–8 (2007)
6. Guye-Vuilleme, A., Thalmann, D.: A high-level architecture for believable social agents. Virtual Reality 5(2), 95–106 (2000)
7. Kats, Y.: Intelligent software design: Challenges and solutions. In: Proc. EEE International Workshop on Intelligent Data Acquisition and Advanced Computing Systems: Technology and Applications, pp. 468–471 (September 2003)
8. Adami, C.: On modeling life. In: Brooks, R., Maes, P. (eds.) Proc. Artificial Life IV, pp. 269–274. MIT Press, Cambridge (1994)
9. Holland, J.H.: Echoing emergence: Objectives, rough definitions, and speculations for echo-class models. In: Cowan, G.A., Pines, D., Meltzer, D. (eds.) Complexity: Metaphors, Models and Reality, pp. 309–342. Addison-Wesley, Reading (1994)
10. Tyrrell, T., Mayhew, J.E.W.: Computer simulation of an animal environment. In: Meyer, J.A., Wilson, S.W. (eds.) From Animals to Animats: Proc. the First International Conference on Simulation of Adaptive Behavior, pp. 263–272. MIT Press, Cambridge (1991)
11. Wilson, S.W.: The animat path to AI. In: Meyer, J.A., Wilson, S.W. (eds.) From Animals to Animats: Proc. the First International Conference on Simulation of Adaptive Behavior, pp. 15–21. MIT Press, Cambridge (1991)
12. Hawick, K.A., James, H.A., Scogings, C.J.: Roles of rule-priority evolution in animat models. In: Proc. Second Australian Conference on Artificial Life (ACAL 2005), Sydney, Australia, pp. 99–116 (December 2005)

13. Scogings, C.J., Hawick, K.A.: Intelligent and adaptive animat resource trading. In: Proc. 2009 International Conference on Artificial Intelligence (ICAI 2009) Las Vegas, USA, July 13-16 (2009)
14. Hawick, K.A., Scogings, C.J.: Spatial pattern growth and emergent animat segregation. Technical Report CSTN-078, Massey University (February 2009)
15. Scogings, C.J., Hawick, K.A., James, H.A.: Tuning growth stability in an animat agent model. In: Proc. the 16th IASTED International Conference in Applied Simulation and Modelling (ASM 2007), August 29-31, pp. 312–317 (2007)
16. Scogings, C.J., Hawick, K.A.: Global constraints and diffusion in a localised animat agent model. In: Proc. IASTED International Conference on Applied Simulation and Modelling, Corfu, Greece, June 23-25, pp. 14–19 (2008)
17. Lotka, A.J.: Elements of Physical Biology. Williams and Wilkins, Baltimore (1925)
18. Volterra, V.: Variazioni e fluttuazioni del numero d'individui in specie animali conviventi. Mem. R. Accad. Naz. dei Lincei, Ser. VI 2 (1926)
19. Gallego, S.: Modelling Population Dynamics of Elephants. PhD thesis, Life and Environmental Sciences, University of Natal, Durban, South Africa (June 2003)
20. James, H.A., Scogings, C.J., Hawick, K.A.: Parallel synchronization issues in simulating artifical life. In: Gonzalez, T. (ed.) Proc. 16th IASTED International Conference on Parallel and Distributed Computing and Systems (PDCS), Cambridge, MA, USA, pp. 815–820 (November 2004)
21. Hawick, K.A., Scogings, C.J., James, H.A.: Defensive spiral emergence in a predator-prey model. Complexity International (msid37), 1–10 (2008)

A Heterogeneous* Particle Swarm

Luke Cartwright and Tim Hendtlass

Faculty of Information & Communication Technologies
Swinburne University of Technology
Melbourne, Australia
cartwrightluke@hotmail.com, thendtlass@swin.edu.au

Abstract. Almost all Particle Swarm Optimisation (PSO) algorithms use a number of identical, interchangeable particles that show the same behaviour throughout an optimisation. This paper describes a PSO algorithm in which the particles, while still identical, have two possible behaviours. Particles are not interchangeable as they make independent decisions when to change between the two possible behaviours. The difference between the two behaviours is that the attraction towards a particle's personal best in one is changed in the other to repulsion from the personal best position. Results from experiments on three standard functions show that the introduction of repulsion enables the swarm to sequentially explore optima in problem space and enables it to outperform a conventional swarm with continuous attraction.

Keywords: Particle Swarm Optimisation, Collective Intelligence, Optimisation.

1 Introduction

The canonical Particle Swarm Algorithm, first introduced in [1], is nature inspired, particularly by the flocking of birds and the schooling of fish. Over the ensuing years it has been developed into a range of robust optimisation algorithms. For more details of the range of algorithms see [2]. Whilst these algorithms differ in many details they almost all assume that the swarm particles are homogeneous and interchangeable.

In this work particles have two different behaviours. Both behaviours involve the conventional swarm components of momentum and attraction to the global best position; in the first behaviour a particle is attracted to its personal best position, in the second the particle is repelled from its personal best position.

A number of strategies for incorporating the two behaviours can be envisioned:
1. Behaviour as a crisp concept, a majority of particles always exhibit the first behaviour while the minority exhibit the second behaviour. No particle ever changes behaviour.
2. Behaviour treated as a fuzzy concept, all particles slowly but smoothly change from full behaviour one (attraction) to full behaviour two (repulsion) as time passes, all simultaneously reverting to full behaviour one as soon as the global best position changes.

---

* Heterogeneous – *adj, :* consisting of dissimilar or diverse ingredients or constituents.

K. Korb, M. Randall, and T. Hendtlass (Eds.): ACAL 2009, LNAI 5865, pp. 201–210, 2009.

3. Particles switch between crisp behaviours independently based on the length of time since they changed their personal best position.

While each of these will result in exploration of the region around a new global best position they will also force exploration away from this position. The first requires a choice for the size of the majority, which experimentation has found to be problem dependent. The second requires a choice of behaviour change rate, which has been found to be problem dependent and hard to predict. The third approach, whilst still introducing a new and slightly problem dependent 'time since last change' threshold parameter, has been found to be the most robust of the three approaches across test problems and is the one focussed on in this paper.

While other approaches such as "Charged Particle Swarm" [3] and "Waves of Swarm Particles" [4] have previously been proposed, these algorithms respectively apply a repulsive or attractive force between particles, rather than a repulsive force between a particle and its personal best point as proposed in this work.

2 Details of the Revised Algorithm

2.1 Formulae

In the standard particle swarm algorithm, the ith particle's vector for the current iteration (at $T + t$) is based the position ($\overline{X}_i$) and velocity ($\overline{V}_i^T$) of this particle last iteration, the positions of two points of particular merit found in this or previous iterations and a set of pre-selected parameters. The two positions are the global best position ($\overline{GB}$), the best position found by any particle so far and which is common to all particles in the swarm, and the personal best position found by this particle ($\overline{PB}_i$). Note that for at least one particle these two positions will be the same.

Pre-selected parameters are the momentum for each particle (M), the weighting factor (G) for the attraction to the global best position ($\overline{GB}$) and the weighting factor (P) for the attraction to the personal best position found by this particle ($\overline{PB}_i$).

From these, the particle's new vector is calculated as:

$$\overline{V}_i^{T+t} = M\overline{V}_i^T + (1-M)(r_1 G(\frac{\overline{GB} - \overline{X}_i}{t}) + r_2 P(\frac{\overline{PB}_i - \overline{X}_i}{t})) \tag{1}$$

Where: r_1 and r_2 are random numbers between zero and one and
       t is the interval since the swarm's last update (typically treated as one).

The modified swarm algorithm utilises the same factors to calculate velocity in the first form of behaviour, but introduces a slightly modified second equation for the second form of behaviour.

$$\overline{V}_i^{T+t} = M\overline{V}_i^T + (1-M)(r_1 G(\frac{\overline{GB} - \overline{X}_i}{t}) - r_2 P(\frac{\overline{PB}_i - \overline{X}_i}{t})) \tag{2}$$

A particle's velocity is calculated using either (1) or (2) depending on the particle's current behaviour.

The new position of the particle is calculated using (3) on the un-biologically plausible hypothesis that the particle's velocity does not alter between iterations.

$$\overline{X}_i^{T+t} = \overline{X}_i^T + \overline{V}_i^{T+t} \tag{3}$$

2.2 Effects of Repulsion on Particle Movement

The movement of particles within a swarm can be viewed in two broad phases: *exploration* of problem space and *exploitation* of a known optimum. In the first phase particles move through the problem space seeking optima and converging towards the best point(s) located. Once a local optimum is located the swarm condenses to better determine the position and quality of this local optimum.

While the swarm is searching the repulsion has little effect on particle movement. Since particles are generally travelling towards the current global best and their personal bests are behind or on their current position the only change is an irregular small increase in velocity (Fig. 1). This allows the swarm to continue to act as it normally would while searching, until an optimum point is found.

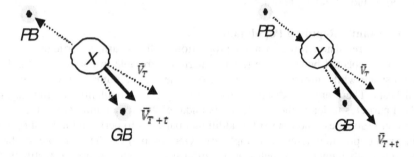

Fig. 1. The differing vector sums for searching particles

Once the swarm has begun exploiting a local optimum the repulsion personal force shows an effect on the particles (Fig. 2). With each particle's personal best tending to approach the global best the repulsive personal force will in effect cancel out the attractive global best force as a particle's personal best converges with the global best. Variances in the random factors will sometimes attract a particle to the converged point and sometimes push it away leading the particle to explore broader arcs of problem space until either a new global best is found or the particle's momentum is entirely spent.

The personal best repulsion force also leads to some particles maintaining a much higher velocity once a global best is found. When the personal and global best are closely grouped on the particle's initial approach to an optimum the repulsed particle maintains, and even increases, its velocity while passing the optimum point. Rather than all the particles settling on a point rapidly, this increased velocity produces a scattering effect, extending the swarm to new areas of the problem.

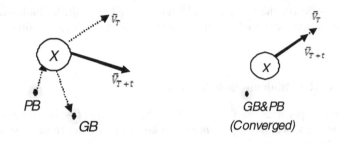

Fig. 2. Vector positions during local optimisation

2.3 Methods of Application

If the repulsion force is applied to all particles at all times the swarm's ability to perform local optimisation is adversely affected; some convergence is required to determine the true quality of a local optimum. Care must be taken as to how the repulsion force is applied to the swarm in order to achieve a balance between exploiting a known point and exploring new areas.

2.3.1 Constantly Repulsed Population

Applying the repulsion force to a set proportion of the swarm's particles provides some of the expected benefits: the repelled particles are able to continue a search of problem space while attracted particles perform the local optimisation of the current global best. There are two disadvantages to using a set population: a tendency for repulsed particles to leave the problem's boundaries as momentum carries them past the current global best and a need for additional particles to perform the local optimisation as the problem rises in complexity. Once a particle has left the problem boundaries it no longer contributes useful information to the swarm, effectively reducing the swarm's size for as long as it remains outside of the boundaries. If too great a proportion of the population has a repelling force applied to it, the ability to perform local optimisation is negatively affected, especially as problems rise in complexity and dimensionality. The appropriate proportion to repel is closely related to the problem being optimised, requiring prior knowledge of the problem to make best use of the swarm.

2.3.2 Trigger Conditions

Allowing the swarm to modify the roles of particles as required during the search removes the need to pre-assign particles as repulsion or attraction, but requires the definition of trigger conditions for a particle state change. Ideally such a definition should require no prior knowledge of the problem.

The main trigger condition used in this paper is to place a threshold on the personal best age of a particle (the time since the last change to the personal best). Particles follow behaviour two (repelled from their personal best) below this threshold age and behaviour one (attracted to personal best) above it (Fig. 3).

In problems with a bounded search space a particle also becomes attracted to its personal best at any time that it leaves the problem's boundaries, limiting a particle's

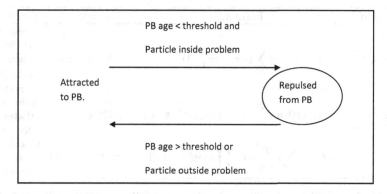

Fig. 3. Particle behaviour state diagram

excursion from problem space as it is attracted to both the global and personal best after crossing the problem's boundary. The boundary condition has the added benefit of prompting the particle to return to the global best point and continue past it in a new direction, again extending the search performed.

The personal best trigger condition leads to the swarm being scattered across a larger area while searching; as some particles are attracted to their personal best and others are repulsed the swarm expands and contracts erratically, providing a broader search pattern. When an optimum is located the particles are able to travel much farther past it than a standard swarm, scattering across a larger area of the problem. If a particle does not find a new personal best before reaching the threshold age, the attraction turns it back towards the global best. Although each particle is switching from repulsion to attraction based on its own personal best age, the swarm is still sharing global best information, maintaining the swarm's collective optimisation capabilities. Over time the swarm does converge, as particles which have settled near the global best remain there unless a new global best is located, however this total convergence takes a much longer time than would be taken by a standard swarm.

There are certain practical limitations on the value range of the personal best age threshold. If the threshold is set too low the swarm will function identically to a normal swarm, the particles will not have the opportunity to travel significantly past the global best and will not maintain the velocity required to search further into problem space. If the threshold is set too high the particles spend the majority of their time repulsed from their personal bests, leading to reduced local search capability. While ideal appropriate threshold values are somewhat related to the problem being optimised, the problem dependence is not strong and certain values have been observed to perform well for all test problems used.

3 Methodology

The modified swarm has been tested on a number of benchmark optimisation problems in varying dimensions. Results for the following problems (equations and limits sourced from [5]) are presented:

Function	Formula	Range	Optimum
Schwefel's	$f(x) = \sum_{i=1}^{n}(-x_i \cdot \sin(\sqrt{\|x_i\|}))$	$-500 \le x_i \le 500$	*Maximum. f(x)* = 412.9829×n where x_i = −412.983 for i=1:n
Rastrigin's	$f(x) = 10 \cdot n + \sum_{i=1}^{n}(x_i^2 - 10 \cdot \cos(2 \cdot \pi \cdot x_i))$	$-5.12 \le x_i \le 5.12$	*Minimum. f(x)* = 0 where x_i 0 for i=1:n
Griewangk's	$f(x) = \sum_{i=1}^{n}(\frac{x_i^2}{4000}) - \prod_{i=1}^{n}(\cos(\frac{x_i}{\sqrt{i}})) + 1$	$-600 \le x_i \le 600$	*Minimum. f(x)* = 0 where x_i 0 for i=1:n

The main challenge in optimising Schwefel's function is the distance between the various peaks – after locating one of the lower peaks the swarm must search across a large distance to find a better alternative. The close clustering of minima across Rastrigin's function raises a different set of requirements to Schwefel's function. Rather than the particles needing to search a large area to find the next optima they are required to perform a concerted local search, overcoming the steepness of peaks between minima. Finally, the extremely large number of local minima in Griewangk's function requires the swarm to perform small incremental improvements, rather than being able to make large positive movements.

Each problem was subjected to 100 independent optimisation runs for each of the swarm configurations, the results are presented as the average obtained for these runs. As each of the test problems is bounded, their maximum value may be used to scale results from 0-100%. For maximisation problems a solution of 100% represents the global maximum being found. For minimisation problems the percentage represents how close to the true minimum the solution is, with 100% being the global minimum.

An optimisation run was judged complete when the global best was unchanged in 100 iterations, using a common change tolerance of ±0.001. The iteration value used for results is the iteration on which the global best was last changed.

4   Results

Schwefel's Function. Especially in higher dimensions it is common for a standard swarm to locate one of the sub-optimal local maxima and settle on this point. The use of a conditional repulsive force shows a marked improvement in result quality. Fig. 4 shows results from two dimensions up to 100 dimensions. In general a PB age threshold of 40 performs best and a standard swarm worst.

Fig. 5 shows the progress of a standard and a modified swarm (personal best age limit of 40 chosen based on experimental results) through a representative optimisation. The modified swarm exhibits a lower value early in the optimisation as the swarm concentrates on exploration over exploitation of known optima. The modified swarm continues to produce new data long after the standard swarm has converged on a solution, resulting in a final result of higher value, although a larger number of iterations are required.

The modified swarm also shows an improvement in ability to locate the problem's global maximum (Table 1). While this does not extend significantly into higher dimensions it does highlight the modified swarm's ability to locate successive optima across the problem.

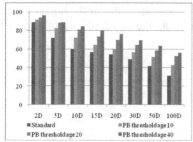

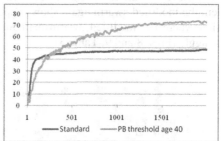

Fig. 4. Average best value found across 100 optimisation runs of Schwefel's function

Fig. 5. Average value of swarm across an entire optimisation run of Schwefel's function in 20 dimensions

Rastrigin's Function. Again the modified swarm shows stronger results than the base swarm, especially as the number of dimensions increases (Figs. 6 and 7). It is also worth noting that the modified swarm performed best with a lower threshold age than Schwefel's function, although the difference is not large. Due to the differing topography of the problems the different threshold values allow the modified swarm to be aimed at a shorter or longer ranged search, although a threshold of 20 appears an acceptable compromise for both problems.

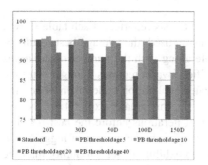

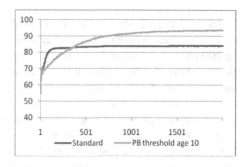

Fig. 6. Average best value found across 100 optimisation runs of Rastrigin's function

Fig. 7. Average value of swarm across an entire optimisation run of Rastrigin's function in 100 dimensions

Similarly the modified swarm is slower in reaching a comparable value to the standard swarm, but continues improving well past the standard swarm.

Comparison of the standard and modified swarm's results for Rastrigin's function (Table 2) serve to highlight the cost and benefit of a repulsive force. As the number of problem dimensions increases the standard swarm shows a reduction in answer quality but little increase in the number of iterations required. Conversely the modified swarm shows significantly less loss in answer quality although with a rapidly climbing number of iterations required.

Griewangk's Function. Although the basic swarm is capable of optimizing this problem the modified swarm again shows superior results, especially in the higher dimension where the modified swarm maintains quality as the standard swarm begins to struggle (Figs. 8 and 9, Table 3).

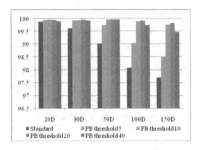   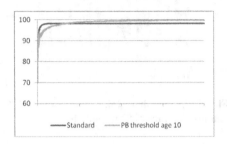

Fig. 8. Average best value found across 100 optimisation runs of Griewangk's function

Fig. 9. Average swarm value across an optimisation run of Griewangk's function in 100 dimensions

Consistent with the other benchmark problems the modified swarm shows a poorer starting but a better finishing performance.

4.1 General Analysis Point

Tables 1 - 3 provide detailed information about the tests reported. Each table shows, for a benchmark problem, the number of dimension used, the average best result for the 100 runs at a percentage of what is possible along with the number of iterations taken to obtain the best result. Clearly the modified swarm requires a longer run time than the standard swarm; the price for the greater exploration undertaken. Since the order of exploration of the optima is unpredictable, this greater exploration also results in a much greater variance in the number of iterations before the best result is obtained. Observation of the modified swarm, along with comparison of logged data, shows that the early stages of the modified swarm exhibit greater variability than the standard swarm. The early repulsion results in many of the particles being widely scattered, producing a chance that the swarm will find a strong optimum early but also slowing the initial optimisation of any point found.

Observation of the modified swarm behaviour also shows a strong capacity to find optima sequentially; that is, first optimising one point then finding another and moving on. This is especially true in higher dimensions where it is difficult for a particle to achieve a positive movement in multiple dimensions at once. The following two images (Fig.10 and Fig. 11) plot all points visited in two arbitrarily chosen dimensions from an optimisation of Schwefel's function in 50 dimensions. Fig. 10 shows the standard swarm, while Fig. 11 shows the modified swarm with a personal best age threshold of 40. This provides a clear demonstration of the more extensive search provided by the modified swarm.

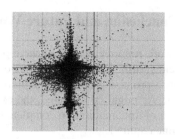

Fig. 10. Standard swarm's search pattern   **Fig. 11.** Modified swarm's search pattern

Table 1. Schwefel's function. Detailed results for a standard swarm and (after the slash) a modified swarm using a personal best age threshold of 40.

Dims	Av Best (%)	Std. Dev.	Global Best Found (%)	Av Iterations	Std. Dev.
2	89.05/96.32	11.16/6.55	42/75	54.45/142.02	14.45/82.89
5	71.90/88.88	10.99/6.45	2/10	95.41/365.35	22.38/157.13
10	59.98/84.57	8.53/6.60	0/1	165.00/763.86	71.37/272.85
15	56.50/80.29	7.59/6.79	0/0	269.54/1119.82	95.83/392.54
20	54.41/76.53	6.17/7.24	0/0	325.73/1511.69	111.93/490.32
30	49.32/69.75	5.89/6.18	0/0	350.06/1965.33	114.12/709.60
50	41.74/63.81	4.97/5.86	0/0	388.69/2950.32	136.39/1234.78
100	31.34/56.36	3.24/7.00	0/0	432.40/4810.63	147.09/2220.19

Table 2. Rastrigin's function. Detailed results for standard swarm and (after the slash) a modified swarm using a personal best age threshold of 10.

Dims	Av Best (%)	Std. Dev.	Av Iterations	Std. Dev.
20	95.29/96.19	1.72/1.35	282.17/873.22	113.70/182.08
30	93.99/95.51	1.42/1.29	293.17/1240.06	76.25/252.92
50	90.89/95.07	1.29/1.24	307.73/1862.21	98.45/340.17
100	86.06/94.86	1.25/1.04	312.53/3629.00	103.26/678.31
150	83.84/94.08	1.08/0.78	284.91/4833.94	88.30/821.51

Table 3. Griewangk's function. Detailed results for standard swarm and (after the slash) a modified swarm using a personal best age threshold of 10.

Dims	Av Best (%)	Std. Dev.	Av Iterations	Std. Dev.
20	99.88/99.96	0.05/0.02	140.21/559.17	11.69/211.53
30	99.61/99.93	0.16/0.06	159.55/312.50	9.19/121.35
50	99.02/99.96	0.27/0.01	166.69/1010.13	6.87/164.04
100	98.07/99.89	0.35/0.05	166.42/1907.17	6.12/287.56
150	97.70/99.76	0.27/0.08	164.32/2618.78	5.81/421.12

5 Conclusion and Future Work

The work reported in this paper represents a proof of concept for a heterogeneous particle swarm. This algorithm has only been tested on a limited set of problems but shows a promising ability to explore a problem to a greater degree than a standard swarm. As might be expected, the extended search comes at a cost of increasing the amount of time required to complete the optimisation along increasing the variance in the time required.

The introduction of a new variable in the personal best age threshold shows some problem dependency, although certain values performed well on all test problems. Further work, in particular tests on a wider range of functions, will be required to establish optimum values for this threshold and to uncover the extent of its problem dependency.

References

1. Kennedy, J., Eberhart, R.C.: Particle Swarm Optimization. In: Proc. IEEE International Conference on Neural Networks, Perth Australia, vol. IV, pp. 1942–1948. IEEE Service Centre, Piscataway (1995)
2. Engelbrecht, A.P.: Fundamentals of Computational Swarm Intelligence. Wiley, West Sussex (2006)
3. Blackwell, T.M., Bentley, P.: Don't push me! Collision-avoiding swarms. In: CEC 2002. Proceedings of the 2002 Congress on Evolutionary Computation, vol. 2, pp. 1691–1696 (2002)
4. Hendtlass, T.: WoSP: a multi-optima particle swarm algorithm. In: CEC 2005 Proceedings of the 2005 Congress on Evolutionary Computation, pp. 727–734 (2005)
5. Pohlheim, H.: GEATbx: Genetic and Evolutionary Algorithm Toolbox for use with MATLAB Documentation (December 2008),
http://www.geatbx.com/docu/index.html

An Analysis of Locust Swarms on Large Scale Global Optimization Problems

Stephen Chen

School of Information Technology, York University
4700 Keele Street, Toronto, Ontario M3J 1P3
sychen@yorku.ca

Abstract. Locust Swarms are a recently-developed multi-optima particle swarm. To test the potential of the new technique, they have been applied to the 1000-dimension optimization problems used in the recent CEC2008 Large Scale Global Optimization competition. The results for Locust Swarms are competitive on these problems, and in particular, much better than other particle swarm-based techniques. An analysis of these results leads to a simple guideline for parameter selection in Locust Swarms that has a broad range of effective performance. Further analysis also demonstrates that "dimension reductions" during the search process are the single largest factor in the performance of Locust Swarms and potentially a key factor in the performance of other search techniques.

1 Introduction

The parameters and optimization problems of the CEC2008 Large Scale Global Optimization (LSGO) competition [12] test two key aspects of heuristic search techniques: scalability and robustness. The complexity of an optimization problem (e.g. the number of local optima) can increase exponentially with dimensionality, so search techniques that are very effective on low-dimension problems may not be efficient enough to handle higher dimension problems. Further, since the performance of a search technique on a given problem may not be representative of its general performance, the LSGO competition uses seven problems with various search space characteristics.

Locust Swarms were specifically developed for non-globally convex search spaces [3]. Due to the diverse locations of the best local optima in these search spaces, it is not possible to "converge" to the global optimum from all parts of the search space. Thus, Locust Swarms explore multiple optima using a "devour and move on" strategy. After a part of the search space is devoured (to find a local optimum), scouts are used to find a promising new area in the search space. Depending on the parameters used, Locust Swarms can display non-convergent and/or non-divergent search behaviours.

Since most of the problems used in the LSGO have globally convex (or unimodal) search spaces, the present implementation of Locust Swarms performs a more convergent/less divergent search. Nonetheless, each swarm still converges to a single

K. Korb, M. Randall, and T. Hendtlass (Eds.): ACAL 2009, LNAI 5865, pp. 211–220, 2009.
© Springer-Verlag Berlin Heidelberg 2009

local optimum, and the necessary balance between exploration and exploitation is achieved by exploring a large number of optima. Exploration (i.e. swarm restarts) is enhanced through the use of "scouts". Scouts perform a preliminary coarse search [5] to identify the most promising areas of a search space for further exploration and exploitation by the subsequent swarms.

Standard particle swarm optimization [8] cannot guarantee convergence to the globally optimal solution in a multi-modal search space, so some form of multi-optima modification can be useful. For example, DMS-PSO [17] reassigns the particles into new swarms which promotes exploration among previously found local optima (like a memetic algorithm), and WoSP [6] occasionally expels particles in unspecified directions to search for new optima. The use of scouts helps to find promising areas where new swarms can be targetted, and this is one of the key ideas behind the development of Locust Swarms [3]. A new analysis shows that the effectiveness of scouts depends greatly on their use of "dimension reductions" to simplify/decompose the search space, and this insight invites a new analysis on the existence and effects of dimension reductions in other search techniques.

To build towards this development, this paper begins with an introduction of Locust Swarms in section 2. Their implementation and performance on the LSGO competition problems are then presented in section 3. A robust guideline for parameter selection in Locust Swarms is then proposed and demonstrated in section 4. The two components of Locust Swarms (i.e. scouts and swarms) are isolated in section 5. Since the performance of the individual components was better than expected, the role of dimension reductions is studied in section 6. The larger than expected effects of dimension reductions is the focus of the discussion in section 7 before the summary occurs in section 8.

2 Locust Swarms

The development of Locust Swarms [3] builds upon particle swarm optimization (PSO) [8] and a multi-optima variation called WoSP (for Waves of Swarm Particles) [6]. The restart mechanism in WoSP is similar to a physics-based short-range force that repulses particles that have converged upon one another due to the attractive effects of the primary long-range forces. The use of discrete time step evaluations can further exaggerate the effects of the short-range force causing particles to not simply repel each other enough to avoid complete convergence, but in fact to expel each other into entirely different areas of the search space.

The restart mechanism for Locust Swarms is based on time. As time is spent "devouring" one area of the search space (i.e. trying to find the exact local optimum), it becomes less and less likely that additional efforts in exploitation will be productive. Thus, there comes a time to "move on" to another area of the search space. This "devour and move on" process is akin to the swarming phase of grasshoppers that become migratory after an intense exploitation of an area's resources.

The two phases of Locust Swarms – migrating and devouring or exploring and exploiting – have been designed and coordinated using a coarse search-greedy search framework. The coarsest coarse search is a single random point, and this leads to a simple random restart of the greedy search technique. A slightly less coarse search technique is a crossover-style operator, and the resulting system might be fashioned into a memetic algorithm [11]. However, memetic algorithms keep a population of (local) optima, and the resulting system tends to converge amongst them.

The coarse search phase of Locust Swarms has been implemented using "smart" start points [5]. In continuous search spaces, the fitness of a random point can be expected to have a correlation to the fitness of its nearby local optima, so random search can help identify promising areas for further exploration and exploitation. The role of this coarse search can be viewed as "scouts" which identify promising new areas for the swarm to exploit. The resulting system of scouts and swarms has three sets of design parameters: scout parameters, swarm parameters, and general parameters.

$$FE = N * (S + L*n) \tag{1}$$

The primary role of the general parameters is to effectively allocate the allowed function evaluations (FE). In (1), N is the number of swarms – each of which explores a (local) optimum, S is the number of scouts, L is the number of locusts, and n is the number of iterations each swarm is allowed to use to find a local optimum.

$$delta = \pm\ range * (gap + \text{abs}(\text{randn}()*spacing)) \tag{2}$$

The scouts perform a coarse search around the previous optima by adding a $delta$ calculated in (2) to a small number of terms in a solution vector. This dimension reduction (DIM_r) allows Locust Swarms to exploit some information from the previous optima. The effect of DIM_r is a problem decomposition – exploration occurs only in the selected DIM_r number of dimensions. Exploration can also be increased by requiring scouts to be a minimum gap from the previous optima (with some additional $spacing$ as appropriate). Note: $range$ is the range for (the given dimension of) the search space and $\text{randn}()$ is the MATLAB® function for a normally distributed random number.

$$velocity_0 = launch*(location_0 - previousoptimum) \tag{3}$$

$$location_{i+1} = location_i + step*velocity_i \tag{4}$$

$$gravityvector_i = globalbestlocation - location_i \tag{5}$$

$$velocity_{i+1} = M*velocity_i + G*gravityvector_i \tag{6}$$

Keeping the L best scouts for the swarm, the initial velocities (3) launch the locusts away from the previous optima with a speed affected by the $launch$ parameter. Each locust moves a $step$ fraction of its velocity during each iteration i (4), and this velocity is updated to reflect a "gravitational" attraction to the "global best" location found by the (current) swarm (5). The rate of convergence is affected by the momentum (M) and gravity (G) parameters in (6). Note: Locust Swarms do not use an attraction to each particle's "local best" (e.g. $pbest$ in [8]) since multiple swarms are used to balance exploration with exploitation.

3 LSGO Competition Results

The competition on Large Scale Global Optimization (LSGO) was held at CEC2008 [12]. The seven benchmark optimization problems used are Sphere (F_1), Schwefel's Problem 2.21 (F_2), Rosenbrock (F_3), Rastrigin (F_4), Griewank (F_5), Ackley (F_6), and "FastFractal" (F_7). Although the competition included results for $D = 100$, 500, and 1000 dimensions, this paper will focus only on the problems at $D = 1000$ dimensions. Competition results were also reported for $50*D$, $500*D$, and $5000*D$ function evaluations (FEs), but this paper will only provide final comparisons at $FE = 5000*D = 5$ million.

To fit this constraint on function evaluations, 1000 function evaluations are used in N = 5000 swarms (each swarm explores an optimum which may or may not be unique). Balancing the computation effort between the scouts and the swarms, the remaining parameters in (1) are set to $S = 500$ scouts, $L = 5$ locusts, and $n = 100$ iterations. The swarm parameters of momentum $M = 0.5$, gravity $G = 0.5$, and step = 0.6 allow a small regional exploration before the swarm converges towards a local optimum at the end of the allowed number of iterations. Due to the globally convex/unimodal landscapes for most of the LSGO problems, the search is concentrated by using small values for the gap (0.0) and spacing (0.1) parameters in (2). Balancing the initial concentration of the scouts, a larger value for the launch (5.0) parameter is used in (3) to cause the swarm to perform some regional search before converging to an optimum. Lastly, each scout changes DIM_r = 3 dimensions/terms from the previous solution – see (2).

Table 1. Results for Locust Swarms on the CEC2008 LSGO problems in competition format. For the 25 independent trials on each problem, the best (1), seventh best (7), median (13), seventh worst (19), worst (25), mean (m), and standard deviation (sd) are reported. For F_1 to F_6, values are error values from the known optimum, and F_7 reports the actual function value since the exact optimum is unknown.

	F_1	F_2	F_3	F_4	F_5	F_6	F_7
1	1.61e-06	4.77e+01	1.03e+03	0.79e-06	0.92e-07	5.43e-05	-1.45e+4
7	1.97e-06	5.02e+01	1.12e+03	1.02e-06	1.18e-07	5.89e-05	-1.45e+4
13	2.23e-06	5.21e+01	1.18e+03	1.26e-06	1.35e-07	6.10e-05	-1.45e+4
19	2.44e-06	5.42e+01	1.23e+03	1.40e-06	1.44e-07	6.65e-05	-1.45e+4
25	3.17e-06	5.93e+01	1.32e+03	1.55e-06	1.73e-07	7.29e-05	-1.44e+4
m	2.25e-06	5.22e+01	1.18e+03	1.22e-06	1.32e-07	6.25e-05	-1.45e+4
sd	0.39e-06	0.27e+01	0.07e+03	0.20e-06	0.24e-07	0.47e-05	34

In Table 1, the results for Locust Swarms on the LSGO problems are given in the LSGO competition format. The total computational time (7 problems, 25 trials each) on a 3 GHz P4 PC running MATLAB® was approximately 250 hours. On an absolute level, the results for Locust Swarms look reasonable with near optimal solutions on four of the problems with known optima (F_1-F_6). For relative performance, the mean results (m) are compared against the other competition methods (i.e. MLCC [15], EPUS-PSO [7], jDEdynNP-F [2], UEP [10], MTS [13], DEwSAcc [16], DMS-PSO [17], and LSEDA-gl [14]) in Table 2.

Table 2. Comparison of mean results for Locust Swarms versus the LSGO competition entrants for $D = 1000$ dimensions. Bolded values indicate the best result for each problem.

	F_1	F_2	F_3	F_4	F_5	F_6	F_7
MLCC	8.46e-13	1.09e+02	1.80e+03	1.37e-10	4.18e-13	1.06e-12	**-1.47e+4**
EPUS-PSO	5.53e+02	4.66e+01	8.37e+05	7.58e+03	5.89e+00	1.89e+01	-6.62e+3
jDEdynNP-F	1.14e-13	1.95e+01	1.31e+03	2.17e-04	3.98e-14	1.47e-11	-1.35e+4
UEP	5.37e-12	1.05e+02	1.96e+03	1.03e+04	8.87e-04	1.99e+01	-1.18e+4
MTS	**0.00e+00**	4.72e-02	**3.41e-04**	**0.00e+00**	**0.00e+00**	1.24e-11	-1.40e+4
DEwSAcc	8.79e-03	9.61e+01	9.15e+03	1.82e+03	3.58e-03	2.30e+00	-1.06e+4
DMS-PSO	**0.00e+00**	9.15e+01	8.98e+09	3.84e+03	**0.00e+00**	7.76e+00	-7.51e+3
LSEDA-gl	3.22e-13	**1.04e-05**	1.73e+03	5.45e+02	1.71e-13	**4.26e-13**	-1.35e+4
LocustSwarm	2.25e-06	5.22e+01	1.18e+03	1.22e-06	1.32e-07	6.25e-05	-1.45e+4

The relative performance of Locust Swarms is generally good with high rankings on problems F_3, F_4, and F_7 (see Table 3), reasonable absolute performance on problems F_1, F_4, F_5, and F_6 (see Table 2), and weak relative and absolute performance on only the highly artificial Schwefel Problem in F_2. After MTS, the clear winner of the competition, Locust Swarms are near the middle tier of entrants (i.e. LSEDA-gl, jDE-dynNP-F, and MLCC), and above the bottom tier of entrants (i.e. DMS-PSO, DEw-SAcc, UEP, and EPUS-PSO). Note: DMS-PSO uses local optimization which is likely responsible for its anomalous performance on F_1 and F_5.

Table 3. Performance rankings (1-9) of results in Table 2

	F_1	F_2	F_3	F_4	F_5	F_6	F_7	mean
MLCC	5	9	5	2	5	2	1	4.1
EPUS-PSO	9	4	8	8	9	8	9	7.9
jDEdynNP-F	3	3	3	4	3	4	4	3.4
UEP	6	8	6	9	7	9	6	7.3
MTS	1	2	1	1	1	3	3	**1.7**
DEwSAcc	8	7	7	6	8	6	7	7.0
DMS-PSO	1	6	9	7	1	7	8	5.6
LSEDA-gl	4	1	4	5	4	1	4	3.3
LocustSwarm	7	5	2	3	6	5	2	4.3

Extending from these results, two additional sets of experiments have been conducted. In one set, Locust Swarms with parameters specified for a non-globally convex search space have been able to find best-overall solutions for the FastFractal problem F_7 – see [4]. The other set described in this paper focuses on analyzing the performance advantage that Locust Swarms have over the other PSO-based methods in the competition (i.e. EPUS-PSO and DMS-PSO).

4   Parameter Analysis

During preliminary parameter selection, results tended to improve with increasing N. This observation led to the hypothesis that the key advantage of Locust Swarms

compared to the other PSO-based techniques was the rapid exploration of many optima. To examine this hypothesis, additional tests with N = 1000, 2500, 10,000, and
20,000 swarms were performed. Note: due to the excessive computational costs for
running 25 trials, 11 trials per problem were performed and the median results are
compared (to reduce the effects of outliers).

Preliminary parameter selection also suggested that the best performance of Locust
Swarms occurs when the computational resources are divided equally between the
scouts and the swarms, and this was achieved with the values for S, L, and n shown in
Table 4. Giving a swarm n iterations to converge, exploration can be increased with
larger M, smaller G, and a larger *step*, and exploitation can be increased with a smaller M, larger G, and a smaller *step*. The best preliminary results were obtained with M
= G = 0.5, so only the *step* parameter was changed to adjust for different values of n
(see Table 4).

Table 4. Parameter values used with various values of N. $FE = N *(S+L*n)$ is held constant at 5
million.

N	S	L	n	M	G	*step*
1,000	2500	5	500	0.5	0.5	0.9
2,500	1000	5	200	0.5	0.5	0.8
5,000	500	5	100	0.5	0.5	0.6
10,000	250	5	50	0.5	0.5	0.5
20,000	125	5	25	0.5	0.5	0.5

The results in Table 5 indicate that Locust Swarms have a broad range of stable
performance with respect to N, the number of optima explored. Therefore, in terms of
the tested hypothesis, these experiments produce a negative result – N is not the key
factor affecting the performance of Locust Swarms. However, the stable performance
of Locust Swarms across all parameter settings suggests a simple guideline for parameter selection. Specifically, N has limited effects, $S = L*n$ and $M = G = 0.5$ work
well, and *step* can be varied with n to ensure swarm convergence.

Table 5. Analysis of the role of N on the performance of Locust Swarms. Median results are
used to reduce the effects of outliers.

N	F_1	F_2	F_3	F_4	F_5	F_6	F_7
1,000	1.83e-07	7.08e+01	1.32e+03	4.25e-07	1.22e-08	4.98e-05	-1.44e+4
2,500	4.12e-10	6.58e+01	1.16e+03	3.79e-10	6.70e-10	1.16e-06	-1.45e+4
5,000	2.23e-06	5.21e+01	1.18e+03	1.26e-06	1.35e-07	6.10e-05	-1.45e+4
10,000	3.78e-07	4.82e+01	1.29e+03	8.08e-08	2.27e-08	7.83e-06	-1.45e+4
20,000	3.73e-02	4.63e+01	1.78e+03	1.92e-02	2.03e-03	8.14e-03	-1.45e+4

5   Component Analysis

Since the best performance of Locust Swarms tends to occur when the scouts and
the swarms have an equal opportunity to make contributions, it was hypothesized
that there may be large synergies between these two components. The following

experiments isolate the performance contributions of the scouts and the swarms. Without the swarms (i.e. $n = 0$), the system behaves like a simplified $(1,\lambda)$-evolution strategy (ES) where $\lambda = S$, the number of scouts. Although an actual evolution strategy would be expected to perform better through features like adaptive mutation rates σ [1], the isolated performance of the scouts can still represent a performance baseline for evolution strategies. Without the scouts (i.e. $S = L$), the system behaves like a simple PSO with regular restarts (and minimal randomizations of particle trajectories and no local best attractors). In both component isolations, $N = 10,000$ is used (versus $N = 5,000$ for Locust Swarms) so that the remaining component can use the function evaluations previously allocated to the eliminated component.

The results in Table 6 indicate that Locust Swarms tend to perform better than either component independently. The two exceptions are F_2 where the scouts perform better independently and F_3 where the swarms perform better independently. Similarly, it can be seen in Table 4 that the performance of Locust Swarms on F_2 improves with increasing N (which emphasizes the role of restarts/scouts) and it tends to improve on F_3 with increasing n (which emphasizes the role of swarms).

Table 6. Isolation of the scout and swarm components in Locust Swarms. All components use $DIM_r = 3$.

	F_1	F_2	F_3	F_4	F_5	F_6	F_7
LocustSwarm	2.23e-06	5.21e+01	1.18e+03	1.26e-06	1.35e-07	6.10e-05	-1.45e+4
Scouts (ES)	2.11e+00	3.24e+01	3.00e+03	1.03e+00	1.20e-01	7.28e-02	-1.45e+4
Swarms	3.56e-01	9.73e+01	1.08e+03	3.32e+02	2.86e-01	2.64e-02	-1.43e+4

In terms of testing the hypothesis that there are large synergies between the scouts and the swarms, the result is essentially negative. Although the combination of the two components tends to improve the performance of Locust Swarms, the performance of each component independently is already quite good. In particular, the isolated performance of the swarms component in Locust Swarms is already better than EPUS-PSO on problems F_1, F_3, F_4, F_5, F_6, and F_7 and DMS-PSO on problems F_2, F_3, F_4, F_6, and F_7 (see Tables 2 and 3).

6   Dimension Reductions

The key difference between the above isolation of the swarms component in Locust Swarms and the PSO-based competition entrants is the use of dimension reductions during the restart process – see (2). To isolate this component, the experiment from section 5 is repeated with $DIM_r = D$ – every term from the previously found local optimum is altered by (2) during scout development. The results for these experiments are presented in Table 7.

Without dimension reductions (i.e. $DIM_r = D$), the performance of Locust Swarms degrades completely. Although some benefits remain from the combination of scouts and swarms, the performance gains realized through their combination pales in comparison to the effects of dimension reductions. It is quite clear that DIM_r is the single most important parameter in Locust Swarms.

Table 7. Analysis of the effects of dimension reductions. $DIM_r = D = 1000$ is used instead of $DIM_r = 3$ for corresponding results in Table 6.

	F_1	F_2	F_3	F_4	F_5	F_6	F_7
LocustSwarm	6.26e+05	1.53e+02	1.79e+11	1.73e+04	5.71e+03	2.13e+01	-7.11e+3
Scouts (ES)	3.17e+06	1.74e+02	2.84e+12	1.98e+04	2.87e+04	2.14e+01	-6.65e+3
Swarms	9.74e+05	1.62e+02	4.93e+11	2.00e+04	8.98e+03	2.13e+01	-7.00e+3

7 Discussion

Many traditional benchmark functions (e.g. Rastrigin F_4 and Ackley F_6) are formed from a one-dimensional equation that is scaled by summing the result of D one-dimensional sub-problems. Since the resulting function is separable, the ability to solve it one dimension at a time causes the function to be both decomposable and to have an axial bias. Decomposability and axial bias can be reduced by rotating the search space, and this is done in the FastFractal problem F_7 [9]. However, the Fast-Fractal problem is built with a two-dimensional base function, so $DIM_r = 3$ may still be able to decompose the large scale problem and solve it one (rotated) two-dimensional problem at a time. The strong performance of all tested variations with $DIM_r = 3$ (e.g. see Table 6) supports this possibility.

Examining several other LSGO competition entrants, it can be seen that MTS includes a one dimensional local search [13], the differential evolution-based jDE-dynNP-F and DEwSAcc have evolvable crossover factors [2][16], and the co-operative co-evolution technique MLCC employs problem decomposition [15]. These search techniques tend to perform much better than the PSO-based DMS-PSO and EPUS-PSO which have no discernable ability to exploit problem decomposition and/or axial bias [7][17]. To learn more about the roles of problem decomposition and axial bias in LSGO performance, it would be useful to know the evolved crossover factors in jDEdynNP-F and DEwSAcc.

Compared to evolvable crossover factors in differential evolution and adaptive mutation rates in evolution strategies [1], dimension reductions offer a direct and explicit means to focus search efforts. This focus is highly dynamic – a different set of DIM_r dimensions can be changed in each scout solution. This probabilistic decomposition can be highly effective on a partially decomposable problem like FastFractal – each selection of dimensions has a chance of matching one of the two-dimensional sub-problems. This probabilistic decomposition also seems to work on a fully-rotated benchmark problem (which should eliminate axial bias and decomposability) [3] and a real-world problem with no known axial bias or axial decompositions [4]. More experimentation on the performance of "dimension reductions" in rotated/non-decomposable search spaces is required.

The value of DIM_r can also be seen as a means to balance exploration and exploitation – large values promote exploration and small values promote exploitation. In Locust Swarms, larger vales for DIM_r can accelerate the early stages of the search process, but smaller values for DIM_r allow more convergence during later stages. The results of section 4 create a simple framework for the specification of the other Locust Swarm parameters. Future work will focus on developing guidelines for DIM_r, testing self-adaptation of the DIM_r parameter, and examining the role of (exploiting) problem decomposition and axial bias in other search techniques.

8 Summary

The performance of Locust Swarms on the CEC2008 LSGO problems is respectable in comparison to the best competition entrants and much better than the PSO-based entrants. The key advantage of Locust Swarms over these other PSO-based entrants appears to be the use of "dimension reductions" which allow the decomposability and the axial bias of the test functions to be exploited. Many of the best LSGO competition entrants also had the ability to exploit these features, so it would be interesting to know how much they actually did. It would also be interesting to know to what extent problem decomposability and axial bias exist and to what extent they can be exploited in real-world problems.

Acknowledgements

This work has received funding support from the Natural Sciences and Engineering Research Council of Canada.

References

1. Beyer, H.-G., Schwefel, H.-P.: Evolution Strategies: A comprehensive introduction. Natural Computing 1, 3–52 (2002)
2. Brest, J., Zamuda, A., Boskovic, B., Maucec, M.S., Zumer, V.: High-Dimensional Real-Parameter Optimization using Self-Adaptive Differential Evolution Algorithm with Population Size Reduction. In: Proceedings of the 2008 IEEE Congress on Evolutionary Computation, pp. 2032–2039. IEEE Press, Los Alamitos (2008)
3. Chen, S.: Locust Swarms – A New Multi-Optima Search Technique. In: Proceedings of the 2009 IEEE Congress on Evolutionary Computation, pp. 1745–1752. IEEE Press, Los Alamitos (2009)
4. Chen, S., Lupien, V.: Optimization in Fractal and Fractured Landscapes using Locust Swarms. In: Korb, K., Randall, M., Hendtlass, T. (eds.) ACAL 2009. LNCS (LNAI), vol. 5865, pp. 211–220. Springer, Heidelberg (2009)
5. Chen, S., Miura, K., Razzaqi, S.: Analyzing the Role of "Smart" Start Points in Coarse Search-Greedy Search. In: Randall, M., Abbass, H.A., Wiles, J. (eds.) ACAL 2007. LNCS (LNAI), vol. 4828, pp. 13–24. Springer, Heidelberg (2007)
6. Hendtlass, T.: WoSP: A Multi-Optima Particle Swarm Algorithm. In: Proceedings of the 2005 IEEE Congress on Evolutionary Computation, vol. 1, pp. 727–734. IEEE Press, Los Alamitos (2005)
7. Hsieh, S.-T., Sun, T.-Y., Liu, C.-C., Tsai, S.-J.: Solving Large Scale Global Optimization Using Improved Particle Swarm Optimizer. In: Proceedings of the 2008 IEEE Congress on Evolutionary Computation, pp. 1777–1784. IEEE Press, Los Alamitos (2008)
8. Kennedy, J., Eberhart, R.C.: Particle Swarm Optimization. In: Proceedings of the IEEE International Conference on Neural Networks, vol. 4, pp. 1942–1948. IEEE Press, Los Alamitos (1995)

9. MacNich, C.: Towards Unbiased Benchmarking of Evolutionary and Hybrid Algorithms for Real-valued Optimisation. Connection Science 19(4), 361–385 (2007)

10. MacNish, C., Yao, X.: Direction Matters in High-Dimensional Optimization. In: Proceedings of the 2008 IEEE Congress on Evolutionary Computation, pp. 2372–2379. IEEE Press, Los Alamitos (2008)

11. Norman, M.G., Moscato, P.: A Competitive and Cooperative Approach to Complex Combinatorial Search, Caltech Concurrent Computation Program, C3P Report 790 (1989)

12. Tang, K., Yao, X., Suganthan, P.N., MacNish, C., Chen, Y.P., Chen, C.M., Yang, Z.: Benchmark Functions for the CEC 2008 Special Session and Competition on Large Scale Global Optimization. Technical Report (2007),
http://www.ntu.edu.sg/home/EPNSugan

13. Tseng, L.-Y., Chen, C.: Multiple Trajectory Search for Large Scale Global Optimization. In: Proceedings of the 2008 IEEE Congress on Evolutionary Computation, pp. 3052–3059. IEEE Press, Los Alamitos (2008)

14. Wang, Y., Li, B.: A Restart Univariate Estimation of Distribution Algorithm: Sampling under Mixed Gaussian and Levy probability Distribution. In: Proceedings of the 2008 IEEE Congress on Evolutionary Computation, pp. 3917–3924. IEEE Press, Los Alamitos (2008)

15. Yang, Z., Tang, K., Yao, X.: Multilevel Cooperative Coevolution for Large Scale Optimization. In: Proceedings of the 2008 IEEE Congress on Evolutionary Computation, pp. 1663–1670. IEEE Press, Los Alamitos (2008)

16. Zamuda, A., Brest, J., Boskovic, B., Zumer, V.: Large Scale Global Optimization using Differential Evolution with Self-adaptation and Cooperative Co-evolution. In: Proceedings of the 2008 IEEE Congress on Evolutionary Computation, pp. 3718–3725. IEEE Press, Los Alamitos (2008)

17. Zhao, S.Z., Liang, J.J., Suganthan, P.N., Tasgetiren, M.F.: Dynamic Multi-Swarm Particle Swarm Optimizer with Local Search for Large Scale Global Optimization. In: Proceedings of the 2008 IEEE Congress on Evolutionary Computation, pp. 3845–3852. IEEE Press, Los Alamitos (2008)

Continuous Swarm Surveillance via Distributed Priority Maps

David Howden

Swinburne University of Technology
dhowden@swin.edu.au

Abstract. With recent and ongoing improvements to unmanned aerial vehicle (UAV) endurance and availability, they are in a unique position to provide long term surveillance in risky environments. This paper presents a swarm intelligence algorithm for executing an exhaustive and persistent search of a non-trivial area of interest using a decentralized UAV swarm without long range communication. The algorithm allows for an environment containing arbitrary arrangements of no-fly zones, non-uniform levels of priority and dynamic priority changes in response to target acquisition or external commands. Performance is quantitatively analysed via comparative simulation with another leading algorithm of its class.

Keywords: UAV, pheromone, swarm intelligence, priority map, exhaustive search.

1 Introduction

UAV use in civilian and military roles for aerial reconnaissance and surveillance has, in recent years, progressed from novelty to given fact for many applications. The current generation of UAVs are mostly monolithic systems controlled by teams of human operators, so while the cost and risk benefits of UAVs over manned vehicles have been realized, the savings in manpower have not [7]. The other major pitfall of these systems is the long-range bandwidth requirements of transmitting large amounts of information between ground station and vehicle [2]. This latter issue has been brought into stark relief during real world deployments where is has been noted that there is a "serious shortfall in long-range communication bandwidth" [1].

The next logical step for UAV systems, which has been increasingly studied over the past decade, is the development of autonomous UAV swarms. The benefits of progressing to a swarm architecture such as scalability, robustness, agent simplicity, communication overhead, risk mitigation, etc., have been exhaustively covered in past work [1][3]. There are currently two main approaches to UAV swarm control. The first is a model where agents have global communication and synchronize their actions to good effect. This is known as a 'consensus level' of autonomy, where agents work as a team to devise actions [4]. As long as communication bandwidth is plentiful and guaranteed, this method is able to

K. Korb, M. Randall, and T. Hendtlass (Eds.): ACAL 2009, LNAI 5865, pp. 221–231, 2009.

produce optimal search patterns. However, some of the main disadvantages of direct control are still present such as lack of scalability and long range bandwidth overheads. The other model, which is the focus of this paper, utilizes local autonomy and is inspired by biological systems; namely ant colonies and pheromone trails. This model translates into each agent having only local knowledge of its environment and planning its own actions based on information gained indirectly via digital stigmergy.

In most cases the missions UAVs are used for, when abstracted, have the unifying goal of searching a bounded problem space. Work in this field has so far been mostly focused on discrete searches, where one or more targets exist within a state-space, and once they are located the search is complete. The discrete search approach is sensible for missions such as search and rescue, mapping a static area such as in agricultural surveying, and for short duration NOTE ISR missions. Continuous state-space exploration, such as would be required for problems such as fire spotting, border surveillance, or long duration ISR missions, is a research area with a small body of published work. This paper will present a new swarm control algorithm, loosely based on the pheromone control model, which is designed explicitly for continuous state-space exploration. It will be shown by quantitative results generated from comparative simulation that it significantly outperforms a leading algorithm of its class in coverage and ability to deal with non-uniform state-spaces.

2   Existing Work

There are two unifying factors in all work published in the field of pheromone based swarm control. The first is in the use of a digital pheromone map to both represent the environmental knowledge of each individual agent and as the main, or only, means of communication between peers. The specific implementation of the pheromone map varies, but can be accurately summarized as overlaying a geographic area with a digital lattice grid and storing pheromone data at the vertices (referred to as cells or nodes), as visualized in Fig 1. The second is that there is no direct communication between agents in the swarm. Communication is in the form of indiscriminate broadcasts of data stored in the individual's pheromone map, and often the spatial coordinates that the communication is being sent from.

Fig. 1.   Pheromone map: Lighter areas have more pheromone and attract agents

Past this, the nature of the pheromone used and the individual search heuristics become varied. Some of the earliest and most widely published class of algorithms are categorized by their clear and direct mapping of biological systems [8] [1]. With this approach the pheromone map starts off with zero pheromone and then digital agents are placed representing areas of interest(AOI). These AOI agents pump

'Interest' pheromone into the environment, which diffuses into neighbouring cells, creating a gradient which can be ascended to locate the source. When a node is visited by a UAV, all pheromone is removed from that cell; when an AOI is visited it stops producing pheromone. Continuous surveillance is accommodated by 'switching on' the AOI agent again after an arbitrary amount of time. This type of algorithm is then further improved by including deterrence pheromones 'Threat' and 'Repulsion' [9]. Repulsion pheromone can be added to the location of AOI agents which have been recently visited to discourage subsequent visits in the short term. It can also be added to the physical location of each agent to stop convergence. Threat pheromone is placed at areas which are actively dangerous, such as directly over a fire storm or around the location of surface-to-air missiles. In all cases, the pheromone placed slowly evaporates over time the same as in the biological model, and for the same reasons as in the standard ACO model [3].

The problem with heuristics based on diffusion is that agents can become stuck in local minima, the diffusion and evaporation rates need to be precisely calibrated, usually using an offline method, to minimize wandering, and most significantly they cannot guarantee exhaustive coverage [7]. A way of getting around these issues is by taking a less literal interpretation of nature and using raw Euclidean distance to cells that need to be observed, rather than pheromone diffusion and evaporation. Using this method cells are either 'explored' or 'unexplored', with explored cells containing the Euclidean distance to the closest unexplored. The heuristic presented is, at its highest level, the same as in the previous methods: greedy hill descent. If there is an adjacent unexplored cell, move to it; if all adjacent cells are explored, move into the one that has the lowest distance to an unexplored cell. While this is a discrete search algorithm, it can be made into a continuous search by changing explored cells to unexplored after a period.

While this approach has been shown to produce excellent results for a single pass of a search area, it is not primarily designed to maintain a persistent search, and is unable to handle state-spaces with non-uniform levels of priority without modification. The standard way of converting these methods to persistent methods is to pick an arbitrary period between when a cell is set to an inert state, and when it becomes active again [6]. To achieve the additional goal of non-uniform state spaces it is necessary to move from the boolean model to one which can be used to differentiate between cells based on time since last visit.

3 The Algorithm

The algorithm presented here uses a time-priority product based pheromone map [10]. Each cell of the map contains two values: the time it was last visited and the priority of that cell. Each cell is initialized to the time at which the surveillance mission began. The quantity of pheromone p at cell C is the product of the cell's *priority* and the cell visit time t_{visit} subtracted from the current time $t_{current}$. This pheromone quantity is reset to zero when flown over by a UAV.

$$p_C = priority_C \times \Delta t \qquad (1)$$

224   D. Howden

Using this map, variations of the traditional value divided by distance heuristic can be applied. Through simulation the best general purpose heuristic h found for use with this map is

$$h = \frac{p_C^2}{d(\boldsymbol{C}, \boldsymbol{P}) + d(\boldsymbol{C}, \boldsymbol{P} + \bar{\mathbf{r}})} \tag{2}$$

where $\boldsymbol{C}$ is the cell being evaluated, d is a distance function, $\boldsymbol{P}$ is the location of the UAV, and $\bar{\mathbf{r}}$ is a repulsion force, calculated as shown in Fig 3.

The repulsion vector is calculated and updated whenever an agent intercepts a pheromone map broadcast from within a pre-set repulsion range. These vectors are stored for an intermediate period of time; if a new repulsion vector is created before the old one has been discarded, it is added to a dequeue and an average is taken when a value is needed. The reason for retaining component vectors, rather than calculating and storing $(\boldsymbol{P} + \bar{\mathbf{r}})$ immediately is twofold.

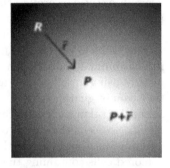

The heuristic is only infrequently updated, and nearby agents often keep their position relative to the other for extended periods of time even though their absolute spatial position is constantly changing. An immediate calculation would mean both UAVs were almost invariably being repulsed from the area they just left, rather than the other agent. The second reason is for continuity as this method of repulsion is relatively light handed and it can sometimes take more than a single 'bump' to gain an effective distance between agents.

Fig. 2. Visualization of the state-space decision surface for a uniform level of pheromone and repulsion being applied; lighter areas indicate fitness. An agent at position $\boldsymbol{P}$ is being subject to repulsion from point $\boldsymbol{R}$. The magnitude of $\bar{\mathbf{r}}$ is set to a constant value and thus only takes its direction from $\boldsymbol{R}$.

In the heuristic, the reason for squaring the pheromone quantity is to help mitigate the distance penalty after all local cells have been exhausted and a long distance decision is required. This power has shown the best average result over the widest range of scenarios, as excessive wandering emerges when it is made any larger, while less leads to local moves being excessively favoured even after they are no longer appropriate. Because long range selections by the heuristic are relatively uncommon and *de facto* bounded by the sheer weight of the distance penalty, a free computation time increase can obtained by limiting the evaluation of cells to the agents local neighbourhood (e.g. with 10-20 steps).

As a final step, a local search heuristic implemented as a point to point pathfinding algorithm is added. This yields a small increase in performance, especially in maps with null priority areas where the agent can intelligently decide between the shortest path, and a longer one that detours over cells with pheromone. The primary reason however is it allows the algorithm to elegantly handle environments with explicit no-fly zones. No-fly zones can accommodate features of the agent's environment which include airports and other prohibited airspace, high density residential areas, sheer cliffs and gullies. The specific implementation used was an A* search, weighted by pheromone, between the agent's position, and the cell chosen by the main heuristic.

4   Comparison Algorithm

Out of the algorithms which were run through a simulator to find a baseline for comparison, the best performing was one published by Erignac [7]. Described briefly in Section 2, Erignac's algorithm uses the Euclidean distance to the closest unexplored cell as its pheromone values. Due to the advantages of this type of pheromone map, and an interesting implementation of state-based behaviours, the search pattern which emerges is, at worst, near optimal. The state-space that Erignac's algorithm was designed for is one with a uniform level of priority, where each cell starts off in the 'unexplored' state, and needs to be visited at least once to change it to 'explored'. To be useful as a comparison, a variation had to be made.

Firstly, the Euclidean distance pheromone map was used side by side with a modified priority pheromone map which said a cell was explored if its pheromone was lower than one, and otherwise it was unexplored. Even when cells were showing as explored, visiting them during a random move or a repulsion move would still reset their pheromone to zero. Consequently, as absolute pheromone values were needed (as opposed to the relative values the map is designed for), the rate of pheromone increase needed to be tweaked offline and tested to optimize the results. The Erignac variant algorithm is referred to as algorithm E.

Fig. 3. An example of emergent contour following observed during the execution of algorithm E, uniform map

As a point of interest, the primary state behaviour, contour following, was found to be largely redundant as simulations show that contour following is an emergent behaviour of both algorithms. An example of the emergent contour following observed is shown in Fig 3.

5   Simulation and Results

The UAV specifications used for the simulation are roughly equivalent to current, off-the-shelf, 3m wingspan vehicles such as the Aerosonde [11]. Key features are a 35 knot cruise speed, 10km LOS communication,[1] and a footprint radius of

---

[1] The algorithm will continue to perform well with a communication range a fraction of this size, as long as the total information flow is past a certain threshold. eg. at 10 minute broadcast intervals, 128 UAVs will perform reasonably with a 500m radius, while 4 UAVs would require 3000m.

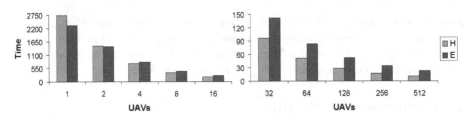

Fig. 4. Uniform Map - Average mean time, in minutes, since cell's last visit

176m (640 × 480 resolution camera, 60 deg FOV with 10× zoom for a 5cm^2 pixel ground resolution). To reduce noise in the data, UAVs were initialized at random locations and set to have unlimited endurance.[2] Agents send out a single broadcast once per minute which includes their individual pheromone map and their spatial coordinates, communicating more frequently than this has no effect on the algorithm.

All scenarios were run in a 50km^2 environment, divided into 20164 (142^2) cells. The cell width is equal to the diameter of the agent's footprint, 352m. During testing, simulated environments of up to 100km^2 with 384^2 cells and as low as 1km^2 were run and the same relative results were observed.

The performance metric used was the length of time since each cell was last visited, averaged for the whole of the map. This measurement was taken 2000 times and then averaged for each scenario, 200 times per 'pass' of the map. A pass was defined as the time taken for each cell to be visited at least once.

5.1   Uniform Map

The first scenario is an exhaustive and persistent search of an area with uniform priority. As can be seen in Fig 4, with the addition of an explored/unexplored mechanism though the priority map, the global-scope Euclidean distance pheromone enables better results for a single UAV, and parity is held until around four agents. After this, with higher agent densities, the H algorithm's emphasis on local seeking of pheromone of any value (not just past a threshold) provides a mean visit time of under half that of the comparison algorithm by 256 agents and onwards.

5.2   Lake Map

The second scenario is a pseudo fire spotting priority map from [10] with three levels of priority, referred to here as the Lake map (Fig 5). Each level is set to twice the level before it, so the highest priority cells are the small white circles which need to be surveyed four times as often as the baseline and the light grey squares need to be surveyed twice as often. Fig 6 shows that, at any density of UAVs on the lake map, algorithm H provides a consistent 25% to 30% decrease in survey times of the high priority survey areas.

---

[2] Tests run with a single launch and refueling point were obviously found to affect the absolute performance, however no significant effect on relative efficiency was observed between algorithms.

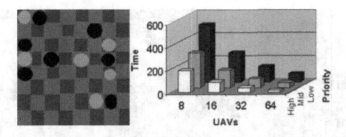

Fig. 5. LEFT: Black circles are null priority lakes. Light grey increases at a x4 rate, middle grey at a x2 rate. RIGHT: Representative sample of visit ratios.

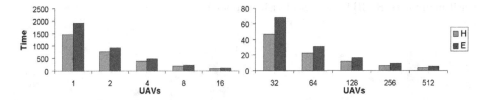

Fig. 6. Lake Map - Mean time since visit for highest priority areas

In a map with multiple priorities, the coverage of the the highest priority area is the key measurement. Lower priority areas will often be visited en route to the high priority areas and thus have an artificially lowered mean time. This can be taken to extremes, however, and the ratio the other cell's visits are still useful as a secondary measurement. Fig 5 shows the visit ratios between priority areas for algorithm H on this map, this ratio is reasonably consistent, especially among higher UAV numbers. Due to the priority map used by both algorithms, these ratios are roughly the same for the comparison algorithm also. The raw data for all three simulations are also presented in the appendix for a more accurate comparison.

5.3   No-Fly Zone Map

The third and most arbitrarily complex environment is the No-Fly Zone map shown in Fig 7. This priority map has the same priority ratios as the fire map, but with the addition of no-fly areas. The environment is also made difficult to optimize by the addition of complex null priority areas in the form of spiralling lane ways.

The results for the No-Fly Zone map, shown in Fig 8, continue the trend seen in the first two environments. The comparison algorithm, with its ability to exploit distant areas of the map, was able to maintain parity for small swarm sizes of one to four, but was unable to compete with larger swarm sizes. By agent count 64, algorithm H is doubling the comparison algorithm's performance. The relative performance between priorities for both algorithms remains similar to

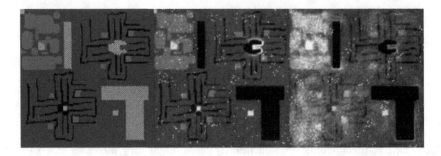

Fig. 7. LEFT: No-Fly Zone Map - Checkered areas are no-fly zones. Black is null priority. Light grey increases at a x4 rate, middle grey at a x2 rate. MIDDLE: Historical visit map for H. RIGHT: Historical visit map for E.

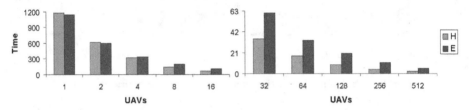

Fig. 8. No-Fly Zone Map - Mean time since visit for highest priority areas

what was shown for the Lake map, and as was previously mentioned the exact numbers can be seen in the appendix.

The reason for the dramatic performance difference at higher agent densities is indicated in Fig 7. While the priority pheromone map allows both algorithms to perform a continuous search with good results, the ability of algorithm H to exploit this data in a continuous, as opposed to binary, form allows it to optimize its moves to a far greater extent. As algorithm E's implementation forces a binary representation to be used, moves to explored cells are essentially random.

6   Analysis of Results

The amount of attention paid to repulsion in description of the algorithm is proportionate to the importance it should play in any swarm search algorithm. Swarm intelligences based on pheromone provably converge and reach a stable state [5], which is an artefact especially relevant to a continuous search which can have a theoretically infinite duration. The quality of the repulsion mechanic is a major component in the performance of this end, stable state.

To its credit, in a search space it was not explicitly designed for, the Erignac variant algorithm used for comparison managed parity when only individuals or pairs of agents were used in two of the three environments. This is due in part to its higher emphasis on the global search space when compared to the paper's algorithm. The more sparse agent coverage is, the larger their decision range needs to be for optimal results. Second, the repulsion method was very light handed: again, with sparse agent density, noise added by repulsion becomes a hindrance rather than an advantage. With repulsion, the rule of thumb should always be to use as little as possible.

In every other scenario, the algorithm presented in this paper provided significant advantages allowing for lower mean visit times, often in the range of 50% or greater. Aside from achieving the primary objective more effectively, there are two other advantages to H over E for continuous surveillance missions. First, H requires no off-line optimization and no adjustment on the fly to accomidate for lost agents. As it works through relative pheromone values, the absolute value is unimportant. For an algorithm which implements the binary abstraction of a priority map, the period between cells switching from explored to unexplored needs to be calibrated off-line as a bad value is nearly insurmountable. A value which is too high, where agents always move to the closest adjacent cell, or too low, where agents spend most of their time using the random move behaviour, leads to results no better than a random search.

The second advantage is computation time. Utilizing an Euclidean distance pheromone map requires that each cell be populated with the distance to the closest unexplored cell. This consequently requires the use of a wave front propagation algorithm every time the map is changed, either via an agent's visit, or through communication of an updated pheromone map through the swarm. This is computationally expensive, and occurs every few seconds in large swarms. Using the priority map algorithm from this paper, only the few nodes on the agent's immediate path need to be checked when new information is received, and only a small and constant sized area of the map needs to be evaluated when a new decision is required. Due to the constant size of the evaluation area, the heuristic's computation time does not increase with map size, opposed to an exponential increase for searching the entire map.

An interesting observation is that while the performance is good, the ratios observed are not the 4/2/1 relationship that was set. Both algorithms are able to maintain an exact relationship if that is the desired result: for algorithm H the heuristic is changed to negate the distance penalty; for algorithm E the period between explored and unexplored is increased. The side effect of these changes is that every area performs worse as the agents spend a disproportionate amount of in transit chasing global maxima. The larger a swarm is, the worse this approach becomes as it becomes rarer that any individual will be the one to reach the target first. Even with the current settings, the ratios approach their 4/2/1 ideal as the agent count increases, often being almost exact by 512.

7   Conclusion

In this paper, an algorithm for performing continuous aerial surveillance of nontrivial search spaces was presented. Through simulation and analysis, it was shown to require less computation time and provide superior coverage to other algorithms in the field. The algorithm, without any off-line calibration between simulations, performed and scaled well through a suite of environments and agent densities. This proven adaptability enables a broad scope for optimization for specific implementations.

References

1. Parunak, H., Purcell, M., O'Connell, R.: Digital pheromones for autonomous coordination of swarming UAVs. In: Proceedings of First AIAA Unmanned Aerospace Vehicles, Systems, Technologies, and Operations Conference, AIAA (2002)
2. Varga, R.: The UAV Reloaded: Placing Video Intelligence On-board. COTS Journal, 72–76 (June 2003)
3. Parunak, H.V.D.: "Go to the Ant": Engineering Principles from Natural Multi-Agent Systems (1999)
4. Martin, C.E., Macfadzean, R.H., Barber, K.S.: Supporting Dynamic Adaptive Autonomy for Agent-based Systems. In: Proceedings of 1996 Artificial Intelligence and Manufacturing Research Planning Workshop, AIAA, pp. 112–120 (1996)
5. Brueckner, S.: Return from the Ant: Synthetic Ecosystems for Manufacturing Control, Thesis at Humboldt University Berlin, Department of Computer Science (2000)
6. Altshuler, Y., Yanovski, V., Wagner, I.A., Bruckstein, A.M.: The Cooperative Hunters - Efficient Cooperative Search For Smart Targets Using UAV Swarms. In: ICINCO-MARS-2005, Barcelona, Spain (2005)
7. Erignac, C.A.: An Exhaustive Swarming Search Strategy Based on Distributed Pheromone Maps. In: Advanced Design Systems M&CT, The Boeing Company (2007)
8. Sauter, J.A., Matthews, R., Parunak, H.V.D., Arbor, A., Brueckner, S.A.: Performance of digital pheromones for swarming vehicle control. In: Proceedings of the fourth international joint conference on Autonomous agents and multiagent systems (2005)
9. Walter, B., Sannier, A., Reiners, D., Oliver, J.: UAV Swarm Control: Calculating Digital Pheromone Fields with the GPU. The Journal of Defense Modeling and Simulation: Applications, Methodology, Technology 3(3), 167–176 (2006)
10. Howden, D., Hendtlass, T.: Collective intelligence and bush fire spotting. In: GECCO 2008: Proceedings of the 10th annual conference on Genetic and evolutionary computation (2008)
11. Aerosonde Pty Ltd, http://www.aerosonde.com/pdfs/aerosonde02-20-09.pdf

Appendix

Table 1. Uniform Map - Mean time, in minutes, since cells were last visited

UAVs	1	2	4	8	16	32	64	128	256	512
H	2719.5	1462.2	729.9	366.9	188.2	96.3	50.5	28.3	16.6	10.9
E	2311.0	1439.2	797.7	432.4	243.5	141.2	83.7	52.0	33.8	23.5

Table 2. Lake Map - Mean time, in minutes, since cells were last visited. Priority levels, in relative terms, are 4:2:1 / High:Mid:Low.

UAVs	1	2	4	8	16	32	64	128	256	512
E Low	3329.6	1783.8	954.1	514.7	281.6	159.0	94.8	57.3	38.4	26.8
E Mid	2199.9	1167.5	574.8	306.6	168.0	93.4	53.6	31.3	19.8	13.4
E High	1905.4	929.3	479.4	231.9	116.6	67.7	30.8	16.2	9.3	5.7
H Low	3707.7	1973.3	997.4	500.6	253.8	131.9	71.0	38.6	21.8	12.7
H Mid	2234.9	1179.1	602.2	299.3	152.3	76.9	41.2	22.0	12.2	6.9
H High	1458.9	768.4	384.3	193.9	93.0	46.7	22.4	11.8	6.1	3.4

Table 3. No-Fly Zone Map - Mean time, in minutes, since cells were last visited. Priority levels, in relative terms, are 4:2:1 / High:Mid:Low.

UAVs	1	2	4	8	16	32	64	128	256	512
E Low	2876.7	1554.1	848.2	457.0	250.1	141.1	79.8	45.6	26.4	15.3
E Mid	1552.1	854.9	464.0	243.2	144.9	76.2	42.9	23.1	12.2	6.0
E High	1144.7	597.4	336.4	198.9	108.2	60.7	33.4	20.2	11.2	4.9
H Low	2491.1	1308.7	661.1	332.3	171.8	89.2	48.3	27.0	15.0	8.7
H Mid	1514.0	853.3	416.4	211.7	99.7	50.2	25.0	13.9	7.8	4.3
H High	1174.2	612.4	319.0	145.7	67.4	34.9	17.6	8.9	4.5	2.3

Optimization in Fractal and Fractured Landscapes Using Locust Swarms

Stephen Chen[1] and Vincent Lupien[2]

[1] School of Information Technology, York University
4700 Keele Street, Toronto, Ontario M3J 1P3
sychen@yorku.ca
[2] Acoustic Ideas Inc.
27 Eaton Street, Wakefield, MA 01880
vincent.lupien@acousticideas.com

Abstract. Locust Swarms are a newly developed multi-optima particle swarm. They were explicitly developed for non-globally convex search spaces, and their non-convergent search behaviours can also be useful for problems with fractal and fractured landscapes. On the 1000-dimensional "FastFractal" problem used in the 2008 CEC competition on Large Scale Global Optimization, Locust Swarms can perform better than all of the methods in the competition. Locust Swarms also perform very well on a real-world optimization problem that has a fractured landscape. The extent and the effects of a fractured landscape are observed with a practical new measurement that is affected by the degree of fracture and the lack of regularity and symmetry in a fitness landscape.

1 Introduction

A fundamental premise of evolutionary computation is that better solutions can be derived from existing solutions. In general, this leads to a concentration of the search process around the best existing/previously found solutions. In globally-convex search spaces (e.g. Rastrigin, Griewank, and Ackley), local optima (i.e. excluding the global optimum) can be expected to have a neighbouring local optimum that has a better fitness and a position in the search space that is between the current local optimum and the global optimum. Since it is possible to converge towards the global optimum in these search spaces, the exploitation of existing solutions can be highly productive.

In non-globally convex search spaces (e.g. W-shaped landscapes like Schwefel's function), it is not possible to converge to the globally optimal solution from all parts of the search space. To escape from "regional" optima (which can consist of many local optima), it is necessary to have non-convergent search behaviours. Locust Swarms [4] are designed to "devour" (i.e. converge on a local optimum) and move on (i.e. employ "scouts" [5] to identify promising new areas of the search space). As a result, Locust Swarms are well-suited for non-globally convex search spaces and other irregular search spaces such as those associated with fractal and fractured landscapes.

K. Korb, M. Randall, and T. Hendtlass (Eds.): ACAL 2009, LNAI 5865, pp. 232–241, 2009.

The first part of this paper focuses on a fractal-based benchmark optimization problem. Due to its exponential computational complexity, the fractal landscape is initially developed in two dimensions [11]. To scale the function up to the sizes used for the Large Scale Global Optimization (LSGO) competition [14] and to avoid problem decomposability, the "FastFractal" variation repeats the two-dimensional version with rotations. On the FastFractal problem, a Locust Swarm parameterized for non-globally convex search spaces performs better than all eight methods that competed in the CEC2008 LSGO competition.

A similarly parameterized Locust Swarm has also been applied to a real-world problem involved with the design of phased array ultrasonic transducers [6][7]. The physics engine used in this problem leads to a function evaluation that is exceptionally noisy – e.g. small changes to the input solution can lead to large changes in the output evaluation. The sharp fractures in the resulting fitness landscape cause gradient descent-based optimization procedures (e.g. fmincon in MATLAB®) to be highly inconsistent and ineffective [7].

The inconsistent behaviour of fmincon is also reflected in the number of function evaluations that it uses to find a solution. Seeking to find a correlation between this observation and the structure of a fitness landscape, the standard deviation in the number of function evaluations used by fmincon from 30 random start points was measured for all seven problems used in the LSGO competition. Since the standard deviation for the transducer design problem exceeds the standard deviations recorded for all LSGO problems and in particular the FastFractal problem, the standard deviation in function evaluations is proposed as a practical measurement (i.e. requiring minimal extraneous computational effort) for estimating the degree of fracture in a fitness landscape.

Overall, the results demonstrate that Locust Swarms are an efficient and effective heuristic search technique for optimization in fractal and fractured fitness landscapes. This demonstration begins with an introduction to the design of Locust Swarms in section 2. The FastFractal problem is introduced in section 3 followed by results for Locust Swarms in section 4. In an attempt to make these results more relevant to real-world problems, a practical measurement to estimate the degree of fracture and the effects of this fracture in arbitrary fitness landscapes is proposed in section 5. Results for Locust Swarms on the fractured landscape involved with the design of phased array ultrasonic transducers are presented in section 6. Lastly, a discussion of the key ideas occurs in section 7 before the paper is summarized in section 8.

2   Locust Swarms

Locust Swarms are a heuristic search technique that have been explicitly designed for non-globally convex search spaces [4]. In globally convex search spaces such as Rastrigin, Griewank, and Ackley, it is possible to progress from any local optimum to the global optimum through a series of progressively better local optima. Search techniques that are convergent in nature (e.g. a $(1+\lambda)$-evolution strategy) can perform well in these search spaces because it is possible to reach the global optimum via a monotonically improving set of local optima that are each in the neighbourhood of the previous local optimum.

To reach the global optimum in non-globally convex search spaces, it can be necessary to pass through a series of worsening local optima and/or to jump to a distant part of the search space (e.g. to explore both halves of a W-shaped fitness landscape like Schwefel's function). The non-convergent search behaviour of Locust Swarms is designed to accommodate both of these possibilities. In the first instance, Locust Swarms store but do not otherwise exploit information from the best previously found local optimum. This allows successive swarms to meander across a non-improving area of the search space. Secondly, Locust Swarms can impose a minimum gap between successive swarms. This allows large jumps from one part of the search space to another.

The development of Locust Swarms [3][4] builds upon particle swarm optimization (PSO) [10] and a multi-optima variation called WoSP (for Waves of Swarm Particles) [8]. The key difference from WoSP is the use of "scouts" [5] to find promising new parts of the search space to explore after the current area has been sufficiently exploited. Overall, this "devour and move on" multi-optima search strategy has three sets of parameters: scout parameters, swarm parameters, and general parameters.

$$FE = N * (S + L*n) \tag{1}$$

The primary role of the general parameters is to effectively allocate the allowed function evaluations (*FE*). In (1), N is the number of swarms – each of which explores a (local) optimum, S is the number of scouts, L is the number of locusts, and n is the number of iterations each swarm is allowed to use to find a local optimum. In general, the best (time constrained) results have been obtained by maximizing N, balancing S and $L*n$, and minimizing n (but not to an extent that disallows the swarm from both performing a small regional search and adequately converging around a local optimum) [3].

$$delta = \pm range * (gap + \text{abs}(\text{randn}()*spacing)) \tag{2}$$

The scouts perform a coarse search around the previous optima by adding a *delta* calculated in (2) to a small number of terms in a solution vector. This dimension reduction (*DIM$_r$*) allows Locust Swarms to partially exploit some information from the previous optima (since it is assumed that a fractal landscape does not lead to a random search space) while still allowing for a non-convergent search behaviour. This non-convergent behaviour is enforced by requiring scouts to be a minimum *gap* from the previous optima (with some additional *spacing* as appropriate). Note: *range* is the range for (the given dimension of) the search space and randn() is the MATLAB® function for a normally distributed random number.

$$velocity_0 = launch*(location_0 - previousoptimum) \tag{3}$$

$$location_{i+1} = location_i + step*velocity_i \tag{4}$$

$$gravityvector_i = globalbestlocation - location_i \tag{5}$$

$$velocity_{i+1} = M*velocity_i + G*gravityvector_i \tag{6}$$

Keeping the L best scouts for the swarm, the initial velocities (3) launch the locusts away from the previous optima with a speed affected by the *launch* parameter. Each locust moves a *step* fraction of its velocity during each iteration i (4), and this velocity is updated to reflect a "gravitational" attraction to the "global best" location found by the (current) swarm (5). The rate of convergence is affected by the momentum (M) and gravity (G) parameters in (6). Note: Locust Swarms do not use an attraction to each particle's "local best" (e.g. *pbest* in [10]) since multiple swarms are used to balance exploration with exploitation.

3 Fractal-Based Benchmark Optimization Problems

The goal of a fractal-based optimization problem is to reduce the biases that can occur in traditional benchmark problems – e.g. central bias, axial and directional bias, scale bias, etc. [11] These biases can inadvertently allow the developers of optimization techniques to "cheat" – e.g. to select a set of parameters that tunes their algorithm's performance to the specific landscape characteristics of the test problem and thus produce results that are unrepresentative for problems in general. It is hypothesized that search techniques that can handle unbiased search spaces can generalize more easily to search spaces that contain (beneficial) biases. Conversely, search techniques that (implicitly or explicitly) exploit a bias may have more difficulty to generalize to a less biased or unbiased search space. [11]

The fitness landscapes of the fractal-based problems are developed by randomly impacting "craters" into initially flat landscapes. The size and number of each type of crater is scaled to replicate the same landscape features at each level of detail. Since the fitness landscapes are created by the superimposed crater impacts, they are naturally fractured, irregular, and non-globally convex. Pseudorandom numbers allow the same fractal landscape to be reproducible as a benchmark. However, the calculation is still exponentially expensive with the number of dimensions, so problems above $D = 2$ dimensions are based on rotated replications of the base problem. [11]

4 FastFractal Results

A separate demonstration of the effectiveness of Locust Swarms for large scale global optimization appears in [3]. However, focusing strictly on the FastFractal problem which is the only non-globally convex problem in the LSGO competition, a different set of parameters is useful. Specifically, the LSGO implementation of Locust Swarms set the *gap* parameter to zero – since better local optima can always be found in the neighbourhood of the most recent local optimum (i.e. assuming that it is not the global optimum), there can be distinct disadvantages to forcing the search process away from previously found local optima.

Table 1. Results for Locust Swarms on the FastFractal problem in CEC2008 LSGO competition format

		$D = 100$	$D = 500$	$D = 1000$
Function evaluations = 50 * D	1 (best)	-1149.7	-5223.4	-9890.8
	7	-1104.0	-5099.8	-9753.1
	13 (median)	-1083.7	-5062.7	-9709.8
	19	-1073.5	-5004.7	-9652.8
	25 (worst)	-1050.1	-4889.7	-9434.1
	mean	-1091.8	-5051.5	-9706.5
	std dev	29.0	74.3	97.2
Function evaluations = 500 * D	1 (best)	-1507.4	-7198.0	-14248
	7	-1501.3	-7183.7	-14179
	13 (median)	-1491.7	-7170.2	-14163
	19	-1488.4	-7162.8	-14151
	25 (worst)	-1478.4	-7134.5	-14107
	mean	-1493.6	-7171.9	-14168
	std dev	7.5	17.5	33.4
Function evaluations = 5000 * D	1 (best)	-1548.9	-7458.0	-14728
	7	-1547.1	-7443.7	-14717
	13 (median)	-1546.1	-7437.8	-14716
	19	-1543.2	-7434.7	-14712
	25 (worst)	-1538.3	-7428.9	-14700
	mean	-1545.0	-7439.4	-14715
	std dev	2.7	7.6	7.8

For these experiments, the swarm parameters are $M = 0.5$, $G = 0.5$, and *step* = 0.6. The number of scouts ($S = 500$), locusts ($L = 5$), and iterations per swarm ($n = 100$) are held constant for dimensions $D = 100$, 500, and 1000, so N is varied as required to reach the indicated function evaluations. The only differences with the LSGO implementation of Locust Swarms [3] are in the scout parameters. Keeping $DIM_r = 3$, the use of *gap* = 0.1 and *spacing* = 0.4 create a more explorative search behaviour as compared to the highly exploitive behaviour achieved with *gap* = 0.0 and *spacing* = 0.1 – the current explorative behaviour is more suitable for non-globally convex search spaces. To keep the magnitudes of the initial velocities similar, *launch* = 0.5 is used (compared to *launch* = 5.0 in [3]).

In Table 1, the results for Locust Swarms on the FastFractal problem are given in the LSGO competition format. In Table 2, the mean result is then compared against the other competition methods (i.e. MLCC [17], EPUS-PSO [9], jDEdynNP-F [2], MTS [15], DEwSAcc [18], DMS-PSO [19], and LSEDA-gl [16] – note: results for UEP [12] were not included in their paper). For 50*D and 500*D function evaluations, MTS is the best competition method, and Locust Swarms are behind at $FE = 50*D$ and ahead at $FE = 500*D$. At $FE = 5000*D$, Locust Swarms find better final solutions on average than all of the methods in the competition. Although Locust Swarms and MLCC both converge at $FE = 5000*D$ to what appears to be near the global optimum, Locust Swarms converge faster and are ahead of MLCC at all recorded time intervals.

Table 2. Comparison of FastFractal results. Bolded values are better than Locust Swarms, or Locust Swarms if it is best.

	Method	$D = 100$	$D = 500$	$D = 1000$
Function evaluations = 50 * D	MLCC	-986.9	-4576.1	-8898
	EPUS-PSO	-812.0	-3410.0	-6490
	jDEdynNP-F	-866.7	-3549.0	-6708
	MTS	**-1277.2**	**-6002.9**	**-11830**
	DEwSAcc	-891.8	-3603.0	-6786
	DMS-PSO	-858.3	-3419.8	-6443
	LSEDA-gl	-804.3	**-5103.1**	-9661
	Locust Swarms	-1091.8	-5051.5	-9707
Function evaluations = 500 * D	MLCC	-1339.3	-6447.3	-12743
	EPUS-PSO	-834.0	-3470.0	-6550
	jDEdynNP-F	-1151.5	-4349.6	-7857
	MTS	-1469.5	-7052.0	-13935
	DEwSAcc	-1123.4	-4329.3	-7855
	DMS-PSO	-977.7	-3711.6	-6707
	LSEDA-gl	-1417.9	-6791.7	-13408
	Locust Swarms	**-1493.6**	**-7171.9**	**-14168**
Function evaluations = 5000 * D	MLCC	-1543.9	-7435.0	-14703
	EPUS-PSO	-855.0	-3510.0	-6620
	jDEdynNP-F	-1476.8	-6881.6	-13491
	MTS	-1486.1	-7081.0	-13999
	DEwSAcc	-1365.0	-5748.7	-10585
	DMS-PSO	-1144.8	-4199.9	-7509
	LSEDA-gl	-1463.8	-6827.6	-13481
	Locust Swarms	**-1545.0**	**-7439.4**	**-14715**

5 Estimating the Degree of Fracture in Fitness Landscapes

Many benchmark optimization problems are based on sinusoidal functions that are repeated in each dimension. For these "traditional" benchmark problems, the resulting fitness landscapes have regularly-shaped areas around each local optimum and a symmetric distribution of local optima around the global optimum. It is hypothesized that a gradient descent procedure will operate with a high level of consistency in a locally smooth search space that exhibits regularity and symmetry. Conversely, the lack of these features in a fractal landscape should lead to large variations in the operation of gradient descent-based search techniques.

Insight into the operation of gradient descent can be gained by observing the number of function evaluations that it requires to find a (locally optimal) solution. In Table 3, the mean and standard deviation for function evaluations used by fmincon to optimize 30 random start points is given for all seven LSGO problems when they are scaled to $D = 5$ dimensions. The low means and standard deviations for the unimodal functions 1 and 2 match their simple fitness landscapes. The standard deviation for the Rosenbrock function drops to 1.9 when two outliers are removed – all solutions outside of its "narrow valley" seem to experience a similar fitness landscape.

Table 3. Function evaluations used by `fmincon` to optimize 30 random start points

Function	Name	Mean	Std Dev
LSGO 1	Sphere	53.2	12.7
LSGO 2	Schwefel 2.21	22.6	3.4
LSGO 3	Rosenbrock	492.2	45.4
LSGO 4	Rastrigin	96.9	22.6
LSGO 5	Griewank	198.2	45.8
LSGO 6	Ackley	81.7	40.5
LSGO 7	FastFractal	190.1	69.4
Real-world	Transducer design	375.3	73.1

Among the "classical" globally-convex functions, the higher standard deviations for the Griewank and Ackley functions may reflect their noisier fitness landscapes caused by having a higher sinusoidal frequency (as compared to Rastrigin). These standard deviations are still much smaller than that recorded for FastFractal – a problem with minimal local smoothness, irregular local optima, and limited global symmetry. It is hypothesized that these search space characteristics may influence the difficulty of certain real-world problems.

For example, as part of the design of phased array ultrasonic transducers in which the number of elements is minimized [7], Probe Designer™ uses a physics engine that receives a candidate real-valued vector representing the probe geometry and determines the number of transducer elements that it takes to steer and/or focus several ultrasonic beams in one or more media. For a given probe geometry, the required element density at each spatial vicinity on the probe surface is driven by one of the ultrasonic beams. As the probe geometry is changed, it is not unusual for the identity of the beams driving the element density to change abruptly. This phenomenon leads to a non-linear search space and great sensitivity in the number of elements as the probe geometry is changed. In Table 3, the higher mean and standard deviation in function evaluations for `fmincon` when it is applied to this problem suggest that this real-world problem has a fractured fitness landscape with limited regularity and symmetry.

6   Locust Swarms for the Design of Phased Array Ultrasonic Transducers

The previous FastFractal results suggest that Locust Swarms may be a promising search technique to apply to the fractured search space of the transducer design problem. The following application uses a limit of 1500 function evaluations due to the high computational costs of the physics engine [7]. The parameters for the developed Locust Swarm are $N = 6$ swarms with $S = 90$ scouts, $L = 4$ locusts, and $n = 40$ iterations per swarm. The scouts have a minimum *gap* of 0.05 and a *spacing* of 0.15 (relative to the range of the search space), and they work with $DIM_r = 2$ dimensions. The initial outward velocity of each locust is equal in magnitude to its distance from the previous optima (i.e. *launch* = 1), and the locust movements are governed by the same swarm parameters used in section 4 (i.e. $M = 0.5$, $G = 0.5$, and *step* = 0.6).

Table 4. Comparison of Locust Swarms to PSO-ES [6]. The previous results with fmincon [7] are added to show how gradient descent methods can perform poorly and inconsistently in a fractured landscape.

	PSO-ES	Locust Swarms	fmincon
Mean	28.47	28.50	76.2
Std Dev	1.31	1.46	70.8
Maximum	27	27	29
Minimum	32	32	263
t-test		46.2%	

The results for the previous single-optima coarse search-greedy search technique (i.e. a particle swarm for the coarse search and an evolution strategy for the greedy search – PSO-ES) [6] and the new implementation of Locust Swarms are shown in Table 4. Under similar test conditions, the results are essentially the same, and a paired t-test confirms that their differences are insignificant. However, preliminary testing across a broader set of problems indicate that Locust Swarms can be more robust and effective than PSO-ES – especially when Probe Designer™ is seeded with a poor initial solution. The advantage of Locust Swarms in fractured landscapes appears related to its ability to traverse broad terrains with many "irrelevant" local optima.

7 Discussion

One of the keys to an efficient, effective, and robust search technique is the ability to automatically adjust parameters to adapt to the characteristics of the current landscape. A simple example is the use of evolving mutation parameters σ in an evolution strategy [1], and similar ideas appear in several of the best performing LSGO competition entrants (e.g. [2][16]). However, these adaptations assume a certain degree of consistency in the search space – e.g. the landscape features that were present while the parameters were learned will continue to be present when the learned parameters are used. When the ability to exploit these biases in the search space is removed (as is the case for a fractal-based fitness landscape [11]), the performance advantage of several LSGO competition entrants is similarly reduced [3].

An explicit goal of the FastFractal problem was indeed to eliminate the opportunity for search techniques to adapt to search space regularity and to use regularly spaced minima as "staging posts" on their path to the global optimum [11]. Another challenge that the FastFractal problem presents to adaptive techniques is the combination of a non-globally convex search space and the elimination of scale bias. Since the "texture" of the fitness landscape is (recursively) similar at all levels of scale (up to 10^{-12} for 64-bit floating point numbers) [11], an adaptive search technique which has reduced its scale to find an exact local optimum will subsequently be ill-suited to begin searching a new region (at a vastly different scale).

The ability to optimize fractal and fractured landscapes is only important if these landscapes exist in the real world. To help identify these landscapes, a new measurement has been introduced to estimate the degree of fracture for an arbitrary problem. A consistent operation of fmincon suggests a similar consistency and smoothness in the fitness landscape, but it is not certain that a high standard deviation in function evaluations is a definitive indicator of a fractured fitness landscape. In particular, the current measure is also affected by variations in the search space (e.g. the irregularities and asymmetries associated with a non-globally convex search space), so it may be better for determining the potential suitability of Locust Swarms than the global degree of fracture (e.g. [13]).

For the transducer design problem, other supportive indicators of its complexity are the high mean in `fmincon` function evaluations and the high variations in `fmincon` solutions. However, `fmincon` requires a large number of function evaluations for the non-fractal Rosenbrock function, and it is difficult to normalize and/or compare variations in solution quality across multiple problem domains. Nonetheless, the use of `fmincon` as a measurement tool can still be useful because it is a popular optimization technique used on many real-world problems (e.g. [7]). If the results for `fmincon` are adequate, then there may be little reason to apply any heuristic search technique. Conversely, if the operation and/or results for `fmincon` are inconsistent (and thus imply a fractured and/or non-globally convex fitness landscape), then the current results suggest that Locust Swarms may be an appropriate search technique to use.

8 Summary

Locust Swarms have been shown to perform effectively in non-globally convex, fractal, and fractured landscapes. The usefulness of these results depends on the prevalence of fractured and other fractal-like landscapes in real-world optimization problems. Future work will focus on measuring the degree of fracture, irregularity, and asymmetry for more real-world problems, improving the measurement of these features, observing the effects of these features on the performance of different heuristic search techniques, and improving the performance of Locust Swarms on these problems.

Acknowledgements

This work has received funding support from the Natural Sciences and Engineering Research Council of Canada.

References

1. Beyer, H.-G., Schwefel, H.-P.: Evolution Strategies: A comprehensive introduction. Natural Computing 1, 3–52 (2002)
2. Brest, J., Zamuda, A., Boskovic, B., Maucec, M.S., Zumer, V.: High-Dimensional Real-Parameter Optimization using Self-Adaptive Differential Evolution Algorithm with Population Size Reduction. In: Proceedings of the 2008 IEEE Congress on Evolutionary Computation, pp. 2032–2039. IEEE Press, Los Alamitos (2008)
3. Chen, S.: An Analysis of Locust Swarms on Large Scale Global Optimization Problems. In: Korb, K., Randall, M., Hendtlass, T. (eds.) ACAL 2009. LNCS, vol. 5865, pp. 232–241. Springer, Heidelberg (2009)
4. Chen, S.: Locust Swarms – A New Multi-Optima Search Technique. In: Proceedings of the 2009 IEEE Congress on Evolutionary Computation, pp. 1745–1752. IEEE Press, Los Alamitos (2009)

5. Chen, S., Miura, K., Razzaqi, S.: Analyzing the Role of "Smart" Start Points in Coarse Search-Greedy Search. In: Randall, M., Abbass, H.A., Wiles, J. (eds.) ACAL 2007. LNCS (LNAI), vol. 4828, pp. 13–24. Springer, Heidelberg (2007)

6. Chen, S., Razzaqi, S., Lupien, V.: Towards the Automated Design of Phased Array Ultrasonic Transducers – Using Particle Swarms to find "Smart" Start Points. In: Okuno, H.G., Moonis, A. (eds.) IEA/AIE 2007. LNCS (LNAI), vol. 4570, pp. 313–323. Springer, Heidelberg (2007)

7. Chen, S., Razzaqi, S., Lupien, V.: An Evolution Strategy for Improving the Design of Phased Array Transducers. In: Proceedings of the 2006 IEEE Congress on Evolutionary Computation, pp. 2859–2863. IEEE Press, Los Alamitos (2006)

8. Hendtlass, T.: WoSP: A Multi-Optima Particle Swarm Algorithm. In: Proceedings of the 2005 IEEE Congress on Evolutionary Computation, pp. 727–734. IEEE Press, Los Alamitos (2005)

9. Hsieh, S.-T., Sun, T.-Y., Liu, C.-C., Tsai, S.-J.: Solving Large Scale Global Optimization Using Improved Particle Swarm Optimizer. In: Proceedings of the 2008 IEEE Congress on Evolutionary Computation, pp. 1777–1784. IEEE Press, Los Alamitos (2008)

10. Kennedy, J., Eberhart, R.C.: Particle Swarm Optimization. In: Proceedings of the IEEE International Conference on Neural Networks, vol. 4, pp. 1942–1948. IEEE Press, Los Alamitos (1995)

11. MacNich, C.: Towards Unbiased Benchmarking of Evolutionary and Hybrid Algorithms for Real-valued Optimisation. Connection Science 19(4), 361–385 (2007)

12. MacNish, C., Yao, X.: Direction Matters in High-Dimensional Optimization. In: Proceedings of the 2008 IEEE Congress on Evolutionary Computation, pp. 2372–2379. IEEE Press, Los Alamitos (2008)

13. Malan, K., Engelbrecht, A.: Quantifying Ruggedness of Continuous Landscapes using Entropy. In: Proceedings of the 2009 IEEE Congress on Evolutionary Computation, pp. 1440–1447. IEEE Press, Los Alamitos (2009)

14. Tang, K., Yao, X., Suganthan, P.N., MacNish, C., Chen, Y.P., Chen, C.M., Yang, Z.: Benchmark Functions for the CEC 2008 Special Session and Competition on Large Scale Global Optimization. Technical Report (2007),
http://www.ntu.edu.sg/home/EPNSugan

15. Tseng, L.-Y., Chen, C.: Multiple Trajectory Search for Large Scale Global Optimization. In: Proceedings of the 2008 IEEE Congress on Evolutionary Computation, pp. 3052–3059. IEEE Press, Los Alamitos (2008)

16. Wang, Y., Li, B.: A Restart Univariate Estimation of Distribution Algorithm: Sampling under Mixed Gaussian and Levy probability Distribution. In: Proceedings of the 2008 IEEE Congress on Evolutionary Computation, pp. 3917–3924. IEEE Press, Los Alamitos (2008)

17. Yang, Z., Tang, K., Yao, X.: Multilevel Cooperative Coevolution for Large Scale Optimization. In: Proceedings of the 2008 IEEE Congress on Evolutionary Computation, pp. 1663–1670. IEEE Press, Los Alamitos (2008)

18. Zamuda, A., Brest, J., Boskovic, B., Zumer, V.: Large Scale Global Optimization using Differential Evolution with Self-adaptation and Cooperative Co-evolution. In: Proceedings of the 2008 IEEE Congress on Evolutionary Computation, pp. 3718–3725. IEEE Press, Los Alamitos (2008)

19. Zhao, S.Z., Liang, J.J., Suganthan, P.N., Tasgetiren, M.F.: Dynamic Multi-Swarm Particle Swarm Optimizer with Local Search for Large Scale Global Optimization. In: Proceedings of the 2008 IEEE Congress on Evolutionary Computation, pp. 3845–3852. IEEE Press, Los Alamitos (2008)

A Hybrid Extremal Optimisation Approach for the Bin Packing Problem

Pedro Gómez-Meneses[1,2] and Marcus Randall[1]

[1] School of Information Technology, Bond University, QLD 4229, Australia
pedgomez@bond.edu.au, mrandall@bond.edu.au
[2] Universidad Católica de la Santísima Concepción, Concepción, Chile
pgomez@ucsc.cl

Abstract. Extremal optimisation (EO) is a simple and effective technique that is influenced by nature and which is especially suitable to solve assignment type problems. EO uses the principle of eliminating the weakest or the least adapted component and replacing it by a random one. This paper presents a new hybrid EO approach that consists of an EO framework with an improved local search for the bin packing problem (BPP). The stochastic nature of the EO framework allows the solution to move between feasible and infeasible spaces. Hence the solution has the possibility of escaping from a stagnant position to explore new feasible regions. The exploration of a feasible space is complemented with an improved local search mechanism developed on the basis of the proposed Falkenauer's technique. The new local search procedure increases the probability of finding better solutions. The results show that the new algorithm is able to obtain optimal and efficient results for large problems when the approach is compared with the best known methods.

1 Introduction

In recent times, a large amount of research has been undertaken to explore novel optimisation meta-heuristics inspired by nature that can solve complex problems. These techniques seek to use the fewest parameters with the smallest use of computational time and memory. Inside the group of evolutionary meta-heuristics, dominated by evolutionary algorithms, there is a simple and effective evolutionary method of optimisation proposed by Boettcher and Percus [1] called extremal optimisation (EO) which is appropriate for solving combinatorial problems. The principle behind EO is to eliminate the weakest or the least adapted component value and replace it by a random one at each iteration without having to tune many parameters.

A hybrid extremal optimisation (HEO) algorithm is presented which finds solutions through the collaborative use of EO with an improved tailor-made local search technique for BPP proposed by Falkenauer [2]. The original technique consists of redistributing a series of items between non-full bins using three different steps. The aim is to reduce the number of necessary bins needed to contain all the items by at least one for each application of the algorithm. In the

K. Korb, M. Randall, and T. Hendtlass (Eds.): ACAL 2009, LNAI 5865, pp. 242–251, 2009.

proposed local search approach, we add a fourth step that achieves improved results. The EO part of this approach allows the solution state to move between feasible and infeasible spaces. This movement permits the exploration of the search space without being stuck in a particular region of the landscape. The results show that the new hybrid algorithm is able to obtain optimal results for large problems. The advantage of this proposal is its simplicity of implementation compared to other techniques [2, 3, 4, 5, 6, 7].

Given this new approach, we test the behaviour of HEO on the BPP using well-known benchmark problems from the OR-Library [8]. The results show that the proposed algorithm is an effective approach and competitive with other methods [2, 3, 6, 7].

The rest of this paper is organised as follows. In Section 2, a summary of EO is given while Section 3 explains the HEO algorithm to solve the BPP. Section 4 presents an analysis of the obtained results. Finally, in Section 5 we conclude and discuss the future work arising from this research.

2   Extremal Optimisation

EO is an evolutionary meta-heuristic proposed by Bak and Sneppen [9] based on a model of co-evolution between species. This model describes species' evolution via extinction events as a self-organised critical (SOC) [10] process. SOC tries to explain the manifestation of complex phenomena in nature such as the formation of sand piles and the occurrences of earthquakes [11]. The main characteristic is that a power law describes the events in the system. Simply put, these systems have some critical points that are configured in a particular way. When an event converges toward one of these points, a critical state is reached which triggers a series of changes over the elements in the vicinity of the critical point. The system is self-organised to reach a new state of equilibrium. Finally, the system evolves in a transient period of stability until then next critical point is reached.

A good way to understand EO's characteristics is to compare it with another well-known method such as the genetic algorithms (GA) [12]. First, a GA commonly has a set of parameters to be tuned for proper operation; however, in EO only one specific parameter must be tuned. Second, in EO the fitness value is not calculated for each structure that represents a solution as in a GA but for each component of the structure. Each component is evaluated according to its contribution in obtaining the best solution. Third, canonical EO works with a single solution instead of a population of solutions as in GA. Last, EO removes the worst components for the next generations; in contrast, GA promotes a group of elite solutions.

There is only one EO specific parameter that is often referred to as τ [13]. This parameter is used probabilistically to choose the component value to be changed at each iteration of the algorithm. The algorithm ranks the components and assigns to them a number from 1 to n using the fitness of each one (where n in the number of components). Therefore, the fitness must be sorted from the worst to the best evaluated. Then a selection method such as roulette wheel

(RWS) is used to choose the component whose value will be changed to a random one according to the probability calculated for each component as is shown in Equation 1. If the new component makes the solution the best found so far, according to the evaluation function, then this solution is saved as X_{best}. The EO process is shown in Algorithm 1.

$$P_i = i^{-\tau} \qquad \forall i \quad 1 \leq i \leq n, \quad \tau \geq 0 \tag{1}$$

where:
  n is the total number of components evaluated and ranked, and
  P_i is the probability that the ith component is chosen.

Algorithm 1. Standard EO pseudo-code for minimisation problem

---

Generate an initial random solution $X=(x_1, x_2, \ldots, x_n)$ and set $X_{\text{best}} = X$;
Generate the probabilities array P according to Equation 1;
for a preset number of iterations **do**
  Evaluate and rank fitness λ_i for each x_i from worst to best, $1 \leq i \leq n$;
  $j =$ Select component based on the probability of its rank P_i using RWS;
  $x_j =$ Generate a random appropriate value that is not equal to x_j;
  $Eva(X) =$ Evaluate the new solution;
  if $Eva(X) < Eva(X_{\text{best}})$ **then** $X_{\text{best}} = X$;
end for
Return X_{best} and $Eva(X_{\text{best}})$;

---

3   HEO for the BPP

This section shows how the HEO approach is implemented to deal with the BPP. Section 3.1 presents a brief and concise definition of the BPP. Section 3.2 describes the improved local search developed for the BPP. Section 3.3 introduces the HEO approach for the BPP using a integrated version of EO with the proposed local search mechanism.

3.1   The Bin Packing Problem

Many production and distribution tasks require that a series of items with different shapes, sizes, or weights be packed into bins or boxes with a limited capacity. The BPP is a known $\mathcal{NP}$-hard combinatorial optimisation problem (COP) [14] and the one-dimensional bin packing problem is the simplest version of this problem and can be formally described as follows.

A infinite set of bins with a one size bin capacity $C > 0$, and a finite set of n items $w_i = \{w_1, w_2, \ldots, w_n\}$ with different weights among $0 < w_i \leq C$ is to be packed. Find the smallest number m of bins needed to contain all of the n items such that the weight of the items packed in each bin must not exceed the bin capacity C, and that one item w_i can be in one and only one bin.

The bin packing problem can be applied to tasks such as to backup of tapes or disks of equal capacity, to allocate data onto blocks memory with the same size, or to assign processes on a system with identical parallel processors. All these tasks follow the idea of minimising the wastage of resources in the bins.

3.2   A Local Search for the BPP

Local Search (LS) is a mechanism used by some variants of evolutionary heuristics to improve the quality of solutions they receive. LS is commonly used as an iterative process that starts with a feasible solution and then this is improved by performing local modifications. This process is repeated until the current solution can not become better.

Preliminary research by Hendlass and Randall [4] showed the benefits of applying a LS at EO to solve 15 BPP instances ranging in size from 120 items to 500 items. Motivated by this promising result, we apply a improved LS to help solve a large set of 80 BPP instances ranging in size from 120 items to 1000 items. We take as a base the LS mechanism proposed by Falkenauer [2] which in turn was inspired by the work of Martello and Toth [7]. This LS mechanism was also applied to an ant colony approach by Levine and Ducatelle [6].

The augmented improved LS implementation can be explained as follows. The procedure begins by moving all the items contained inside the two least full bins into a (temporary) free bin. Four sequential steps are then applied for each bin that is not full. These steps consist of interchanging one or two items between the current bin and the bin with free items so that the current bin could be refilled up to the limit. The first step swaps two current items by two free items, the second step swaps two current items by one free item and the third step swaps one current item by one free item. The new fourth step swaps one current item by two free items. Next, free items are reinserted into bins provided that the latter have enough space to contain it. Finally, the remaining items in the free bin are then put into a new bin.

This local search process is repeated while new solutions are feasible. The aim of this local search is to reduce by one the number of necessary bins to contain every item.

3.3   A HEO Implementation for the BPP

One of the main characteristics of EO is its ability to locate a possible optimal solution stochastically in search space. This characteristic enables EO to avoid being trapped in a local optimum; however, this characteristic also hinders EO from being able to refine the search when a solution is near to the optimal result. Recent research has reported the achievement of better results on some optimisation problems when a local search technique complements EO [4, 15, 16, 17]. Therefore, HEO is made up of an EO framework and a local search procedure that is executed every time a feasible solution is found.

HEO starts by generating an initial solution using the Best Fit Decreasing (BFD) strategy [18]. BFD sorts the items in decreasing order according to its weight and then trys to put an item into a bin that has minimal free space.

BFD guarantees to use no more than $\frac{11}{9}B + 1$ bins [18], where B is the optimum number of bins. The reason for generating an initial solution using BFD, rather than purely at random, is based on our interest in minimising the initial number of bins necessary to contain all the items. That is done with the goal of making the process more efficient.

As solutions in EO can move between infeasible and feasible space (see Randall [17]), the next step is to carry out an appropriate selection process for both cases. These processes follow the EO scheme but with a slight modification.

When the current solution is feasible, the bin with the largest free space available is selected since the aim of BPP is to minimise the numbers of bins. One item is picked randomly from the selected bin to be moved into other available bin which is chosen according to the new EO rules. Hence, these bins are ranked by a fitness value which is defined as follows:

$$\lambda_j = \frac{R_j}{N_j} \qquad \forall j \text{ such that } 0 < R_j < C \tag{2}$$

where:
  R_j is the current weight of the jth bin,
  N_j is the number of items in the jth bin, and
  λ_j is the fitness of the jth bin.

The idea of this selection process is to try to put the item in a bin where the relation $used\_space/number\_of\_items$ is the smallest. This means that between two bins with the same free space to receive a new item, it is preferable to use the bin with more items of small size. This is because if the bin gets overloaded then it is easier to adjust several small items than a few big items within the bin. This helps the infeasible solution procedure to obtain a feasible solution again.

On the other hand, when the current solution is infeasible, the most overloaded bin is selected since this bin degrades the solution. A light item is chosen with a greater probability than a heavy item from the selected bin as it is less likely to overload another bin. Next, the bin to which the item will be added is chosen between the non-full bins according to the EO rules. Thus, these bins are ranked by a fitness value using Equation 2.

Finally, the new solution is accepted and evaluated to see if it is feasible or not. In the case that a better feasible solution is obtained, the best solution found so far is updated and the local search process is invoked. Algorithm 2 shows the steps to solve the BPP using the HEO approach.

4   Computational Experiments

The proposed HEO method was coded in the C language and compiled with **gcc** version 4.3.0. The computing platform used to perform the tests has a 1.86 GHz Intel Core2 CPU and 917 MB of memory, and runs under Linux.

The set test used in this paper comes from the OR-Library [8]. This test was contributed by Falkenauer [2] and consists of 80 problems which are divided into four groups of 20 problems. Each of these has 120, 250, 500 and 1000 items. The items' weights are spread uniformly between 20 and 100 and the capacity for all bins is 150. The problem's name is presented using the nomenclature uNNN_PP

Algorithm 2. Pseudocode of the HEO model for the BPP

Initialise a feasible solution S using the BFD method and set $S_{best} \leftarrow S$;
Generate the probabilities array P according to Equation 1;
$S_{new} = LocalSearch(S)$;
Evaluate the new solution $Eva(S_{new})$;
if $Eva(S_{new}) < Eva(S_{best})$ **then** $S_{best} \leftarrow S_{new}$;
for a preset number of iterations **do**
   if the current solution is feasible **then**
      Select the emptiest bin and choose a random item;
   else
      Select the most overloaded bin and choose a light item;
   Evaluate and rank the fitness λ_i for the remaining bins with free space using Eq. 2;

   Choose a bin based on the probability of its rank P_i using RWS;
   Put the selected item into the chosen bin;
   Evaluate the new solution $Eva(S_{new})$;
   if S_{new} is feasible **then**
      if $Eva(S_{new}) < Eva(S_{best})$ **then** $S_{best} \leftarrow S_{new}$;
      $S_{new} = LocalSearch(S_{new})$;
      if $Eva(S_{new}) < Eva(S_{best})$ **then** $S_{best} \leftarrow S_{new}$;
end for

---

where NNN is the number of items and PP is the problem's identification. Each problem is accompanied by its theoretical optimal result.

All results are presented as a percentage gap determined using the theoretical optimal solution and the solution obtained by each analysed method. It is defined as $\%gap = \frac{b-a}{b} \times 100$, where a is the cost of the obtained solution and b is the cost of the theoretical optimal solution. A value of zero means that the method found the optimal.

The first part of the experiments describes the results when a local search mechanism is applied. Here, the local search method proposed by Falkenauer [2] is compared with the improved version proposed in this paper when one new step is added. Table 1 shows the results of the BSD method, the three step Falkenauer method and the four step proposed method. The latter two were tested using the BSD result as the initial solution. As can be seen, the improvement made in the new version, with respect to Falkenauer's local search method, was better for 52 out of 74 problems where the initial solution is not optimal. Also, the average percentage gap for the four groups decreased from around 1.1% to under 0.67%. This means that for the four groups, the improved local search method was able to find closer results to the theoretical solution. Thus, the improved local search method was chosen to be used in the next part when the HEO approach is implemented.

In the second part, HEO is configured to run 10 test trials. The number of iterations to complete a HEO process is of 100000. The τ parameter is set at 1.4 which has been reported in previous research [4, 17, 19] as a good value to obtain efficient solutions.

The results generated by the HEO approach, shown in Table 2, are compared with those reported by Martello and Toth (MTP) [7], Alvim et al. (LS-BPP) [3],

Table 1. Results for the initial BSD solution, Falkenauer local search (FLS) and the proposed local search (PLS). Note that the number of bins and the % gap are shown.

Problem name	Theo Opt	BFD	% gap	FLS	% gap	PLS	% gap	Problem name	Theo Opt	BFD	% gap	FLS	% gap	PLS	% gap
U120_00	48	49	2.08	49	2.08	48	0	U500_00	198	201	1.52	201	1.52	200	1.01
U120_01	49	49	0	49	0	49	0	U500_01	201	204	1.49	203	1	202	0.5
U120_02	46	47	2.17	46	0	46	0	U500_02	202	205	1.49	204	0.99	203	0.5
U120_03	49	50	2.04	50	2.04	49	0	U500_03	204	207	1.47	207	1.47	206	0.98
U120_04	50	50	0	50	0	50	0	U500_04	206	209	1.46	209	1.46	208	0.97
U120_05	48	49	2.08	48	0	48	0	U500_05	206	207	0.49	207	0.49	206	0
U120_06	48	49	2.08	49	2.08	49	2.08	U500_06	207	210	1.45	210	1.45	209	0.97
U120_07	49	50	2.04	50	2.04	49	0	U500_07	204	207	1.47	207	1.47	206	0.98
U120_08	50	51	2	51	2	51	2	U500_08	196	199	1.53	198	1.02	198	1.02
U120_09	46	47	2.17	47	2.17	47	2.17	U500_09	200	204	0.99	203	0.5	202	0
U120_10	52	52	0	52	0	52	0	U500_10	200	202	1	202	1	201	0.5
U120_11	49	50	2.04	50	2.04	49	0	U500_11	200	203	1.5	203	1.5	202	1
U120_12	48	49	2.08	49	2.08	49	2.08	U500_12	199	202	1.51	202	1.51	201	1.01
U120_13	49	49	0	49	0	49	0	U500_13	196	198	1.02	198	1.02	196	0
U120_14	50	50	0	50	0	50	0	U500_14	204	206	0.98	206	0.98	204	0
U120_15	48	49	2.08	49	2.08	48	0	U500_15	201	204	1.49	203	1	202	0.5
U120_16	52	52	0	52	0	52	0	U500_16	202	205	1.49	204	0.99	203	0.5
U120_17	52	53	1.92	53	1.92	53	1.92	U500_17	198	201	1.52	201	1.52	200	1.01
U120_18	49	50	2.04	49	0	49	0	U500_18	202	205	1.49	204	0.99	203	0.5
U120_19	49	50	2.04	50	2.04	50	2.04	U500_19	196	199	1.53	198	1.02	199	1.53
Average			1.44		1.13		**0.62**	Average			1.34		1.14		**0.67**

Problem name	Theo Opt	BFD	% gap	FLS	% gap	PLS	% gap	Problem name	Theo Opt	BFD	% gap	FLS	% gap	PLS	% gap
U250_00	99	100	1.01	100	1.01	99	0	U1000_00	399	403	1	403	1	401	0.5
U250_01	100	101	1	101	1	100	0	U1000_01	406	411	1.23	410	0.99	407	0.25
U250_02	102	104	1.96	103	0.98	103	0.98	U1000_02	411	416	1.22	414	0.73	415	0.97
U250_03	100	101	1	101	1	100	0	U1000_03	411	416	1.22	416	1.22	415	0.97
U250_04	101	102	0.99	102	0.99	102	0.99	U1000_04	397	402	1.26	400	0.76	399	0.5
U250_05	101	104	2.97	103	1.98	102	0.99	U1000_05	399	404	1.25	404	1.25	403	1
U250_06	102	103	0.98	103	0.98	102	0	U1000_06	395	399	1.01	399	1.01	396	0.25
U250_07	103	105	1.94	104	0.97	104	0.97	U1000_07	404	408	0.99	408	0.99	406	0.5
U250_08	105	107	1.9	106	0.95	106	0.95	U1000_08	399	404	1.25	402	0.75	403	1
U250_09	101	102	0.99	102	0.99	102	0.99	U1000_09	397	404	1.76	402	1.26	403	1.51
U250_10	105	106	0.95	106	0.95	106	0.95	U1000_10	400	404	1	404	1	401	0.25
U250_11	101	103	1.98	102	0.99	102	0.99	U1000_11	401	405	1	405	1	403	0.5
U250_12	105	107	1.9	107	1.9	106	0.95	U1000_12	393	398	1.27	397	1.02	395	0.51
U250_13	102	104	1.96	104	1.96	103	0.98	U1000_13	396	401	1.26	401	1.26	397	0.25
U250_14	100	101	1	101	1	100	0	U1000_14	394	400	1.52	399	1.27	397	0.76
U250_15	105	107	1.9	107	1.9	106	0.95	U1000_15	402	408	1.49	407	1.24	406	1
U250_16	97	99	2.06	98	1.03	98	1.03	U1000_16	404	407	0.74	407	0.74	405	0.25
U250_17	100	101	1	101	1	100	0	U1000_17	404	409	1.24	408	0.99	407	0.74
U250_18	100	102	2	102	2	101	1	U1000_18	399	403	1	402	0.75	402	0.75
U250_19	102	103	0.98	103	0.98	102	0	U1000_19	400	406	1.5	406	1.5	403	0.75
Average			1.52		1.23		**0.64**	Average			1.21		1.04		**0.66**

Falkenauer (HGGA) [2], Levine and Ducatelle (HACO) [6], and Randall, Hendtlass and Lewis (SEO, PEO)[1] [20]. For the four methods that do not use EO, the results are presented only using one column that shows the percentage gap with respect to the theoretical solution. For the remaining three methods that use EO, the results are presented by three columns, "mi%", "me%", and "ma%". These denote the minimum, median and maximum percentage gap respectively.

HEO is confirmed to be an efficient and competitive approach to solve the BPP. In 77 out of 80 problems, the theoretical optimal result was found, only one less than HACO which found 78. The result of the three remaining bins (u250_07, u250_12, u250_13) is only one bin from the theoretical optimal solution. In fact, the problem u250_13 has no solution for the theoretical optimal as is reported in Levine and Ducatelle [6].

---

[1] SEO means EO with a single solution. PEO means EO with a population of solutions.

Table 2. Comparative HEO test results. Note that blank spaces are present because these problems were not solved by the authors.

Problem name	Theo Opt	MTP %	LS-BPP %	HGGA %	HACO %	SEO mi%	SEO me%	SEO ma%	PEO mi%	PEO me%	PEO ma%	HEO mi%	HEO me%	HEO ma%
U250.00	99	1.01	0	0	0	0	0.51	1.01	0	1.01	1.01	0	0	0
U250.01	100	0	0	0	0	0	0	0	0	0	0	0	0	0
U250.02	102	0	0.98	0	0	0	0.49	0.98	0	0.98	0.98	0	0	0
U250.03	100	0	0	0	0	0	0	0	0	0	1	0	0	0
U250.04	101	0	0	0	0	0	0	0.99	0	0.99	0.99	0	0	0
U250.05	101	1.98	0.99	0	0	0	0.99	1.98	0.99	0.99	1.98	0	0	0
U250.06	102	0	0.97	0	0	0	0	0	0	0	0	0	0	0.99
U250.07	103	0.97	0.95	0.97	0	0	0	3.85	0	0	1.9	0.97	0.97	0.97
U250.08	105	0	0	0	0	0.95	0.95	0.95	0.95	0.95	0.95	0	0	0
U250.09	101	0.99	0	0	0	0.99	0.99	2.97	0.99	0.99	1.9	0	0	0
U250.10	105	0.95	0.99	0	0	0	0.99	1.9	0	0.95	0.95	0	0	0
U250.11	101	0	0.95	0	0	0.99	0.99	0.99	0.99	0.99	0.99	0.95	0.95	0.95
U250.12	105	0.95	0.98	0.95	0.95	0	0	0.94	0	0	0.94	0.95	0.98	0.98
U250.13	102	0.98	0	0.98	0.98	0	0	0.97	0	0	0.97	0.98	0.98	0.98
U250.14	100	0	0.95	0	0	0	1	1	0	1	1	0	0	0
U250.15	105	0.95	1.03	0	0	0.95	0.95	3.81	0.95	1.9	1.9	0.95	0.95	0.95
U250.16	97	1.03	0	0	0	0	0	3.09	1.03	1.03	1.03	0	0	0
U250.17	100	0	0	0	0	0	0	1	1	1	1	0	0	0
U250.18	100	0	0	0	0	1	1	1	1	1	1	0	0	0
U250.19	102	0	0	0	0	0	0	0.98	0	0	0.98	0	0	0
Average		0.59	0.44	0.15	0.1	0.19	0.34	1.47	0.35	0.64	0.98	0.15	0.19	0.24

Problem name	Theo Opt	MTP %	LS-BPP %	HGGA %	HACO %	SEO mi%	SEO me%	SEO ma%	PEO mi%	PEO me%	PEO ma%	HEO mi%	HEO me%	HEO ma%
U1000.00	399	1	0	0	0							0	0	0.25
U1000.01	406	0.99	0	0	0							0	0	0.25
U1000.02	411	1.22	0	0	0							0	0	0
U1000.03	411	1.22	0.49	0	0							0	0.24	0.49
U1000.04	397	1.01	0.25	0	0							0	0.25	0.5
U1000.05	399	0.75	0.25	0	0							0	0.25	0.25
U1000.06	395	0.76	0	0	0							0	0	0
U1000.07	404	0.5	0	0	0							0	0	0.25
U1000.08	399	0.75	0.5	0	0							0	0	0.25
U1000.09	397	1.26	0	0	0							0	0	0.25
U1000.10	400	1	0	0	0							0	0	0
U1000.11	401	0.75	0.25	0	0							0	0	0.25
U1000.12	393	0.76	0	0	0							0	0	0.25
U1000.13	396	1.26	0	0	0							0	0.25	0.25
U1000.14	394	1.27	0.51	0	0							0	0.13	0.25
U1000.15	402	1.24	0.25	0	0							0	0	0.25
U1000.16	404	0.74	0	0	0							0	0	0.25
U1000.17	404	0.74	0.25	0	0							0	0	0.25
U1000.18	399	1	0	0	0							0	0	0
U1000.19	400	1.25	0	0	0							0	0	0.25
Average		0.97	0.14	0	0							0	0.04	0.22

Problem name	Theo Opt	MTP %	LS-BPP %	HGGA %	HACO %	SEO mi%	SEO me%	SEO ma%	PEO mi%	PEO me%	PEO ma%	HEO mi%	HEO me%	HEO ma%
U120.00	48	0	0	0	0	0	0	2.08	0	0	0	0	0	0
U120.01	49	0	0	0	0	0	0	0	0	0	0	0	0	0
U120.02	46	0	0	0	0	0	2.04	4.08	2.04	2.04	2.04	0	0	0
U120.03	49	0	0	0	0	0	0	0	0	0	0	0	0	0
U120.04	50	0	0	0	0	0	0	0	0	0	0	0	0	0
U120.05	48	0	0	0	0	0	0	2.08	0	0	2.08	0	0	0
U120.06	48	0	0	0	0	0	0	2.08	0	0	2.08	0	0	0
U120.07	49	0	2	0	0	0	0	0	0	0	0	0	0	0
U120.08	50	2	0	2	0	0	0	1.96	0	0	0	0	0	0
U120.09	46	0	2.17	0	0	0	1.09	2.17	0	2.17	2.17	0	0	0
U120.10	52	0	0	0	0	0	0	0	0	0	0	0	0	0
U120.11	49	0	0	0	0	0	0	2.04	0	0	0	0	0	0
U120.12	48	0	0	0	0	0	2.08	2.08	0	2.08	2.08	0	0	0
U120.13	49	0	0	0	0	0	0	0	0	0	0	0	0	0
U120.14	50	0	0	0	0	0	0	0	0	0	0	0	0	0
U120.15	48	0	0	0	0	0	0	2.08	0	0	0	0	0	0
U120.16	52	0	0	0	0	0	0	1.92	0	0	0	0	0	0
U120.17	52	0	0	0	0	0	0	5.77	0	1.92	1.92	0	0	0
U120.18	49	0	0	0	0	0	0	2	0	0	0	0	0	0
U120.19	49	2.04	2.04	2.04	0	0	0	0	0	0	0	0	0	0
Average		0.2	0.31	0.2	0	0	0.26	1.41	0.05	0.41	0.62	0	0	0

Problem name	Theo Opt	MTP %	LS-BPP %	HGGA %	HACO %	SEO mi%	SEO me%	SEO ma%	PEO mi%	PEO me%	PEO ma%	HEO mi%	HEO me%	HEO ma%
U500.00	198	1.52	0.51	0	0	0.51	0.51	2.53	0.51	1.01	1.01	0	0	0.25
U500.01	201	0.5	0.5	0	0	0.5	0.5	0.5	0.5	1	1	0	0	0.5
U500.02	202	0.99	0	0	0	0.5	0.5	0.99	0.5	0.74	0.99	0	0	0
U500.03	204	0.98	0.49	0	0	0.49	0.49	1.47	0.98	0.98	0.98	0	0	0.49
U500.04	206	1.46	0	0	0	0.49	0.49	1.94	0.49	0.49	0.97	0	0	0
U500.05	206	0.49	0	0	0	0.97	0.97	2.91	0.49	0.73	0.97	0	0	0
U500.06	207	1.45	0.48	0	0	0.48	0.72	1.45	0.97	0.97	1.45	0	0	0.48
U500.07	204	1.47	0.49	0	0	0.49	1.23	3.43	0.98	1.47	1.96	0	0.25	0.49
U500.08	196	1.02	0.51	2	0	0.26	1.02		0.51	0.51	1.02	0	0.25	0.49
U500.09	202	0.99	0	0	0	0	0.5	0.5	0.5	0.5	0.99	0	0	0.51
U500.10	200	1	0.5	0	0	0.5	0.5	0.5	0.5	0.5	1	0	0	0
U500.11	200	1	0.5	0	0	0.5	0.5	3	0.5	0.5	1.5	0	0	0.5
U500.12	199	1.51	0.5	0	0	0.5	0.5	1.01	1.01	1.01	1.01	0	0	0.5
U500.13	196	0.51	0	0	0	0.51	0.51	1.53	0.51	0.51	0.51	0	0	0
U500.14	204	0.49	0	0	0	0.49	0.49	8.33	0.49	0.49	0.98	0	0	0
U500.15	201	1	0	0	0	0.5	0.5	0.99	0.5	0.5	1	0	0	0
U500.16	202	0.99	0	0	0	0.51	0.51	0.99	0.51	1.01	1.01	0	0	0.5
U500.17	198	1.52	0	0	0	0.51	0.51	1.01	0.51	1.01	1.01	0	0	0
U500.18	202	1.49	0	0	0	0	0.5	1.98	0.99	1.49	1.49	0	0	0.51
U500.19	196	1.53	0.51	2.04	0	0.51	0.51	1.02	0.51	1.02	1.53	0	0	0
Average		1.09	0.25	0.2	0	0.27	0.51	1.81	0.52	0.82	1.09	0.01	0.01	0.2

Table 3. Average time, in seconds, for each group test

Problem name	MTP time	LS-BPP time	HGGA time	HACO time	HEO time
u120	370	0.2	381	1	1
u250	1516	6.7	1337	52	1.2
u500	1535	37.25	1015	50	1.2
u1000	9393	143.55	7059	147	1.8

Despite the good results obtained by the previous two works where EO is applied (SEO and PEO), we can infer from the development of HEO that the better results are due to the use of a good initial solution, the improved local search mechanism and the modified EO selection process.

In Table 3 we can observe the excellent runtime of HEO in relation to the other methods. Indeed, the difference in the average time amongst the four test groups is quite similar. This shows a balanced performance in relation to the number of items to be packed.

5   Conclusions

This paper has described a Hybrid EO approach for the BPP. EO is a recent form of search for which there exists a relatively small amount of work that is applicable to combinatorial problems. Some of the research has been in combining EO with a local search mechanism to obtain better solutions. The strength of HEO, which is EO and improved local search, lies in exploiting the stochastic nature of EO as a coarse grain solver to move solutions from stagnant positions to explore new regions toward the optimal result. The improved local search mechanism works in a fine grain way to find better solutions from the last found feasible solution.

Results obtained for the uniformly distributed test set with HEO have been encouraging. This approach has been able to find the optimal solutions for 77 out of 80 problems. For the three remaining problems, the result was only one item from the theoretical optimal solution. Note that for one of these three problems, it is not possible to find the optimal solution.

When the proposed HEO is compared to the existing evolutionary approaches, we can see that HEO is able to obtain quality solutions as good as the best of them. The advantage that our approach has are the low requirements of runtime, memory, parametrisation and implementation.

The proposed approach could be adapted to solve other problems such as the cutting stock problem, the multiprocessor scheduling problem, and the graph colouring problem. Also, we are looking to adapt the algorithm presented to solve the multi-objective version of BPP.

References

[1] Boettcher, S., Percus, A.G.: Evolutionary strategies extremal optimization: Methods derived from co-evolution. In: GECCO 1999: Proceedings of the Genetic and Evolutionary Computation Conference, pp. 825–832 (1999)

[2] Falkenauer, E.: A hybrid grouping genetic algorithm for bin packing. Journal of Heuristics 2(1), 5–30 (1996)

[3] Alvim, A., Glover, F., Ribeiro, C., Aloise, D.: Local search for the bin packing problem. In: Extended Abstracts of the III Metaheuristics International Conference (MIC 1999), Angra dos Reis, Brazil, pp. 7–12 (1999)

[4] Hendtlass, T., Randall, M.: Extremal optimisation and bin packing. In: Li, X., Kirley, M., Zhang, M., Green, D., Ciesielski, V., Abbass, H.A., Michalewicz, Z., Hendtlass, T., Deb, K., Tan, K.C., Branke, J., Shi, Y. (eds.) SEAL 2008. LNCS, vol. 5361, pp. 220–228. Springer, Heidelberg (2008)

[5] Karmarkar, N., Karp, R.M.: The differencing method of set partitioning. Technical Report UCB/CSD-83-113, EECS Department, University of California, Berkeley, CA, USA (1983)

[6] Levine, J., Ducatelle, F.: Ant colony optimisation and local search for bin packing and cutting stock problems. Journal of the Operational Research Society 55(7), 705–716 (2004)

[7] Martello, S., Toth, P.: Lower bounds and reduction procedures for the bin packing problem. Discrete Applied Mathematics 28(1), 59–70 (1990)

[8] Beasley, J.E.: OR-Library: Distributing test problems by electronic mail. Journal of the Operational Research Society 41(11), 1069–1072 (1990)

[9] Bak, P., Sneppen, K.: Punctuated equilibrium and criticality in a simple model of evolution. Physical Review Letters 71(24), 4083–4086 (1993)

[10] Bak, P., Tang, C., Wiesenfeld, K.: Self-organized criticality: An explanation of the 1/f noise. Physical Review Letters 59(4), 381–384 (1987)

[11] Bak, P.: How nature works. Springer-Verlag New York Inc., Heidelberg (1996)

[12] Goldberg, D.E.: Genetic Algorithms in Search, Optimization and Machine Learning. Addison-Wesley Longman Publishing Co., Boston (1989)

[13] Boettcher, S.: Extremal optimization: Heuristics via co-evolutionary avalanches. Computing in Science and Engineering 2, 75–82 (2000)

[14] Garey, M.R., Johnson, D.S.: Computers and Intractability: A Guide to the Theory of NP-Completeness. Series of Books in the Mathematical Sciences. W. H. Freeman & Co., New York (1979)

[15] Huang, W., Liu, J.: Extremal optimization with local search for the circular packing problem. In: Proceedings of the Third International Conference on Natural Computation, ICNC 2007, vol. 5, pp. 19–23. IEEE Computer Society, Los Alamitos (2007)

[16] Gómez-Meneses, P., Randall, M.: Extremal optimisation with a penalty approach for the multidimensional knapsack problem. In: Li, X., Kirley, M., Zhang, M., Green, D., Ciesielski, V., Abbass, H.A., Michalewicz, Z., Hendtlass, T., Deb, K., Tan, K.C., Branke, J., Shi, Y. (eds.) SEAL 2008. LNCS, vol. 5361, pp. 229–238. Springer, Heidelberg (2008)

[17] Randall, M.: Enhancements to extremal optimisation for generalised assignment. In: Randall, M., Abbass, H.A., Wiles, J. (eds.) ACAL 2007. LNCS (LNAI), vol. 4828, pp. 369–380. Springer, Heidelberg (2007)

[18] Martello, S., Toth, P.: Knapsack problems: Algorithms and computer implementations. John Wiley & Sons, Inc., New York (1990)

[19] Boettcher, S., Percus, A.G.: Extremal optimization for graph partitioning. Physical Review E 64, 026114 (2001)

[20] Randall, M., Hendtlass, T., Lewis, A.: Extremal optimisation for assignment type problems. In: Biologically-Inspired Optimisation Methods: Parallel Algorithms, Systems and Applications. Studies in Computational Intelligence, vol. 210, pp. 139–164. Springer, Heidelberg (2009)

Feedback of Delayed Rewards in XCS for Environments with Aliasing States

Kuang-Yuan Chen and Peter A. Lindsay

ARC Centre for Complex Systems, The University of Queensland
{k.chen3,p.lindsay}@uq.edu.au

Abstract. Wilson [13] showed how delayed reward feedback can be used to solve many multi-step problems for the widely used XCS learning classifier system. However, Wilson's method – based on back-propagation with discounting from Q-learning – runs into difficulties in environments with aliasing states, since the local reward function often does not converge. This paper describes a different approach to reward feedback, in which a layered reward scheme for XCS classifiers is learnt during training. We show that, with a relatively minor modification to XCS feedback, the approach not only solves problems such as Woods1 but can also solve aliasing states problems such as Littman57, Miyazaki A and MazeB.

Keywords: XCS, Learning Classifier Systems, aliasing states problem, credit assignment, maze problems.

1 Introduction

Learning classifier systems such as XCS have proven very successful in solving maze problems and single step problems such as Multiplexer [13]. During training, XCS agents develop classifier sets using a clever combination of a genetic algorithm and feedback of local rewards (Fig. 1 - see Section 2 for details). Wilson adapted XCS to multi-step problems by introducing a method similar to discounting in Q-learning, whereby the strength of the feedback is adjusted according to the "distance" of the classifier from the goal [15]. This proved effective for finding efficient solutions for many multi-step problems, such as maze problems [17]. However, in partially observable environments with aliasing states (i.e., where there are two or more states with the same input conditions but different distance from the goal) Wilson's approach often gives suboptimal solutions, since the local reward function often does not converge [11, 18].

Fig. 2 gives an example of such a problem, called Littman57 [8]. In this example the agent cannot tell the difference between cells 1a, 1b and 1c since they all have the same input conditions. But since the distance to the food F is different in the three cases, and the same classifier is used at each corresponding step in the solution, the discounted reward will be different for each application of the classifier. Littman et al [9] showed that Q-learning systems could not solve this problem. Although XCS is more general than Q-learning since its update procedure is based on the accuracy of classifier predictions rather than on predictions of individual states, it still suffers

K. Korb, M. Randall, and T. Hendtlass (Eds.): ACAL 2009, LNAI 5865, pp. 252–261, 2009.
© Springer-Verlag Berlin Heidelberg 2009

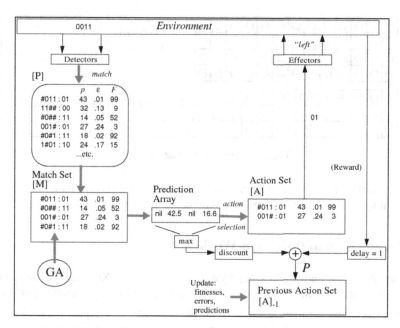

Fig. 1. XCS classifier matching, selection and feedback schematic [14]

from the aliasing states problem, as will be shown in Section 6.2. Another example of an aliasing states problem that is known to cause difficulties for XCS is the intertwined spirals problem [6]. And the problem exists even in the case of two-step problems, as shown by Tang and Jarvis [11].

In this paper we explore a new approach to back-propagating delayed rewards in XCS by replacing the Q-learning-like mechanism by one that uses credit assignment rather than discounting. In our approach, the classifiers used in the Action Set A are tracked at each step in the training epoch, together with the maximum SNP value from the Prediction Array generated at that step. (In the example shown in Fig. 1, the action selected is 01 and its SNP value is 42.5) At the end of the epoch the global reward is fed back to each of the classifiers used in the epoch weighted according to how the SNP value of the classifier compares with the maximum SNP value achieved in the epoch. If the global reward was 1000, for example, and the highest SNP in the epoch was 100, then the two classifiers in the Action Set A in Fig. 1 would receive payoff 425. This payoff is then used to update the classifiers' predicted reward. The details are given in Section 3 below.

We show that in many cases the resulting predicted reward structure in the XCS agent converges to a layered reward scheme [2, 3] which solves the problem efficiently. This approach overcomes the aforesaid aliasing states problem because now the predicted local reward has a better chance of converging, instead of bouncing between multiple different values.

The rest of this paper explains the approach (which we call XCS-GR) in detail and demonstrates its performance on some standard benchmark problems, including Woods1, Littman57 and MiyazakiA. We have also shown that the approach works on

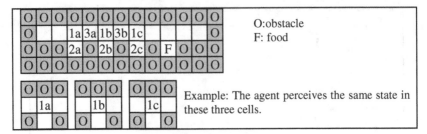

Fig. 2. The Littman57 maze problem

MazeB, another Type-I aliasing maze problem, and a large class of path-finding problems, but space does not permit treatment here.

2 Background and Related Work

XCS stands for eXtended Classifier System. A classifier in XCS consists of a condition, action and three main parameters: A prediction p of the estimated payoff, a prediction error ε, and a fitness estimate F of the accuracy of the payoff prediction [4] (see Fig. 1). For example in Fig. 1, the first classifier in the XCS population [P] has condition #011 and action 01. # is a wildcard that can match any value.

At each step in an epoch, the XCS agent detects the state of its environment and forms a match set [M] of classifiers from [P] whose condition matches the current state. A coverage Genetic Algorithm (GA) is used to create new classifiers in the case that [M] is empty in Fig. 1.) For each possible action a, XCS calculates the *System Net Prediction* (SNP), which is the fitness-weighted average of all classifiers from [M] that use action a. The results are put into a Prediction Array. During exploitation XCS selection the action with the maximum value of SNP; during exploration the choice is random. The classifiers from [M] with the selected action become the Action Set [A]. The agent takes the action and gets external reward R from the environment.

As explained by Wilson [14], XCS uses an adaptation of the Q-learning approach to back-propagation of delayed rewards in which the predicted rewards p_j of classifiers in [A] are updated using the standard Widrow-Hoff delta rule as follows:

$$p_j \leftarrow p_j + \alpha[P - p_j]$$
$$P = R + \gamma V$$

$\qquad$ (1)

where α is the learning rate, R is the external reward, γ is a discount factor, V is the maximum SNP in the Prediction Array in the next step, and P is the resulting payoff. With this, XCS can efficiently handle many single-step problems such as Multiplexer [13] and multi-step maze problems such as Woods1 [8, 14].

In Mitchell's terminology [10], the Woods1 and Woods2 environments are *deterministic*, in the sense that they do not have aliasing states: given the same input conditions, it is sufficient to choose the same action each time. In such cases, back propagation of the global reward in the above manner will eventually cause convergence to a unique Q-value [17]. The mechanism leading to convergence is that the value at each state is changed according to the discount factor γ (i.e. γ times the value of the best move from the next state), so the values will all reflect the minimal distance to food. The environments and the corridor problem [1] in Fig. 3 are classified to be *spatial-oriented* environments [11].

On the other hand, in *temporal-oriented* environments, the sequence of states traversed is *nondeterministic*. In this case, a set of states can lead to the same reward regardless the sequence of these states visited by the agent. So, the identical state found in different distance to the food should receive the same reward instead of discounted reward. The problem was also classified as *aliasing Type-I* in [18]. For example, Littman57 contains the aliasing cells 1a, 1b, 1c, 2a, 2b, 2c, 3a, and 3b.

In the research of Zatuchna [18], mazes with aliasing states were classified into four types based on two major factors having a significant influence on learning process: minimal distance to food d and correct direction to food a. When d_3 and d_5 (of cell-3 and cell-5 respectively) are different but a_3 and a_5 should be the same (such as Littman57) the aliasing states are said to be *Type I*. The average steps to the food are 3.71 and the maximum steps to the food are 8. Their self-adjusting and gradual AgentP systems found solutions for Littman57 with an average of 5.24 and 4.82 steps to the food, respectively. However, this performance is still suboptimal.

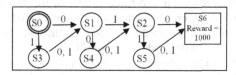

Fig. 3. Corridor problem from [1]: the reward is given when S6 is reached

In Tang and Jarvis's experiment [11], the intertwined spirals problem consists of 194 points, as shown in Fig. 4. The goal of this problem is to learn to classify two sets of points belonging to two distinct spirals in the x-y plane. The reward was deferred for one step only. That is, a reward in the range between 2.0 and -2.0 was given, after two spiral points were classified. The result showed that the prediction of the classifier failed to converge. The problem was that the actions have nondeterministic outcomes. It was "a form of the aliasing state problem: the existence of more than one identical state, which required the same action but returned with different rewards" [11]. Therefore, they concluded that the reason XCS could work well was because those experimental environments such as Woods2 and Corridor environments were "spatially dependent", while their classification problem was "temporally dependent".

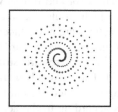

Fig. 4. The intertwined spirals problem in [11]

In [12], Tangamchit et al. showed that two Q-learning robots could be effective at action level, but did not achieve cooperation at the task level (i.e., purposeful division of labour), due to reward discounting. To illustrate the phenomenon, the authors used an example of two robots with a sequential task. The task consists of two parts in strict order. Part 2 starts only after Part 1 finishes. A reward is given to the robots when Part 2 is finished. In their experiment only Robot 1 could do Part 1 but both robots could do Part 2. They gave reward 10 if Robot 1 did Part 1 and Robot 2 did Part 2, but only 8 if Robot 1 did both parts. With Q-learning robots they showed that Robot 1 learnt to do Part 2 rather than leave it to Robot 2, because it could then receive the immediate reward which was higher than the propagated discounted reward when it chose Part-1 and Robot-2 chose Part-2. However, the problem did not occur when they used a Monte-Carlo learning approach with average reward.

These examples demonstrated some of the issues of the discounted reward approach of Q-learning and standard XCS. Our approach, called *XCS with Global Reward* (XCS-GR), was developed to get around these issues.

3   Details of the Approach

Instead of updating the prediction, error and fitness of the classifiers every step (as Fig. 1), our approach is to record the action set and corresponding SNP from its Prediction Array at each step and update the classifiers at the end of each epoch, based on a Global Reward Function (GRF) which measures the effectiveness of the solution. The result is called XCS-GR. The two key components comprised in our approach, the GRF and the method for credit assignment, are described below.

Generally a Global Reward Function would be defined based on the performance of the agent, with suitable penalties deducted. In what follows we used a maximum global reward value of 1,000, as is common in the XCS literature. In this paper we applied a penalty according to the number of steps taken to reach the food. Without the penalty for the number of steps, all states and actions along the path will have the same Q-value, as long as the agent can reach the food eventually.

Our approach to back-propagating the global reward ρ is similar to Wilson's (see equation (1) above) except that the payoff function P is replaced by the following

$$P = \rho \times (V / Vm) + LR \qquad (2)$$

where V is the SNP in Prediction Array for the classifiers used in the selected action set, Vm is the maximum SNP used in the epoch, and LR is an offset parameter used to prevent the payoff becoming lower than the initial value assigned to the predicted reward (we use $LR=15$ below). The error and fitness values of classifiers are updated in the standard way. For actions which would take the agent into a cell containing an obstacle, P is set to zero, so such actions are never taken in exploration or exploitation.

In what follows we show that the predicted reward function converges to a layered reward structure in many cases. Layered reward structures are often used to solve multi-step problems in XCS [2, 3]. They take advantage of one of the key differences between XCS and other Learning Classifier System approaches, whereby during exploration XCS selects classifiers according to the accuracy of their predicted reward rather than simply selecting the classifier with the highest predicted reward.

In the presence of aliasing states, the advantage of this new approach over the standard XCS approach now becomes apparent: since the payoff no longer depends on the distance to the goal, different applications of a classifier in an epoch result in the same payoff in each instance, instead of different payoffs because of discounting. We use partial credit assignment (V/Vm) to ensure that the predicted reward of over-general classifiers gets reduced below the maximum available ($\rho + LR$) in order to allow more specific classifiers that contribute to optimal paths to get higher rewards.

In what follows we show that the new approach gives solutions which are at least as good as standard XCS on certain benchmark problems, and substantially better on certain benchmark problems involving aliasing states. We have also found it to give significantly better results on certain path-finding problems, but space does not permit their treatment here.

4 Case Studies

Three benchmark problems are studied and tested here, to illustrate the generality of our approach. The first is Woods1, a typical spatial-oriented maze environment. The other two are in Littman57 and MiyazakiA, which are typical Type-I aliasing maze environments from machine learning.

4.1 Case Study I: Woods1

Woods1 [8] is a typical sequential environment for testing the learning system/agent (Fig. 5 left). In the environment, the agent is expected to take no more than two steps to reach the food (F), no matter where the agent is located initially. The agent will not change its position if it chooses an action to move a cell with the obstacle inside, though one time-step still elapses. The optimal average path from start to the food is 1.6875 steps. Since the position of the goal is fixed and no aliasing state exists in the environment, thus it is a spatial-oriented reward problem. It has shown that Learning Classifier Systems are capable of solving this problem, such as ZCS in [16], although the average steps to the food converged to four steps. Thus, it will be used to demonstrate that with our approach XCS-GR can solve the general maze problem.

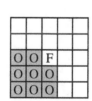

O	O	O	O	O	O	O	O
O		O		O	O	O	
O	3a					O	
O	O	2a	1a	4a	O	F	O
O		S	3b			O	
O	4b	O	2b	1b		O	
O						O	
O	O	O	O	O	O	O	O

Fig. 5. (left) The Woods1 maze problem; (right) The MiyazakiA maze problem

4.2 Case Study II: Littman57

As shown in Fig. 2, the environment contains eight aliasing cells. The agent needs to learn to find one shortest path to the food/goal from any cell. The agent will stay on the current cell, if it chooses to move to a cell with obstacle 'O' inside, but one step still elapses. The optimal average path to the food is 3.71 steps.

4.3 Case Study III: MiyazakiA

As shown in Fig. 5 (right), MiyazakiA, another Type-I aliasing maze listed in [17], contains eight cells with aliasing states. Due to many cells without aliasing states existing in the environment, the improvement will not be obvious if the agent is placed in a random cell. Therefore, one better way for analysing the difference of the results should be placing the agent in a cell (such as S) close to aliasing cells or in an aliasing cell (such as 4b). The results demonstrated in the paper are the results when the agent is placed at the cell with '4b', so the shortest path to the food is 4 steps.

5   Experimental Setup

5.1   Experimental Configuration

In the maze problems, the experiment typically proceeded as follows. The agent was placed into a randomly chosen empty cell. Then it moved around the environment avoiding obstacles until either it reached the food or had taken 30 steps, at which point the epoch ended (as done in [7]). For Case Study I and II, the agent was initially placed at a random cell. For Case Study III, the agent was placed at an aliasing cell, 4b. The results reported below are the moving average of 50 epochs and 20 seeds.

5.2   Structure of Classifiers

The following classifier structure was used for the XCS agents in the experiments: Each classifier had 16 binary nodes in the condition field: three nodes for each of the 8 neighbouring cells, with 00 representing the condition that the cell is empty, 11 that it contains food (F), and 01 that it is an obstacle. The following XCS parameter values were used: $N = 3200$, $\alpha=.1$, $\beta=.2$, $\gamma=.71$, $\theta_{GA} = 25$, $\chi=.8$, $\mu=.01$, $\delta=.1$, $\theta_{del}=20$, $v=5$, $p_I =10$, $F_I =.01$, $\varepsilon_I = 0$.

5.3   Global Reward Function (GRF)

The following Global Reward Function was used:

$$\rho = MaxR - TotalSteps \times pnty \qquad (3)$$

where
MaxR: is the maximum reward attainable. It was set to 1000 in the experiment.
TotalSteps: The number of steps in the epoch.
pnty: The penalty for each step. It was set to 100 in the experiment.

6 Results

6.1 Results for Woods1

Fig. 6 shows the comparison between the results of applying XCS and XCS-GR to the Woods1 environment. In both cases the results converged to the optimal solution, about 1.6875. In this case study XCS converged a lot faster than XCS-GR, since it updates its classifiers at each step whereas XCS-GR needs to wait until the end of the epoch to do so. But the main point is that XCS-GR is able to solve a standard maze problem such as Woods1.

6.2 Results for Littman57

Fig. 7 shows the results for Littman57. For this case study the classifier sets found by XCS were far from optimal. The average number of steps for the classifier systems AgentP with Self-adjusting and Gradual described in [18] were 5.24 and 4.82 steps respectively. XCS-GR performance quickly converged to an average between 3.67 and 3.9 steps to the food, which is very close to the optimal solution 3.71.

6.3 Results for MiyazakiA

Fig. 8 shows the results for MiyazakiA. XCS performance converged to an average of 4.25 steps to find food, while XCS-GR converges to 4 steps after 500 training epochs,

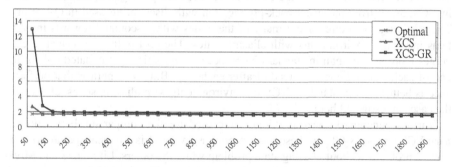

Fig. 6. The result of applying XCS-GR to Woods1

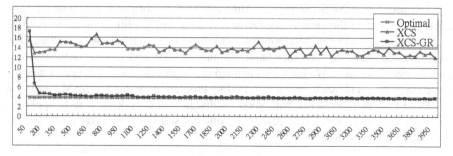

Fig. 7. The result of applying XCS-GR to Littman57

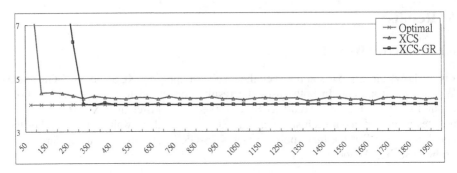

Fig. 8. The results of applying XCS-GR to MiyazakiA

which is optimal. If instead we measure the steps from a random cell, XCS-GR gets 3.4, which is not as good as Zatuchna[18], which gets 3.05. The reason XCS-GR performs less well seems because the GA generates rules with too many wildcards.

7 Conclusion

The paper demonstrated a modification (XCS-GR) to the mechanism for how delayed rewards are assigned in XCS, using credit assignment based on the System Net Prediction (SNP) calculated as part of the Prediction Array in standard XCS. This involved a relatively minor change to the delayed reward propagation mechanism used in XCS, away from the usual back-propagation-with-discounting used in Q-learning. The result seems to circumvent many of the issues with discounted rewards in problems involving environments with aliasing states. The cost is that performance converges more slowly than in standard XCS, because classifiers get updated only at the end of each training epoch instead of after each step. But the benefit is that performance is better than in standard XCS for environments with aliasing states, at least for the cases investigated here.

Various benchmark maze problems were used for the comparison between ordinary XCS and XCS-GR. In these maze problems, Woods1 is a typical spatial-oriented reward problem without aliasing state problem while the others, Littman57 and MiyazakiA, contain several cells with Type-I aliasing states. We showed that our approach finds near-optimal or optimal solutions for those maze problems, and showed how much improvement XCS-GR makes. Although XCS performance can sometimes be improved by tuning the discounting factor γ in some cases, a little thought shows that such an approach will typically not generalise from one environment to another. The XCS-GR approach, on the other hand, is quite general and can be used with any global reward function. We have also applied the approach to another Type-I aliasing maze environment (MazeB) and a path finding problem from Air Traffic Management similar to the one used in [5], with good results. In future work we plan to apply the approach to more complex path planning problems, starting with larger environments with the Type-I aliasing states.

References

[1] Barry, A.M., Down, C.: Limits in Long Path Learning with XCS. In: Cantú-Paz, E., Foster, J.A., Deb, K., Davis, L., Roy, R., O'Reilly, U.-M., Beyer, H.-G., Kendall, G., Wilson, S.W., Harman, M., Wegener, J., Dasgupta, D., Potter, M.A., Schultz, A., Dowsland, K.A., Jonoska, N., Miller, J., Standish, R.K. (eds.) GECCO 2003. LNCS, vol. 2724, pp. 1832–1843. Springer, Heidelberg (2003)

[2] Butz, M.V., Kaloxylos, R., Liu, J., Chou, P.H., Bagherzadeh, N., Kurdahi, F.J.: Analysis and Improvement of Fitness Exploitation in XCS: Bounding Models, Tournament Selection, and Bilateral Accuracy. Evolutionary Computation 11, 239–277 (2003)

[3] Butz, M.V., Kovacs, T., Lanzi, P.L., Wilson, S.W.: How XCS Evolves Accurate Classifiers. In: Proceedings of the Genetic and Evolutionary Computation Conference (GECCO 2001), pp. 927–934 (2001)

[4] Butz, M.V., Kovacs, T., Lanzi, P.L., Wilson, S.W.: Toward a theory of generalization and learning in XCS. Evolutionary Computation 8, 28–46 (2004)

[5] Chen, K.-Y., Dam, H.H., Lindsay, P.A., Abbass, H.A.: Biasing XCS with Domain Knowledge for Planning Flight Trajectories in a Moving Sector Free Flight Environment. In: IEEE Symposium on Proceedings of Artificial Life, 2007. ALIFE 2007, Honolulu, HI, USA, pp. 456–462 (2007)

[6] Lang, K.J., Witbrock, M.J.: Learning to Tell Two Spirals Apart. In: Proceedings of the 1988 Connectionist Models Summer School, San Mateo, CA, pp. 52–59 (1988)

[7] Lanzi, P.L.: A Model of the Environment to Avoid Local Learning with XCS, Dipartimento di Elettronica e Informazione, Politecnico do Milano, IT, Tech. Rep. N. 97. 46 (1997)

[8] Lanzi, P.L.: Solving Problems in Partially Observable Environments with Classifier Systems, Dipartimento di Elettronica e Informazione, Politecnico do Milano, IT, Tech. Rep. N. 97. 45 (1997)

[9] Littman, M.L., Cassandra, A.R., Kaelbling, L.P.: Learning Policies for Partially Observable Environments: Scaling up. In: Proceedings of the Twelfth International Conference on Machine Learning (1995)

[10] Mitchell, T.M.: Machine Learning. McGraw-Hill, New York (1997)

[11] Tang, K.W., Jarvis, R.A.: Is XCS Suitable for Problems with Temporal Rewards. In: Proceedings of International Conference on Computational Intelligence for Modelling, Control and Automation (CIMCA 2005), Vienna, Austria, pp. 258–264 (2005)

[12] Tangamchit, P., Dolan, J.M., Khosla, P.K.: The Necessity of Average Rewards in Cooperative Multirobot Learning. In: Proceedings of the 2002 IEEE International Conference on Robotics & Automation, Washington, DC USA, pp. 1296–1301 (2002)

[13] Wilson, S.W.: Classifier Fitness Based on Accuracy. Evolutionary Computation 3, 149–175 (1995)

[14] Wilson, S.W.: Generalization in the XCS classifier system. In: Proceedings of Genetic Programming 1998: Proceedings of the Third Annual Conference San Francisco, CA, pp. 665–674 (1998)

[15] Wilson, S.W.: Structure and Function of the XCS Classifier System. In: Cantú-Paz, E., Foster, J.A., Deb, K., Davis, L., Roy, R., O'Reilly, U.-M., Beyer, H.-G., Kendall, G., Wilson, S.W., Harman, M., Wegener, J., Dasgupta, D., Potter, M.A., Schultz, A., Dowsland, K.A., Jonoska, N., Miller, J., Standish, R.K. (eds.) GECCO 2003. LNCS, vol. 2724, pp. 1857–1869. Springer, Heidelberg (2003)

[16] Wilson, S.W.: ZCS: A Zeroth Level Classifier System. Evolutionary Computation 2, 1–18 (1994)

[17] Zatuchna, Z.V.: AgentP: a Learning Classifier System with Associative Perception in Maze Environments, University of East Anglia, PhD thesis (2005)

[18] Zatuchna, Z.V., Bagnall, A.: Learning Mazes with Aliasing States: An LCS Algorithm with Associative Perception. Adaptive Behavior - Animals, Animats, Software Agents, Robots, Adaptive Systems 17, 28–57 (2009)

Memetic Approaches for Optimizing Hidden Markov Models: A Case Study in Time Series Prediction

Lam Thu Bui[1] and Michael Barlow[2]

[1] School of Computer Science, University of Adelaide
[2] School of SEIT and DSARC, UNSW@ADFA, University of New South Wales
lam.bui07@gmail.com, m.barlow@adfa.edu.au

Abstract. We propose a methodology for employing memetics (local search) within the framework of evolutionary algorithms to optimize parameters of hidden markov models. With this proposal, the rate and frequency of using local search are automatically changed over time either at a population or individual level. At the population level, we allow the rate of using local search to decay over time to zero (at the final generation). At the individual level, each individual is equipped with information of when it will do local search and for how long. This information evolves over time alongside the main elements of the chromosome representing the individual.

We investigated the performance of different local search schemes with different rates and frequencies as well as the two newly proposed strategies. Four time series of the exchange rate were used to test the performance. The results showed the inconsistent behaviour of the approaches that used manual settings on local search's parameters, while showing the good performance of adaptive and self-adaptive strategies.

1 Introduction

The likelihood function of an observation sequence obtained by a hidden Markov model (HMM) possesses a multi-modal property in which there are many local optima. This is a challenge for optimizing the HMM's structure; the popular algorithm Baum-Welch (BW) usually falls into a local optimum. Our motivation for the work comes from this challenge in which we desire to design a global optimization algorithm for optimizing HMMs.

In this paper, we propose to investigate the use of evolutionary algorithms for this task. In many works [9,8,4], it has been shown that the hybridization of Evolutionary Algorithms (EAs) and the Baum-Welch algorithm (as a local search (memetic) tool) produced a much better performance in comparison with EAs or BW alone. However, the challenge for these hybridization schemes is that there is no consistent rule for defining parameters for local search such as the rate and the frequency of using local search. So far, the authors manually set the parameters for employing BW within their evolutionary algorithms.

K. Korb, M. Randall, and T. Hendtlass (Eds.): ACAL 2009, LNAI 5865, pp. 262–271, 2009.

Our hypothesis is that instead of using fixed values for these parameters, we should adapt them during the optimization process. In particular, we propose two strategies for memetics: (1) **Adaptive strategy**: The rate of using local search (defined as how many percentage of the population will receive local search) is set at the maximal rate at the beginning of optimization process and it will be reduced slightly over time to zero at the final generation. (2) **Self-adaptive strategy**: there will be no concept of rate for this strategy. Each individual will have added two pieces of information : should this individual be allowed to do local search? and if so, how many steps of local search it should take? This information is also evolved over time. In order to validate our proposal, we used several time series prediction problems including forecasting exchange rates for four currencies against the Australian dollar (AUD): New Zealand dollar (NZD), British pound (POUND), US dollar (USD), and Japanese yen (YEN). With a case study, we found that the results strongly support our proposal in which the local search schemes with predefined and fixed rates and frequency showed inconsistent performance over the set of testing data, while the proposed adaptive and self-adaptive schemes always give competitive results against the best cases using manual settings.

The paper is organized as follows: the second section is given for background information on hidden markov models and time series prediction. The third section covers the proposed methodology, while a case study is given in the fourth section. Section five is a concluding discussion.

2   Background

2.1   Hidden Markov Models

A HMM can be considered as the combination of three components $\lambda = \{A, B, \pi\}$ where A is the state transition probability distribution $A = \{a_{ij}\}$, $1 \leq i, j \leq N$, and N is the number of states, B is the probability distribution of the observation in a state denoted as $B = \{b_j(k)\}$, $1 \leq j \leq N$ and $1 \leq k \leq M$, M is the number of distinct observations (or symbols) per state, and π is the initial state distribution $\pi = \{\pi_i\}$ $1 \leq i \leq N$ [6]. There are three important questions to be resolved in addressing HMMs: (1) Given a HMM λ and a sequence of observations $O = O_1, O_2, ..., O_T$ - T is the length of the sequence, how do we compute the likelihood of the observation sequence $P(O|\lambda)$? (2) Given a HMM λ and a sequence of observations O, how do we generate a sequence of states $Q = q_1, q_2, ..., q_T$ which is optimal in some meaningful sense (i.e the one that is the most suitable for the sequence of observations)? (3) Given a sequence of observations O, how do we determine λ?

While the first and the second questions are about utilizing the model, the third question concerns optimizing the model based on the observation. For the third question, the most popular local search training technique is the Baum-Welch (BW) algorithm where we are always sure that the current re-estimated model is better than the initial model ($P(O|\overline{\lambda_t}) \geq P(O|\lambda_0)$). The re-estimation process will stop if there is no further improvement on the likelihood function. For this algorithm, the parameters for re-estimation are $\overline{\pi}_i$, $\overline{a}_{ij}$, $\overline{b}_j(k)$. They are calculated as follows:

$$\overline{\pi}_i = \gamma_1(i), \quad \overline{a}_{ij} = \frac{\sum_{t=1}^{T-1} \xi_t(i,j)}{\sum_{t=1}^{T-1} \gamma_t(j)}, \quad \text{and} \quad \gamma_t(i) = \sum_{j=1}^{N} \xi_t(i,j) \tag{1}$$

$$\xi_t(i,j) = \frac{\alpha_t(i)a_{ij}b_j(O_{t+1})\beta_{t+1}(j)}{P(O|\lambda)} = \frac{\alpha_t(i)a_{ij}b_j(O_{t+1})\beta_{t+1}(j)}{\sum_{j=1}^{N} \alpha_T(j)} \tag{2}$$

$$\alpha_t(i) = \begin{cases} \pi_i b_i(O_1), & t=1, 1 \leq i \leq N; \\ \{\sum_{i=1}^{N} \alpha_{t-1}(i)a_{ij}\}b_j(O_t), & 1 < t \leq T, 1 \leq i \leq N. \end{cases} \tag{3}$$

$$\beta_t(i) = \begin{cases} 1, & t=T, 1 \leq i \leq N; \\ \sum_{i=1}^{N} a_{ij}b_j(O_{t+1})\beta_{t+1}(j), & 1 \leq t < T, 1 \leq i \leq N. \end{cases} \tag{4}$$

Note that $P(O|\lambda) = \sum_{j=1}^{N} \alpha_T(j)$; and for a continuous HMM, B will be in the form of a mixture of distributions:

$$b_j(O) = \sum_{l=1}^{L} c_{jl}G(O, \mu_{jl}, U_{jl}) \tag{5}$$

where L is the number of distributions, G is the member distribution (usually a Gaussian), c_{jl} is the mixture coefficient for the lth mixture in state j, μ_{jl} is the mean vector and U_{jl} is the covariance matrix for the lth mixture in state j.

$$\overline{c_{jk}} = \frac{\sum_{t=1}^{T} \gamma_t(j,k)}{\sum_{t=1}^{T} \sum_{k=1}^{M} \gamma_t(j,k)}, \quad \overline{\mu_{jk}} = \frac{\sum_{t=1}^{T} \gamma_t(j,k).O_t}{\sum_{t=1}^{T} \gamma_t(j,k)} \tag{6}$$

$$\overline{U_{jk}} = \frac{\sum_{t=1}^{T} \gamma_t(j,k).(O_t - \mu_{jk})(O_t - \mu_{jk})'}{\sum_{t=1}^{T} \gamma_t(j,k)} \tag{7}$$

$$\gamma_t(j,k) = \gamma_t(j) * \frac{c_{jk}G(O_t, \mu_{jk}, U_{jk})}{\sum_{m=1}^{M} c_{jm}G(O_t, \mu_{jm}, U_{jm})} \tag{8}$$

2.2   Using Evolutionary Algorithms to Optimize HMMs

Most of the published work on using EAs to optimize HMMs employ a combination of EAs and BW in which an EA is used to search for a population of HMMs and BW will act as a local fine-tuner (see [10] for an example of not using local search). Note that here we use the terms "local search" and "memetics" interchangeably. In [4], the authors used BW after an interval of 10 generations. Each individual in the population will go through BW for 10 iterations. After every 3 generations, they allows BW to fine-tune the offspring as well. They reported a much better performance of the algorithm in comparison to BW itself. They call these 'hybrid operations' alongside the genetic operations. The likelihood value of the HMM encoded by an individual was used as the fitness for the individual. Meanwhile, in [9], the authors used BW for every generation on the training set, but the fitness value of an individual will be determined based on the likelihood on the validation set. In this way, the purpose of the evolution is to improve the generalization of BW. Another approach reported in [8] is to allow the evolution to find the best solution and then let this solution to undergo the local fine-tuning process by BW.

So far in the literature, the use of BW within the process of evolutionary computation has been done manually by pre-defining parameters such as rate and frequency. As we know that the balance between exploration and exploitation is a very important factor in an evolutionary process. If we focus too much on using BW for the local search, the process will be exploitation-oriented and result in premature convergence and the advantage of EAs over BW itself will not be employed to maximum effect. The question is: how can we adapt the parameters of hybridization over the evolution process?

2.3 Time Series Prediction

Since HMMs are suitable for sequential data, the problem of time series prediction (or forecasting) can be conveniently used to provide a test environment for optimizing HMMs. Prediction of time series has been intensively analyzed by the means of statistics such as moving average (MA), or the auto regressive method (AR) [5]. Prediction of financial time series is a particular challenging task. It is because the financial market is one of the most complex markets with so many factors that affect the fluctuation of the series of financial values such as politics, securities and interest rates. This makes it difficult for conventional statistical approaches, and hence there is a need to discover more efficient methods with less pre-assumptions such as the ones in artificial intelligence (AI) and machine learning. To date, much work has been dedicated to this area, such as using artificial neural networks [11], fuzzy-based systems [7], support vector machines [12], and HMMs [3].

3   Methodology

This section will address the issues of when and how to apply BW during the evolution process without any manual procedures of pre-defining hybrid parameters. A population of some individuals (HMMs) is maintained over time. The representation of an individual is similar to the one used in some previous works [4]. The structure of a chromosome C_i is comprised by the following components: a matrix of transition probabilities A, an array of the initial distribution π, a matrix of mean values for the mixture of distributions $\mu = \{\mu_{ij}\}$ with $1 \leq i \leq N$ and $1 \leq j \leq L$, and a matrix of (co-)variance values for the mixture of distributions $var = \{v_{ij}\}$ with $1 \leq i \leq N$ and $1 \leq j \leq L$.

To evolve the populations of HMMs, we employ a version of Differential Evolution [2]. Local search is applied to each individual in one of the two strategies reported in Sections 3.1 and 3.2.

3.1   Adaptive Reduction of the Local Search Rate

It is a very hard question to determine the most suitable parameters for employing local search in the literature of evolutionary computation. These include the rate (how many individuals to receive local search) and the frequency (the interval of generations to receive local search). Also, the above literature analysis indicates that there is no common method to incorporate BW with EAs for optimizing HMMs. From an evolutionary computation point of view, if all solutions go through the local search procedure, they might give the evolution process a quick convergence, but it might also quickly converge to the same area and therefore lead to the premature convergence. On the other hand, too few solutions doing local search might have a limited advantage.

In order to overcome this challenge, here we propose to have the rate of local search reducing over time. The motivation is that at the earlier stage there should be a necessity to have more solutions being able to search around their newly discovered areas. However, at the later time when the optimization process converges, the use of local search might not be important and therefore there is no need to keep a high rate of local search. The rate function $R(t)$ is defined as $R_{max}(1-\frac{t}{G})$ where G is the maximal number of generations, and $t = \{1, 2, ..., G\}$. At the beginning, the local search process is started with maximal rate (R_{max}). It will be reduced over time to zero at the final generation.

3.2   Evolving Hybrid Parameters

The above procedure is used to adapt the local search rate over time. However, it can not be adjusted upon a change in the problem. In order to do that, we propose to evolve both the rate and the degree of doing local search (how many steps for each local search). A part from the main structure of a chromosome C_i, we add two pieces of information: (1) a single bit to indicate if the individual is allowed to do local search or not; and (2) a single integer value to present the number of iterations when applying local search.

Note that with this strategy, there is no concept of frequency, since it works at the individual level. Each individual will determine for itself when to perform local search.

3.3   Individual Evaluation

Each individual is associated with an objective value (that is used to define fitness of the individual). In this study, we use the log likelihood function $log(P(O|\lambda))$ for evaluation. For each individual, the values of A, B, and π are obtained from the chromosome. The forward procedure is triggered to calculate α and the fitness F is calculated as $log(P(O|\lambda))$ or equivalently $log(\sum_{j=1}^{N} \alpha_T(j))$.

4   A Case Study

Since HMMs are particular suitable for sequential data, we propose to test them on several time series prediction problems. In order to make HMMs become a prediction tool, we follow the suggestion from [3] where the likelihood is calculated for every sub series from the start to the current time. Whenever a prediction is required, a procedure is triggered to search for a time slot that has the closest likelihood with the current one's. The behaviors of the time series at the current time and the found time slots will be considered as similar. The difference of the value between the found time slot with its next one will be used to calculate the next value from the current time slot.

We selected four time series for the exchange rate (to the Australian dollar): NZD, POUND, USD, and YEN. These data were obtained by the Reserved Bank of Australia [1] and they contain the monthly rates from 1970 up to now. We divided these data into different subsets for training and testing. The training set has 400 records while the test set has 268 records. Differential Evolution (DE) was used as the underlying evolutionary algorithm for optimizing HMMs since DE has been widely applied to solve non-linear optimization problems. The following approaches were tested: DE without local search - **DE1**, with the local search rate of 0.1 - **DE2**, with the local search rate of 0.5 - **DE3**, with a full rate - **DE4**, with an adaptive rate - **DE5**, and with a self-adaptation -**DE6**.

Each algorithm was tested 30 runs on each problem and with different random seed. Population size is 30 and they were allowed to run for about 2000 evaluations as in [4]. For HMMs, there were 5 states and 3 distributions to create the mixture. BW was allowed to run for 10 iterations in manual-setting local search.

4.1 Effects of Local Search Rates and Frequencies

We start with an investigation on the performance of DE with different rates of local search. These rates included 0, 0.1, 0.5 and 1.0. We measured the average of the best log-likelihood values among 30 runs for each test. The *mean* and *standard error* from 30 runs are reported in Table 1 with frequency of 10 generations and Table 2 for doing local search after every generation.

Table 1. The log likelihood values for all approaches with an interval of ten generations

	NZ	POUND	USD	YEN
DE1	264.263±7.741	449.707±7.793	62.416 ±7.085	146.509±14.692
DE2	780.493±1.970	913.226±1.765	645.318±2.450	816.615±3.991
DE3	788.589±1.025	914.992±0.724	652.469±1.336	830.871±0.973
DE4	790.005±0.467	915.081±0.319	650.910±1.160	831.992±0.386

From Table 1, it is clear that without local search, DE (or here we denoted it as DE1) worked less effectively on optimizing HMMs. The highly multi-modal function of the log-likelihood posed a real challenge for DE1 with a budget of 2000 evaluations. However, with the use of local search, the performance of DE versions (DE2, DE3, and DE4) was greatly improved. These approaches obtained much higher value of the log likelihood. This is also similar when we did not use the frequency (or the frequency of 1 generation).

Table 2. The log likelihood values for all approaches with an interval of 1 generation

	NZ	POUND	USD	YEN
DE1	264.263±7.741	449.707±7.793	62.416 ±7.085	146.509±14.692
DE2	789.600±1.093	915.911±0.160	656.342±1.444	831.667±0.931
DE3	790.258±0.744	915.541±0.121	659.139±1.186	833.252±0.352
DE4	790.423±0.301	915.230±0.071	657.507±1.177	832.687±0.229

In terms of frequency, it seems that if we allow local search after every generation (rather than 10 generations), the performance of the DE versions was better. However, for the rate of local search, it is a different matter. In the case of 10-generations frequency, the use of a full rate (1.0) seemed to be the best option (although it was bad with the USD series). When we allowed local search after every generation, the full rate did not give the best results and the winner was the rate of 0.5. This clearly supports our previous observation that the use of local search with a manual predefined rate brings up a question of how to find the appropriate rate for a time series. It is worthwhile to find an automatic approach to determine a rate that can provide competitive results to the above options.

Although it is not the purpose of this paper to test the generalization of approaches since it a completely different matter from optimizing HMMs, we still provide some results of memetic approaches on testing data sets. Fore this, we tested approaches on the testing data without any generalization techniques. We use the mean absolute percentage error (MAPE) [3] to measure the performance of approaches on testing data sets.

Table 3. MAPE values obtained by approaches with a frequency of 10 generations

	NZ	POUND	USD	YEN
DE1	1.938±0.036	2.302±0.046	3.304±0.029	3.152±0.003
DE2	2.095±0.007	2.263±0.019	2.935±0.024	2.913±0.013
DE3	2.096±0.004	2.226±0.008	3.007±0.032	2.894±0.000
DE4	2.102±0.003	2.238±0.009	2.967±0.033	2.894±0.000

In both cases of frequencies, DE1 still showed the inferior results in comparison to others DE versions using local search (except the case of NZD series). Once more, it indicates the usefulness of incorporating local search to the evolution algorithms during the optimization process of HMMs. These results also indicate the need for a consistent means of defining the rate of local search, since the use of a full rate or a small rate does not show better results over other strategy (such as the rate of 0.5). In both Tables 3 and 4, DE4 was better DE3 only in the case of USD, while DE2 was worse than DE3 and DE4 in most of the cases. Note that in all cases, the standard errors were very small in comparison with the mean; hence the comparison results are reliable.

Table 4. MAPE values obtained by approaches with a frequency of 1 generation

	NZ	POUND	USD	YEN
DE1	1.938±0.036	2.302±0.046	3.304±0.029	3.152±0.003
DE2	2.079±0.007	2.229±0.006	3.019±0.029	2.894±0.000
DE3	2.097±0.005	2.225±0.000	3.001±0.028	2.894±0.000
DE4	2.101±0.004	2.225±0.000	2.938±0.029	2.894±0.000

4.2 Effects of Adaption and Self-adaptation

As stated in the previous sections, our hypothesis is that we can adapt the
local search parameters automatically while still maintaining the performance
of the algorithms at least equally to that of the original algorithms. Firstly, we
also measured the values of the log likelihood for our adaptive and self-adaptive
approaches (see table 5). Clearly, these approaches provided very competitive
results over testing data sets of time series. The adaptive one seemed to be better
with the frequency of 1 generation (DE5.1), while DE6 showed an overall better
performance. In comparison to other cases of using manually defined rates, it
was better than DE4 and DE5 with POUND and YEN, but worse with NZD and
USD. Note that in these cases, the standard error was very small in comparison
with the mean except the case of DE6 (with YEN). A closer look is given in
this case where there was one outlier that made the value very high while the
average value of the remaining 29 runs is about 830 (the standard error ≈ 1).

Table 5. The values of log-likelihood obtained by DE5 and DE6

	DE5.10	DE5.1	DE6
NZ	786.333±1.440	790.129±0.739	790.181±1.055
POUND	915.286±0.390	915.424±0.121	915.682±0.249
USD	648.181±1.184	657.994±1.301	653.794±1.853
YEN	829.550±1.270	833.177±0.352	1045.022± 214.794

In terms of the performance on testing data, the adaptive and self-adaptive
approaches showed promising results. DE5.1 and DE6 were competitive with
each other and with DE3 and DE4. DE5.1 was better than DE3 on USD, but
worse on NZD and equal on POUND and YEN, while DE4 was better than DE5.1
on USD, worse on NZD and equal on the rest. It is also worthwhile to note that
DE5.1 seemed to give better performance on testing data but worse on training
data in comparison with DE6. This is understandable since it is common that if
an algorithm focuses too much on training, it is likely to be in a overfitting trap.
There have been many generalization techniques to be introduced to overcome
this trap. This is one thread of future work in terms of extending the technique
discussed here.

To get a better view on the prediction of time series, we visualized the prediction results from DE5.1 (not shown due to the space limitation) and DE6 over

Table 6. MAPE values obtained by DE5 and DE6

	DE5.10	DE5.1	DE6
NZ	2.087±0.006	2.098±0.005	2.079±0.006
POUND	2.240±0.008	2.225±0.000	2.235±0.007
USD	2.952±0.029	2.973±0.027	2.950±0.030
YEN	2.894±0.000	2.894±0.000	2.903±0.008

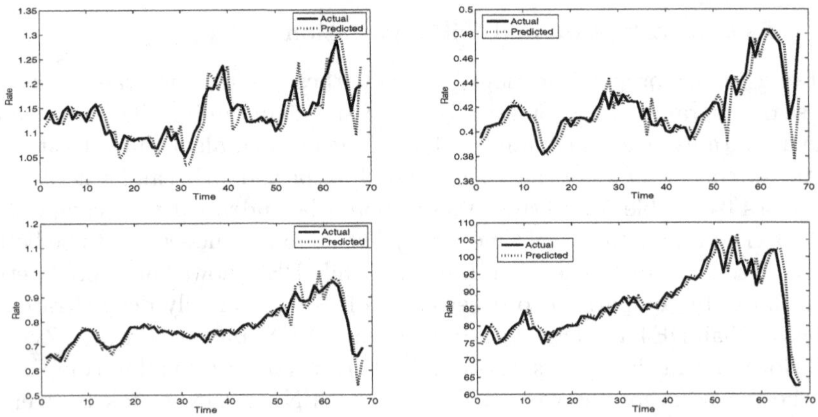

Fig. 1. Actual and predicted exchange rates using self-adaptive strategy DE6: NZD, POUND, USD, and YEN

all testing data sets for all four time series in Figure 1. We can see from the figures that our proposed approaches captured quite well the pattern of the all four time series. They might not always predict exact values, but the patterns of peaks and troughs were presented quite precisely.

5   Conclusion and Future Work

We introduced a methodology for employing memetics within a framework of evolution algorithms to optimize parameters of HMMs. The rate and frequency of using local search are automatically changed over time. For the adaptive strategy, we allow the rate of using local search to decay over time to zero at the final generation. For the self-adaptive strategy, each individual is equipped with information of when it will do local search and for how long. This information evolves over time alongside the main elements of the chromosome representing the individual.

We investigated the performance of different local search schemes with different rates and frequencies as well as the two newly proposed strategies. Four time series of the exchange rate were used to test the performance. The results clearly showed the inconsistent behaviour of the approaches using manual settings for the local search's parameters, while showing the good performance of adaptive and self-adaptive strategies. For the future work, we intend to incorporate our memetic strategies with generalization techniques (such as ensemble learning) to obtain better performance from HMMs.

Acknowledgements

This work was carried out when the first author was with UNSW@ADFA under funding of the ARC Discovery Grant No. DP0667123.

References

1. Reserve bank of Australia. Exchange rate (2009),
   http://www.rba.gov.au/Statistics/HistoricalExchangeRates/index.html
   (April 16, 2009)
2. Corne, D., Dorigo, M., Glover, F., Dasgupta, D., Moscato, P., Moscato, P., Poli, R., Price, K.V., Price, K.V.: New ideas in optimization (1999)
3. Hassan, M.R., Nath, B.: Stock market forecasting using hidden Markov model: a new approach. In: Proc. of ISDA 2005, pp. 192–196 (2005)
4. Kwong, S., Chau, C.W., Man, K.F., Tang, K.S.: Optimisation of HMM topology and its model parameters by genetic algorithms. Patt. Recog. 34(2), 509–522 (2001)
5. Montgomery, D.C., Jennings, C.L., Kulahci, M.: Introduction to time series analysis and forecasting. Wiley, Chichester (2008)
6. Rabiner, L.R.: A tutorial on hidden Markov models and selected applications in speech recognition. Proceedings of the IEEE 77(2), 257–286 (1989)
7. Romahi, Y., Shen, Q.: Dynamic financial forecasting with automatically induced fuzzyassociations. In: The Ninth FUZZ IEEE 2000, vol. 1 (2000)
8. Thomsen, R.: Evolving the topology of hidden markov models using evolutionary algorithms. LNCS, pp. 861–870 (2003)
9. Won, K.J., Prugel-Bennett, A., Krogh, A.: Evolving the structure of hidden Markov models. IEEE Trans. on Evol. Comp. 10(1), 39–49 (2006)
10. Yada, T., Ishikawa, M., Tanaka, H., Asai, K.: DNA sequence analysis using hidden Markov model and genetic algorithm. ICOT Technical Memorandom TM-1314. In: Institute for New Generation Computer Technology (1994)
11. Zhang, G., Eddy Patuwo, B., Hu, M.Y.: Forecasting with artificial neural networks: The state of the art. Int. J. of Forecasting 14(1), 35–62 (1998)
12. Zhang, Z., Shi, C., Zhang, S., Shi, Z.: Stock Time Series Forecasting Using Support Vector Machines Employing Analyst Recommendations. In: Wang, J., Yi, Z., Żurada, J.M., Lu, B.-L., Yin, H. (eds.) ISNN 2006. LNCS, vol. 3973, pp. 452–457. Springer, Heidelberg (2006)

The Effects of Different Kinds of Move in Differential Evolution Searches

James Montgomery

Complex Intelligent Systems Laboratory
Faculty of Information & Communication Technologies
Swinburne University of Technology
Melbourne, Australia
jmontgomery@swin.edu.au

Abstract. In the commonly used DE/rand/1 variant of differential evolution the primary mechanism of generating new solutions is the perturbation of a randomly selected point by a difference vector. The newly selected point may, if good enough, then replace a solution from the current generation. As the replaced solution is not the one perturbed to create the new, candidate solution, when the population has divided into isolated clusters large moves by solutions are the result of small difference vectors applied within different clusters. Previous work on two- and 10-dimensional problems suggests that these are the main vehicle for movement between clusters and that the quality improvements they yield can be significant. This study examines the existence of such non-intuitive moves in problems with a greater number of dimensions and their contribution to the search—changes in solution quality and impact on population diversity—over the course of the algorithm's run. Results suggest that, while they frequently contribute solutions of higher quality than genuine large moves, they contribute to population convergence and, therefore, may be harmful.

1 Introduction

Differential evolution [8] is a population-based algorithm for optimisation in continuous domains. The numerous studies examining its performance on different problems attest to its robustness [5] and its relative insensitivity to the values of its governing parameters [7]. However, there are relatively few analyses of the mechanisms underlying its operation, which makes prediction of its actual search behaviour difficult [2]. Such prediction is complicated by the indirect way in which solutions are perturbed in the most commonly used variant of the algorithm: DE/rand/1/bin.[1] At each iteration, each member of the population of solutions is considered in turn as a *target* for replacement in the subsequent generation. The new candidate solution is generated by adding the weighted difference between two randomly chosen population members (hereafter referred to as x_{r2} and x_{r3}, neither of which is the *target*) to a third, randomly-chosen

---

[1] See [6] for a description of variants of the algorithm and their respective names.

K. Korb, M. Randall, and T. Hendtlass (Eds.): ACAL 2009, LNAI 5865, pp. 272–281, 2009.

population member, referred to here as the *base*. Uniform crossover is then performed between the *target* and candidate; this further modified candidate is the one that may replace the *target*.[2] The weighting factor is denoted F, while the probability of using a vector component from the newly generated point instead of the *target* is given by Cr. Typically, the candidate replaces the *target* if it is as good or better. As the *target* solution is not the one perturbed to generate the candidate point, the magnitude of the *observed* displacement of solutions— replacement of *target* solutions by improved candidates—may not be a reflection of the magnitude of the difference vector employed.

The relationship between the magnitude of difference vectors and moves was previously explored by Montgomery [3,4]. That work classified difference vectors and *target* displacement as either *small* or *large*, yielding four combinations: *real small*, *real large*, *fake small* and *fake large*, where *fake* moves are those with a disparity between the magnitudes of the move and difference vector employed. Fig. 1 reproduces the classification scheme used, while an expanded description is given in Section 2 below. The relative frequency of these four move classes was examined across a range of two- and 10-dimensional problems. The impetus for that study was the observation that, when multiple optima exist and the population splits into separate clusters around these, movement between the separate clusters takes place. The standard explanation for this behaviour is that the large difference vectors that exist between members of different clusters allow movement between them [6,9], which is certainly part of the explanation. However, it was found that small difference vectors producing candidate solutions in one cluster were frequently the cause of *target* solutions from other clusters being replaced so that they appear to move the large distance between clusters, so-called *fake large* moves. In problems with competing optima, such moves tend to account for a greater proportion of large moves as a DE search progresses than do those caused by large difference vectors. Additionally, the average improvement in solution quality that they produce is greater than genuine large moves, a trend that apparently increases with the dimensions of the problem in question.

This paper examines three issues raised by that previous work. First, do such *fake large* moves exist in problems of more than 10 dimensions? Second, do such moves produce improvements in solution quality comparable to or better than other move types and does this increase with problem dimensions? Last, given that such moves necessarily place a solution near to the *base* solution used to generate the candidate, do they lead to the algorithm's convergence? Thus, even if they are shown to produce moves of high quality, it may be that such moves do not assist in the exploration of the search space.

1.1 Other Analyses of DE's Emergent Behaviour

Price, Storn and Lampinen [6] provide a quite extensive analysis of the performance of different DE variants with different values for the key parameters F and Cr (and others in the pertinent variants). Results are presented as phase

---

[2] At a minimum, one randomly chosen vector component will always come from the initially generated candidate point.

portraits, with successful combinations plotted on 2D maps of the space of values for F and Cr. While of undoubted benefit in demonstrating the algorithm's insensitivity (in some cases) to the values of these parameters and in providing a guide to suitable values for different kinds of problem, such portraits cannot explain *why* the algorithm is successful or not. In a similar vein, Zaharie [10] analysed the effect of the value of F on the convergence behaviour of DE, concluding that higher values work against premature convergence; what is an appropriately high value, however, is somewhat problem-dependent.

Recently, Dasgupta, Biswas, Das and Abraham [2] developed a mathematical model of DE's behaviour and used this to predict its convergence behaviour when the population is located near an optimum. Their model is highly constrained, however, restricted to a one-dimensional problem space with the population located on a gradual slope leading to the optimum. Ali [1] derives a probability density function of the location of new solutions. This is used in a simulation to show the likely distribution of generated solutions over time when DE is applied to a particular problem, chiefly to illustrate the large number of generated solutions that lie outside the problem's bounds. Ali characterises DE's search as one of initial expansion (constrained by the bounds of the problem) followed by a period of contraction. In essence, this paper examines the alternative mechanisms of that contraction and their contribution to the search.

2   Classifying Moves

In this study a *move*, whether accepted by the algorithm or not, is considered to be from the *target* solution to the new candidate solution, as this is the only observable change in DE's population of solutions with each iteration [3]. In the DE/rand/1 algorithm the size of each move is primarily a function of the scaling factor F (typically high), the magnitude and direction of the difference vector $||x_{r2} - x_{r3}||$, the location of the *base* solution and amount of crossover performed (controlled by Cr). As in [3] the potentially confounding effects of crossover are largely ignored, as Cr is typically set to a high value and hence the position of candidate solutions is more a function of the difference vector used and location of the *base* solution than of crossover. While the effects of crossover will not be considered further, setting Cr to a value a little less than 1 (as opposed to exactly 1) has been observed to offer some advantage to the search, so in the experiments described in Section 3 it is set to the commonly used value of 0.9 so that the typical evolution of the population can be observed.

Difference vectors and moves will be classified as either *small* or *large*, with *large* moves considered to be those greater than 20% the magnitude of the space-diagonal of the solution space. Moves are then classified according to the combination of their magnitude and the magnitude of the difference vector employed in their creation: moves and difference vectors of similar magnitude (i.e., both *small* or both *large*) are labelled "real" and those with dissimilar magnitudes are labelled "fake", as shown in Fig. 1.[3] A *fake small* move can result from a

---

[3] Negative connotations of the word "fake" are *not* implied.

$$||target - candidate|| \begin{cases} small \begin{cases} ||x_{r2} - x_{r3}|| \text{ is } small \;\rightarrow real\ small \\ ||x_{r2} - x_{r3}|| \text{ is } large \;\rightarrow fake\ small \end{cases} \\ large \begin{cases} ||x_{r2} - x_{r3}|| \text{ is } large \;\rightarrow real\ large \\ ||x_{r2} - x_{r3}|| \text{ is } small \;\rightarrow fake\ large \end{cases} \end{cases}$$

Fig. 1. Characterising exploration by DE/rand/1 in terms of distances between solutions involved in generating candidate solutions

distant *base* solution perturbed by a large difference vector, when the vector points towards the *target*'s location.[4] A *fake large* move is the result of a small difference vector applied to a *base* solution that is far from the *target*.

3 Experiment Design, Results and Discussion

Previous work [3] examined the relative proportion of different classes of move in a mix of two- and 10-dimensional problems, finding that *fake large* moves account for many of the candidate and accepted moves made by the algorithm when the population splits into separate clusters. Measures of the average quality (magnitude of solution improvement) of accepted moves from each class suggested that the contribution of *fake large* moves is non-negligible and, in several cases, better than other move classes (see Montgomery [4] for corrected results from [3]). However, taking the average quality across an entire run is potentially quite misleading as the scope for improvement in solutions diminishes as the search progresses. As the population converges, *real small* moves will dominate, with many accepted moves of low absolute value; the average quality of such moves may thus be very low, even though earlier in the algorithm's run they made valuable contributions.

This study has three objectives: to determine if all move classes are represented in higher dimensional problems; to examine the relative contribution to the search made by different moves; and to examine the possible relationships between different moves and contraction or expansion of the population. These are dealt with in the three sections that follow.

A standard DE/rand/1/bin algorithm was used with $F = 0.8$ and $Cr = 0.9$, values which are commonly used and frequently effective [6,8]. The problems to which the algorithm was applied are given in Table 1 together with the number of solution evaluations (consequently, number of iterations) allowed. These limits were chosen so that, in the runs for all problems except Langerman, the population has converged to a single location by the end of a run. In the case of Langerman, considerable progress is made after the population has converged to the point where only *real small* moves are made. To ensure the results are informative—the algorithm's *behaviour* is of interest, not the quality of its final result—those runs were halted at this point, which was consistent across problem sizes.

---

[4] *Fake small* moves may also result from very low values of Cr or the truncation of difference vectors that extend outside the search space's bounds.

Table 1. Problems used, allowed function evaluations and resultant number of iterations given population size of 50

Problem	n	Optima	Evaluations	Iterations
Fletcher-Powell	30	1	100000	2000
Langerman	30, 50, 100	1	120000	2400
Modified Shekel's Foxholes	30	1	50000	1000
Shubert	30	many	1.2×10^6	24000

In all cases a population of 50 solutions was employed; even though this is far below the oft-suggested size of $10 \cdot n$, experimentally it performed well and allowed the population to converge in a reasonable amount of time. Examination of runs using this larger population size and increased number of solution evaluations exhibited the same patterns reported in the following sections, but spread over a considerably longer time. For instance, with the 100D Langerman instance, a population of 1000 solutions requires four orders of magnitude more solution evaluations to have progressed to the same point as when using 50 solutions.

3.1   Relative Proportions of Successful Moves

Fig. 2 shows the relative proportions of accepted moves from each class over successive fifths of the algorithm's run. As the exact figures for individual runs vary—the general shape is frequently the same—the data were smoothed by taking the mean values at each iteration.

As has been observed previously on smaller instances, *real large* moves account for a high proportion of successful moves during early iterations, while *fake large* moves take over as the mechanism for large movement in later stages. As the population converges, *real small* moves take over as the primary, and frequently sole, driver of population changes.

3.2   Contributions to Solution Quality by Move Class

Figs. 3 and 4 show the minimum, mean and maximum observed *improvement* in solution quality resulting from each of the four kinds of move across the problems examined. The extrema and mean are averages of those measures from 100 runs and are divided into the values observed during each fifth of the algorithm's run. The bubbles at the bottom of each chart show the relative proportion of each kind of move within each block of execution time. As the magnitude of solution quality varies considerably across problems and as the search progresses, some of the charts use logarithmic scales.

In the Modified Shekel's Foxholes, Fletcher-Powell and Shubert functions, although *real small* moves become the primary source of accepted moves, the small number of *fake large* moves in later stages of the search contribute good improvements. In later iterations where successful *real large* moves still exist, the improvement from *fake large* moves is greater. A similar pattern is observed in the three Langerman instances, with *fake large* moves contributing greater

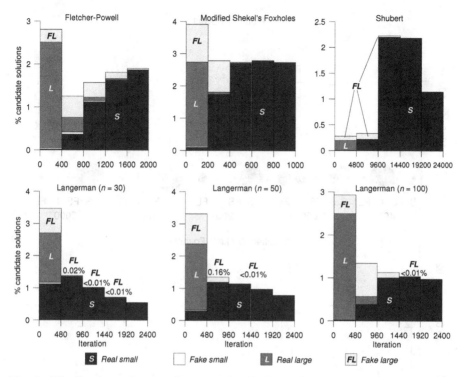

Fig. 2. Distributions of accepted moves of each class. Classes in each figure are, from bottom, *real small*, *fake small* (often absent), *real large*, *fake large*.

improvements than *real large* moves, although the quality of *real small* moves is in all cases typically many orders of magnitude better. A possible explanation for the greater improvement provided by *fake* rather than *real large* moves is that *fake large* moves will place the modified *target* near the *base*, which may already be located in a good region, while *real large* moves are more exploratory and may find new areas which are not as good as some of those already found.

3.3 Impact on Population Diversity

Population diversity or spread was measured at each iteration by taking the upper quartile of distances between all pairs of solutions. To assess the impact of different kinds of move on the population spread, the correlation between the change in this measure at the end of an iteration and the absolute number of accepted moves of each kind (when greater than zero) was calculated. Table 2 reports the mean and standard deviation of the observed correlation coefficients taken across 100 runs.[5] Although correlation does not imply causation, an

---

[5] The distributions of these measures were mostly normal; in those instances where they were not the median and interquartile range were found to agree strongly with the mean and standard deviation.

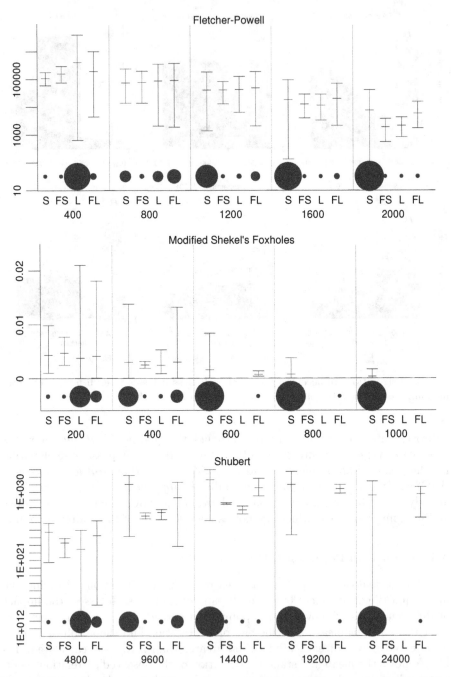

Fig. 3. Minimum, mean and maximum improvement from different move classes within each fifth of the algorithm's run; the last iteration of each block is shown underneath. The relative proportions of each accepted move class are indicated by the bubbles at the base of each chart. S = *small*, FS = *fake small*, L = *large*, FL = *fake large*.

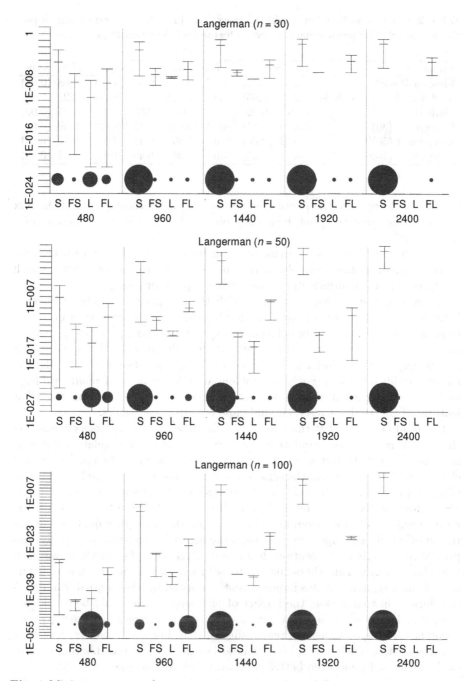

Fig. 4. Minimum, mean and maximum improvement from different move classes within each fifth of the algorithm's run; the last iteration of each block is shown underneath. The relative proportions of each accepted move class are indicated by the bubbles at the base of each chart. S = *small*, FS = *fake small*, L = *large*, FL = *fake large*.

Table 2. Correlation between number of moves within each class and change in population spread. The mean value is given with standard deviation in parentheses.

Problem	Small	Fake small	Large	Fake large
Fletcher-Powell	-0.03 (0.06)	0.02 (0.15)	0.53 (0.08)	-0.08 (0.04)
Modified Shekel's Foxholes	-0.26 (0.05)	-0.02 (0.12)	-0.36 (0.08)	-0.25 (0.04)
Shubert	0 (0.02)	0 (0)	0.65 (0.06)	-0.05 (0.02)
Langerman (30)	-0.14 (0.11)	-0.01 (0.18)	0.41 (0.12)	-0.09 (0.06)
Langerman (50)	-0.21 (0.06)	0 (0.06)	0.41 (0.1)	-0.10 (0.04)
Langerman (100)	-0.16 (0.05)	0 (0.04)	0.49 (0.1)	-0.11 (0.02)

increase or decrease in the population spread is a direct consequence of the mix of moves that were accepted, hence any observed correlation is suggestive of a causal link.[6]

Small moves show a very weak to weak negative correlation with changes in diversity, suggesting they play a role in the population's convergence, which would be expected intuitively as they drive population change in later stages of the search when it does converge. With the exception of Modified Shekel's Foxholes, *real large* moves show a moderate positive correlation with diversity; large moves thus help the population to explore. On most problems *fake large* moves show a very weak negative correlation with diversity. Thus, *fake large* moves *may* be somewhat related to decreases in population diversity. Results for the Modified Shekel's Foxholes problem are interesting: *real small*, *real large* and *fake large* moves all show a moderate negative correlation. It is unclear what features of this problem's search landscape lead to these results.

It should be noted that the strength of the observed correlations depends on the chosen measure of population diversity; a precise linear relationship may not exist even if there does exist a relationship between particular moves and changes in diversity. Thus, the observed correlations are moderately suggestive of an exploratory effect of *real large* moves and somewhat suggestive of a convergence effect of *fake large* moves. Taken with the results from Section 3.2, if the recruitment of solutions from one cluster to another of higher quality—an apparent effect of *fake large* moves—causes the algorithm to converge rapidly, and prematurely, to that one location then it is deleterious to the search. However, if much improvement can still be made after solutions have been drawn in to a particular small region, and this improvement is assisted by having greater numbers of solutions in that region, the impact of *fake large* moves could be considered positive. The considerable improvements observed in Langerman once the population has converged to a small region suggest that, in those instances, *fake large* moves may serve such a useful function, although it is also possible that greater exploration could find even better areas on which to converge.

---

[6] An alternative interpretation of any correlation is that certain stages of the search afford more opportunities for a reduction (or expansion) of diversity as well as supporting different kinds of moves.

4   Conclusions

This study confirms that *fake large* moves—large moves caused by small difference vectors—exist when DE is applied to benchmark problems with many dimensions and that they become the primary vehicle for large movement of solutions as the search progresses. Also as the search progresses the quality of such moves is often greater than that of *real large* moves. Weak correlations were observed between both *real* and *fake large* moves and population diversity, suggesting *real large* moves assist in exploration while *fake large* cause the population to converge. Thus, the large improvements in solution quality that *fake large* moves provide may not indicate high utility, subject to the alternative interpretations of convergence and its role in the search of high-dimensional problems outlined in the previous section. Further study of this issue may assist in determining whether such moves should be avoided or encouraged.

References

1. Ali, M.M.: Differential evolution with preferential crossover. Eur. J. Oper. Res. 181, 1137–1147 (2007)
2. Dasgupta, S., Biswas, A., Das, S., Abraham, A.: On stability and convergence of the population-dynamics in differential evolution. AI Commun. 22, 1–20 (2009)
3. Montgomery, J.: Differential evolution: Difference vectors and movement in solution space. In: IEEE CEC 2009, Trondheim, Norway, pp. 2833–2840. IEEE, Los Alamitos (2009)
4. Montgomery, J.: Erratum: Differential evolution: Difference vectors and movement in solution space. Technical Report SUTICT-TR2009.02, Swinburne University of Technology (2009)
5. Paterlini, S., Krink, T.: High performance clustering with differential evolution. In: IEEE CEC 2004, vol. 2, pp. 2004–2011. IEEE, Los Alamitos (2004)
6. Price, K., Storn, R., Lampinen, J.: Differential Evolution: A Practical Approach to Global Optimization. Springer, Berlin (2005)
7. Salman, A., Engelbrecht, A.P., Omran, M.G.H.: Empirical analysis of self-adaptive differential evolution. Eur. J. Oper. Res. 183, 785–804 (2007)
8. Storn, R., Price, K.: Differential evolution – a simple and efficient heuristic for global optimization over continuous spaces. J. Global Optim. 11, 341–359 (1997)
9. Storn, R.: Differential evolution research – trends and open questions. In: Chakraborty, U.K. (ed.) Advances in Differential Evolution. SCI, vol. 143, pp. 1–31. Springer, Berlin (2008)
10. Zaharie, D.: Critical values for the control parameters of differential evolution algorithms. In: Matoušek, R., Ošmera, P. (eds.) MENDEL 2002, 8th International Conference on Soft Computing, Brno, Czech Republic, pp. 62–67 (2002)

Author Index